Sustainable Technology Transfer

Sustainable Technology Transfer

A Guide to Global Aid & Trade Development

Edited by

Hans Henrik Lidgard

Jeffery Atik

Tu Thanh Nguyen

Wolters Kluwer
Law & Business

Published by:
Kluwer Law International
PO Box 316
2400 AH Alphen aan den Rijn
The Netherlands
Website: www.kluwerlaw.com

Sold and distributed in North, Central and South America by:
Aspen Publishers, Inc.
7201 McKinney Circle
Frederick, MD 21704
United States of America
Email: customer.service@aspenpublishers.com

Sold and distributed in all other countries by:
Turpin Distribution Services Ltd.
Stratton Business Park
Pegasus Drive, Biggleswade
Bedfordshire SG18 8TQ
United Kingdom
Email: kluwerlaw@turpin-distribution.com

Printed on acid-free paper.

ISBN 978-90-411-3448-6

Printed and Bound by CPI Group (UK) Ltd, Croydon, CR0 4YY.

STRENGTHENING
LEGAL EDUCATION
IN VIETNAM
A SIDA SUPPORTED PROJECT

SUFFOLK UNIVERSITY LAW SCHOOL

The Editors

Hans Henrik Lidgard is Professor of Private Law, Lund University, Sweden, specializing in Innovation Market related problems. Professor Lidgard holds a Licencié en Droit Européenne, Université Libre, Brussels, 1974, a Juris Doctor, Lund University, 1977. He has been visiting professor, McGeorge School of Law, Sacramento (1984/85) and Suffolk University Law School (2000). He was a co-founder and partner in Advokatfirman Vinge and has worked with legal and strategic matters within Pharmacia AB. His legal writing is focused on European competition law and intellectual property rights. Professor Lidgard has published numerous articles and books on intellectual property, competition law and international trade issues and has lectured internationally on these subjects. (hhl@jur.lu.se).

Jeffery Atik is Professor of Law and Sayre Macneil Fellow at Loyola Law School – Los Angeles. He holds a A.B. with distinction from the University of California at Berkeley, a J.D. from the Yale Law School, a Ph.D. from the Universidad Autonoma de Madrid and a Juris Dr (honoris causa) from Lund University. He has also taught at Berkeley (Boalt Hall), Boston College, Indiana-Bloomington, Lund, Suffolk, and UCLA law schools. He has served on three NAFTA bi-national panels, including the review in Softwood Lumber from Canada. He has practiced law at Shearman & Sterling in New York City and later with firms in Boston, Milan and St. Louis. He writes on a range of international trade, international finance and international intellectual property topics. (Jeffery.Atik@lls.edu)

Tu Thanh Nguyen, Doctor of Laws, LL.L., LL.M., Associate Professor, Faculty of Law, Lund University, Sweden; Deputy Director General of the Department of International Law, Ministry of Justice of Vietnam. He was a lecturer at the Ho Chi Minh City School of Law (Vietnam), trainee at the Intellectual Property Division of the WTO (Geneva, Switzerland), scholarship holder at the Max Planck Institute for Intellectual Property and Competition Law (Munich, Germany), visiting scholar at the Suffolk University School of Law (Boston, USA), and visiting researcher at the Faculty of Law, Copenhagen University (Denmark). He has written approximately 50 publications in both English and Vietnamese, including 'Competition Law, Technology Transfer and the TRIPS Agreement: Implications for Developing Countries' (Edward Elgar, 2010) (tu.nguyen@jur.lu.se).

List of Contributors

Tran Viet Dung, PhD, Law Faculty, National University of Singapore (NUS), LL.B, Ho Chi Minh City University of Law (HCMUL). He also read international studies in the Peoples' Friendship University of Russia and international trade policy in Harvard Kennedy School. He lectures law at HCMUL, was visiting lecturer at universities in Singapore and Hong Kong, scholar of the Education Forum of the US-Vietnam Trade Council and Asian Research Institute of NUS, and legal consultant of Ho Chi Minh City Information Technology Project Management Office. His research is primary in the areas of world trade law, investment law, anti-dumping law and competition law. He is the author of 'Anti-Dumping Policy and Law of Vietnam: A Critical Analysis from Integration and Competition Policy Perspectives' (2011) and co-author of 'International Trade Law' (2005) (email: tvdzung@gmail.com).

Steven Ferrey, Professor of Law, Suffolk University Law School where he lectures on environmental, energy and contract law. Professor Ferrey holds a B.A. in Economics, a J.D. from the University of California, Berkeley, a Masters Degree in Regional Planning, and was a post-doctoral Fulbright Fellow at the University of London on energy redevelopment. Professor Ferrey has been working with the Government of Vietnam since 2002, regarding systemic legal and regulatory barriers and incentives to renewable energy development within the Vietnam legal system and the new energy regulatory authority of Vietnam, ERAV in launching a renewable energy program to develop independent small power projects. Professor Ferrey has served over the past 18 years as a legal advisor for the World Bank and various Asian and African countries on renewable energy and climate control and has authored 7 books and 75 articles on energy policy and the environment (email: sferrey@suffolk.edu).

Christopher Gibson, Associate Dean and Professor of Law, Suffolk University Law School. He holds a B.A. from the University of Chicago, a M.P.P from the

Kennedy School of Government, Harvard University, and J.D. from Boalt Hall Law School, University of California, Berkeley. Following law school he clerked for a United States District Court judge in San Francisco, served as a Legal Assistant at the Iran-United States Claim Tribunal in The Hague, as a Senior Legal Officer for the United Nations Compensation Commission in Geneva (Head of the 'C' Claims Division), and as Head of the Electronic Commerce Law Section of the World Intellectual Property Organization in Geneva. Dean Gibson has practiced law in California, London and Washington, D.C. He writes in the areas of intellectual property, trade, foreign direct investment and international dispute resolution (email: cgibson@suffolk.edu).

Peter Gottschalk, L.L.M. Lund University Faculty of Law 1995, Doctoral candidate in public international law at the Faculty of Law, Lund University, Sweden. Gottschalk's current research for his dissertation focuses on the negotiations within the framework of the Convention on Biological Diversity leading up to the adoption of the Nagoya Protocol on Access to Genetic Resources and the Fair and Equitable Sharing of Benefits Arising from their Utilization. He also teaches international environmental law, intellectual property law, human rights law, world trade law, investment law and procedural law at the graduate and Master's level. Gottschalk also served as law clerk at the District Court of Malmö and at the Scania and Blekinge Courts of Appeal, and practises business law, specializing in intellectual property law (email: peter.gottschalk@jur.lu.se).

Xavier Groussot, Associate Professor of EU Law, Lund University and guest Professor at the Collège Européen de Paris, University of Panthéon-Assas (Paris II). He received his PhD in 2005 from Lund University. During his time as PhD student he was also Marie Curie Fellow at the Europa Institut, Amsterdam Center of International law. He has studied international law and EU law both at Lund University and Lyon (Université Jean Moulin, Lyon III) (email: Xavier.Groussot@jur.lu.se).

Keith E. Maskus, Professor of Economics and Associate Dean for Social Sciences at the University of Colorado, Boulder, USA. Professor Maskus was a Lead Economist in the Development Research Group at the World Bank. He is a Research Fellow at the Peterson Institute for International Economics, a Fellow at the Kiel Institute for World Economics, and an Adjunct Professor at the University of Adelaide. He was a visiting professor at the University of Adelaide and the University of Bocconi, and a visiting scholar at the CES-Ifo Institute at the University of Munich and the China Center for Economic Research at Peking University. He serves also as a consultant for the World Bank, the World Health Organization, and the World Intellectual Property Organization (email: keith.maskus@colorado.edu).

Hari M. Osofsky is an associate professor with tenure at the University of Minnesota Law School and the Associate Director of Law, Geography & Environment of the University of Minnesota's Consortium on Law and Values in Health, Environment & the Life Sciences. She received her B.A. and J.D. from Yale University.

She is currently a Ph.D. student in the Department of Geography at the University of Oregon. Her recent and forthcoming publications include articles in leading law reviews, a co-edited book on climate change litigation with Cambridge University Press, and a casebook on climate change law and policy forthcoming with Aspen Publishers. Professor Osofsky is i.a. a member of the Climate Legacy Initiative's Consultants Working Group, the International Law Association's Committee on the Legal Principles of Climate Change and the Journal of Environmental Law and Litigation's National Advisory Board (email: hosofsky@umn.edu).

Le Thi Nam Giang, PhD candidate of Hochiminh University of Law, LLM (Ho Chi Minh City University of Law, 2002). She is concurrently a lecturer and a Head of International Private Law – Comparative Law Section, International Law Faculty, Ho Chi Minh City University of Law. She specializes in Private International Law and Intellectual Property Law. Her publication approximately consists of 20 articles in Vietnamese on those fields (email: lenamgiang@gmail.com).

Nguyen Nhu Quynh, Doctor of Laws (the Doctoral Joint Training Program between Sweden and Vietnam), LL.B (Hanoi Law University, Vietnam), LL.M (Faculty of Law, Lund University, Sweden). Dr Quynh is Deputy Head of the Intellectual Property Law Centre of Hanoi Law University and specializes in Intellectual Property Law as well as Civil Law. She was a visiting scholar at Suffolk University School of Law (Boston, MA, USA) and an internship at WIPO (Geneva, Switzerland). She was also a researcher at WIPO for Project on Intellectual Property and Competition Policy (Geneva, Switzerland). She has published 20 articles in both Vietnamese and English, co-authored several books, presented papers at national and international IPR related conferences. Dr Quynh has taken part in drafting some Vietnamese legal normative documents on intellectual property (email: quynhnguyen.hlu@gmail.com).

Thu-Lang Tran-Wasescha is currently Counsellor in the Intellectual Property Division of the World Trade Organization and, *inter alia*, serves as Secretary to the Special Session of the Council for TRIPS. Before joining the WTO Secretariat in 2000, she was Director of the Department of International Affairs of the Swiss Intellectual Property Institute. In that capacity, she represented Switzerland in IP/TRIPS negotiations and discussions in GATT/WTO, WIPO, WHO, FAO, CBD and EFTA. She was deputy head of the Swiss TRIPS negotiating team in the Uruguay Round and responsible for the implementation of the TRIPS provisions into Swiss domestic law. From 1976 to 1986, she has served as Principal Legal Officer in the Industrial Property Division of the World Intellectual Property Organization, dealing with special projects such as the revision or adoption of treaties (email: Thu-Lang.TranWasescha@wto.org).

Christoffer Wong M.A. (St And.), B.Phil. (Oxon.), LL.M., LL.D. Lund University Faculty of Law. Christoffer is a senior lecturer in criminal law and procedural law at the Faculty of Law, Lund University. His research interests lie in the fields of the general part of criminal law, international and European criminal law and cooperation in criminal matters (email: Christoffer.Wong@jur.lu.se).

Summary of Contents

Table of Contents

List of Abbreviations

ACTA	Anti-Counterfeiting Trade Agreement
AFEC	Applicants for Africa Enterprise Challenge Fund
AHTEG	Ad Hoc Technical Experts Group on Technology Transfer and Scientific and Technological Cooperation
ANSI	American National Standards Institute's
ARV	Anti-retroviral Medicine
ASEAN	1967 Association of Southeast Asian Nations
BTA	Bilateral Trade Agreement with the United States
BI	Boehringer Ingelheim
BIT	Bilateral Investment Treaty
BRIC	Brazil, Russia, India, and China
CAFOs	Concentrated Animal Feeding Operations
CBD	1992 Convention on Biological Diversity, Adopted Rio de Janeiro 5 June 1992, entered into force 29 December 1993, 1760 U.N.T.S. 79
CDIP	WIPO Committee on Development and Intellectual Property
CDM	Clean Development Mechanism
CDR	Carbon Dioxide Removal
CDM	Clean Development Management
CDM EB	CDM Executive Board
CER	Certified Emission Reductions
CFC	Chlorofluorocarbon
CGIAR	Consultative Group on International Agricultural Research
CJEU	Court of Justice of the European Union
CL	Compulsory License
COP	Council of Parties

DRM	Digital Rights Management
DRAM	Dynamic Random Access Memory
DSB	Dispute Settlement Body
DSU	Dispute Settlement Understanding
ECJ	European Court of Justice
EGAT	Electricity Generation Authority of Thailand
EIF	Enhanced Integrated Framework
ERAV	Energy Regulatory Authority of Vietnam
EU	European Union
EVN	Electricite de Vietnam
FAO	Food and Agriculture Organisation
FDI	Foreign Direct Investment
FTA	Free Trade Agreement
FTC	US Federal Trade Commission
GATT	1947 General Agreement on Tariffs and Trade 55 U.N.T.S 194
GDA	Global Development Alliance
GDC	German Development Corporation, Deutsche Gesellschaft für Internationale Zusammenarbeit (GIZ) GmbH
GEF	Global Environmental Facility
GHG	Greenhouse Gas
GSK	GlaxoSmithKline
HTTP	Hypertext Transfer Protocol
ICL	International Common Law
ICTs	Information and Communications Technologies
IEC	International Electrotechnical Committee
IFAD	International Fund for Agricultural Development
ILA	International Law Association
ILO	International Labour Organization
IPCC	Intergovernmental Panel on Climate Change
IP	Intellectual Property
IPR	Intellectual Property Rights
ITC	International Trade Commission
ITU	International Telecommunication Union
JPS	Japanese Patent System
LDC	Least Developed Country
MDG	8th UN Millennium Development Goal, UN, General Assembly, Draft Res. A/65/L.1, 17 September 2010
MEA	Multilateral Environmental Agreement
MNE	Multi-National Enterprise
MOIT	Ministry of Industry and Trade
MONRE	Ministry of Natural Resources and Environment of Vietnam
NGO	Non-Governmental Agency
NIC	Newly Industrialised Country
NIEO	United Nations Declaration on the Establishment of a New International Economic Order, Resolution 3201(S-VI) of 1 May 1974, A/RES/S-6/3201

NOIP	National Office of Intellectual Property
NSSPP	Non-Negotiable Standardized Small Power Program
NSPPA	Non-Negotiable Standardized Power Producer Agreements
OECD	Organisation for Economic Co-operation and Development
ODA	Official Development Assistance
PCT	Patent Cooperation Treaty
PDD	Project Designed Document
PPA	Power Purchase Agreement
PPP	Public Private Partnership
R&D	Research and Development
RESPP	Renewable Energy Small Power Program
SAAP	State Aids Action Program
SACC	South African Competition Commission
SBSTTA	Subsidiary Body on Scientific, Technical and Technological Advice
SIDA	The Swedish International Development Cooperation Agency (Styrelsen för Internationellt Utvecklingssamarbete)
SME	Small and Medium Sized Enterprises
SPP	Small Power Producer
SRM	Solar Radiation Management
SSO	Standard-Setting Organization
SSP	Sub-Saharan Publishers
TAC	Treatment Action Campaign
TBT Agreement	Agreement on Technical Barriers to Trade
TFEU	Treaty on the Functioning of the European Union
TFP	Total Factor Productivity
TFTC	Fair Trade Commission of Taiwan
TIPO	Intellectual Property Office of Taiwan
ToT Code	International Code of Conduct on the Transfer of Technology
TPP	Trans Pacific Strategic Economic Partnership Agreement among Brunei Darussalam, the Republic of Chile, New Zealand and the Republic of Singapore, entered into force May 2006
TRIPS Agreement	Trade-related Aspects of Intellectual Property Rights
TT	Technology Transfer
UNCCD	1996 United Nations Convention to Combat Desertification
UNCED	United Nations Conference on the Environment and Development
UNCLOS	United Nations Convention on the Law of the Sea, 16 November 1994, 21 ILM 1245 (1982)
UNCTAD	1964 United Nations Conference on Trade and Development
UNDP	United Nations Development Programme
UNEP	United Nation Environment Programme
UNFCCC	United Nations Framework Convention on Climate Change
UNIDO	United Nations Industrial Development Organization
USAID	US Agency for International Development
USTR	United States Trade Representative

WCT	WIPO Copyright Treaty, 20 December1996, 36 I.L.M. 65 (1997)
WGABS	Working Group on Access and Benefit-sharing
WHO	World Health Organisation
WIPO	1967 World Intellectual Property Organization
WTO	1994 World Trade Organization
WPPT	WIPO Performances and Phonograms Treaty 20 December 1996, 36 I.L.M. 76 (1997)

Institutional Definitions and Abbreviations

ACTA	Anti-Counterfeiting Trade Agreement dated 15 November 2010 among Australia, Austria, Belgium, Bulgaria, Canada, Cyprus, the Czech Republic, Denmark, Estonia, Finland, France, Germany, Greece, Hungary, Ireland, Italy, Japan, the Republic of Korea, Latvia, Lithuania, Luxembourg, Malta, Morocco, the Netherlands, New Zealand, Poland, Portugal, Romania, Singapore, Slovakia, Slovenia, Spain, Sweden, Switzerland, the United Kingdom, the United Mexican States, the United States, and the European Union
Bayh-Dole Act	Pub. L. No. 96-517 (1980) codified at 35 USC. § 200-12
Berne Convention	Berne Convention for the Protection of Literary and Artistic Works September 9, 1886, as last revised at Paris on July 24, 1971, 1161 U.N.T.S. 30
Copenhagen Accord	Copenhagen Accord of 18 December 2009
Cotonou Agreement	Partnership Agreement between the Members of the African, Caribbean and Pacific Group of States on the One Part and the European Community and its Member States on the Other Part, Signed in Cotonou Benin on 23 June 2000, entered into force 1 April 2003
Doha Declaration	WTO, Ministerial Conference Fourth Session Doha, 9 – 14 November 2001, Decision of 14 November 2001, Implementation-Related Issues and Concerns, WT/MIN (01)/17, DEC/1, 41 I.L.M. (2002)

Kyoto Protocol	Kyoto Protocol to the United Nations Framework Convention on Climate Change, 10 December 1997, 37 I.L.M. 22 (1998)
NAFTA	1994 North American Free Trade Agreement 32 I.L.M.289 (Parts 1-3) and 32 I.L.M. 612 (Parts 4-8)
Paris Convention	Paris Convention for the Protection of Industrial Property of March 20, 1883, as revised, 828 U.N.T.S. 305
The 1948 Universal Declaration of Human Rights	UN, General Assembly Res. 217 (III), Universal Declaration of Human Rights of 10 December 1948
TRIPS	The Agreement on Trade-Related Aspects of Property Rights Annex 1C to the Marrakesh Agreement Establishing the World Trade Organization, done at Marrakesh 15 April 1994, 33 I.L.M. 1125
UNFCCC	United Nations Framework Convention on Climate Change United Nations Framework Convention on Climate Change, May 1992, S. TREATY DOC. NO. 102-38, 1771 U.N.T.S. 164, 166, 170

Preface

Interdependence is a term that captures the essence of modern times. We all share in the world's scarce natural resources and we are affected by cross border measures. The world cannot remain indifferent to poverty and natural disasters in other countries. Collaboration and sharing of knowledge is both demanded by reality and required by international law.

Based on overriding humanitarian considerations, richer states support the least developed countries (LDCs) with aid programs to address acute emergencies and with longer-term measures to assist in building viable infrastructures. Several international conventions require that such measures be undertaken in a specific and concrete way. For example, the Agreement on Trade Related Aspects of Intellectual Property Rights (TRIPS) requires developed countries to provide incentives to their national industries in order to foster concrete transfer of useful technologies.

Problems inevitably arise when good intentions are put to the test. Rights to the ownership of intellectual property assets create problems for developing countries. When actors in developing countries copy trademark protected products or items covered by copyright, the reaction from the developed countries is often swift, and one which sometimes disregards the overall goal of developed country support. On the other hand some claim that it is only when IPRs are respected that sustainable support can be expected.

In October 2010 five universities from three different continents held a legally-oriented conference in Vietnam on the subject of Sustainable Technology Transfer. The event was one of the final elements of a twelve yearlong project financed by the Swedish Development Agency (SIDA) under the title 'Strengthening Legal Education in Vietnam' and carried out by the Faculty of Law of Lund University. Over the course of its existence, the project has supported Vietnamese teachers studying educational and research methodology with a view to promoting international understanding. More than 200 teachers have undergone such training in Vietnam and abroad. In addition some 80 younger teachers earned a master's degree in International Law and a small group out of these is in the process of completing their PhD dissertations – all now learning from international experience and applying their findings to the Vietnamese context.

The project is an example of what might be characterized as a technical assistance project. The Conference, which assembled distinguished scholars, international and national agency representatives and Vietnamese teachers, was focused on Sustainable Technology Transfer – from developed countries to developing countries – primarily LDCs. A review of past developments indicated their shortcomings and provided a lesson for the future. The more difficult task was to find ways to improve the outcome.

The purpose of this book is to summarize the most important findings in a coherent manner. What is required to satisfy the existing international obligations and to what extent may developing countries use flexibilities in international conventions to advance their own development? Our focus is on the legal problems which Sustainable Technology Transfer may give rise to and how they may be addressed. Along with technical legal discussion, the book focuses on health, environment, energy and climate change as examples of the more problematic areas. The intention is less to find a one-size-fits-all solution or to give a full presentation of these problems, but rather to look for clues on how to approach them in the future.

Firstly, the book draws a global map of the territory we are entering providing an overall context. To that end it sketches out the major issues, purposes, institutions, and players. The book then reviews legislative enactments and seeks to establish their real value. A factual account of actual technology transfer follows. Subsequent chapter discuss to what extent competition rules offer developing countries a way to escape the more severe consequences of the undertakings they have accepted. With this background, the presentations turn to the health care sector, discussing the impact of compulsory licensing and the recently agreed ACTA agreement. It also deals with the transfer of technology in the energy sector, the climate change area and the specific conditions pertaining to conservation of biodiversity in separate studies. Finally the situation in Vietnam is examined by way of example.

Having penetrated all these areas as a way to expand on real life situations, the book concludes with forward looking statements in which the experience that has been studied is used to provide recommendations for concrete actions.

Many have contributed to the result, but the editors wish to gratefully acknowledge the valuable support provided by Robert Schwartz, who served as research assistant throughout the project.

Chapter 1
Framing the Issues

Hans Henrik Lidgard, Jeffery Atik & Tu Thanh Nguyen

1 DEFINING SUSTAINABLE TECHNOLOGY TRANSFER

The overall aim of the 2010 Conference on Sustainable Technology Transfer in Vietnam and accordingly also this book was to discover what obligations developed countries have to support developing countries in line with the UN Millennium Development Goals to eradicate poverty and improve living conditions by closing the technology gap. This is done by investigating the obligations and flexibilities expressed in numerous international conventions – including the Trade-related Aspects of Intellectual Property Rights (TRIPS Agreement), the Convention on Biological Diversity (CBD), and the UN Framework Convention on Climate Change – as well as how these obligations and flexibilities function in practice.

We wish to emphasize that we are addressing technology transfer in the development context – i.e. transfers of technology from developed to developing countries. We are not concerned here with transfers between developed countries, although of course this is an important topic that may have important implications for developing countries. Nor are we concerned with the interfirm characteristics of technology transfer – this is not a book on the private law or public law of licensing. Rather our focus is the role that technology transfer might play, if properly deployed, in the development of the disadvantaged countries of the world.

Hans Henrik Lidgard, Jeffery Atik & Tu Thanh Nguyen, *Sustainable Technology Transfer*, pp. 1–12.

This first chapter will draw a global map of the territory of technology transfer by providing an overall context. It sketches out central definitions, the major issues, purposes, institutions, and players. The objective is to return to these central issues and discuss them in the conclusion – precisely as was done during the discussions in Vietnam – and try to advance both overriding and practical solutions to the problems discussed in the subsequent chapters.

The title 'Sustainable Technology Transfer' is neither self-explanatory nor precise enough to indicate who has what obligation and therefore is in need of some preliminary observations. We begin our treatment with an examination of this concept, building from its elements.

1.1 TECHNOLOGY TRANSFER AND TECHNICAL ASSISTANCE

The notion of 'technology transfer' is found in Article 16 of the CBD,[1] in various provisions of the UN Framework Convention on Climate Change[2] and, perhaps most importantly, in Article 66.2 of the TRIPS Agreement.[3] Yet it has no clearly defined meaning. In the broadest sense it covers any type of advice that helps recipient countries (which in the development context is often confined to LDCs) address problems having a technical character – from simple applications, such as drilling a water well, to, at the other extreme, developing and producing a specific biopharmaceutical medicine. Posited too broadly or too narrowly, the notion of technology transfer begs the question of whether it embraces all types of development aid, or if it only concerns highly technical information permitting the production of modern commodities for both the local and the export markets (production technology).

The transfer of production technology is to be distinguished from technical assistance. The latter concerns assistance in the preparation of laws and regulations on the protection and enforcement of IPRs. It also includes support regarding the establishment of relevant agencies, including the training of personnel. Technical assistance aims to create an environment that can receive and successfully adopt production technology.[4]

---

1. The Convention on Biological Diversity (CBD), known informally as the Biodiversity Convention, is a legally binding treaty. The Convention was opened for signature at the Earth Summit in Rio de Janeiro on 5 Jun. 1992 and entered into force on 29 Dec. 1993. <www.cbd.int/>.
2. United Nations Framework Convention on Climate Change, FCCC/INFORMAL/84, opened for signature 9 May 1992, entered into force 21 Mar. 1994 (FCCC). <http://unfccc.int/resource/docs/convkp/conveng.pdf>.
3. TRIPS Art. 66.2 provides:

   > Developed country Members shall provide incentives to enterprises and institutions in their territories for the purpose of promoting and encouraging technology transfer to least-developed country Members in order to enable them to create a sound and viable technological base.

4. Keith E. Maskus, *Encouraging International Technology Transfer.* UNCTAD-ICTSD Programmes on IPRs and Sustainable Development, Issue Paper No.7, (2004) International Centre

The European Union (EU) has addressed the hurdles encountered with the definition of technology transfer at several meetings of the World Trade Organization (WTO) Council for Trade Related Aspects of Intellectual Property (TRIPS Council).[5] The EU emphasized that the private and commercial sector has clearly been the main source of production technology. Governments do not normally own rights to existing production technologies, nor can they compel the private sector to transfer their technologies. Governments can only provide incentives in the form of encouragement, promotion, and facilitation of the most fruitful projects. Such incentives are only one of many factors relevant to companies' decisions to enter into technology transfer arrangements. Transfer of production technology is one component of a complex project, where macroeconomic stability, a safe legal framework, and a well-functioning government administration are also important drivers.[6] These more general efforts at providing technical assistance should then be coupled to the notion of technology transfer, although it seems clear that the notion should not encompass the even-broader concept of addressing emergencies and survival in a broad sense through ordinary development aid.[7]

This developed-country position has not been well-received by the developing world block. The concern is that if the notion of technology transfer becomes too broad, encompassing all forms of technical assistance, then the only 'technology transfer' effected would be the provision of technical assistance, and more concrete

---

for Trade and Sustainable Development, Geneva <www.iprsonline.org/unctadictsd/projectoutputs.htm>. As an alternative Maskus proposes a distinction between market-based forms of Technology Transfer (including FDI, technologies embodied in imported goods, licensing, exports by firms in the recipient country) and non-market-based forms of Technology Transfer (including imitation and reverse engineering, learning from patents filed by foreign inventors, learning from scientific publications, movement of workers and students and other types of spillovers). According to Maskus, certain market-based channels have increased in importance over time.

5. Copies of all documents hereinafter referred to of Minutes of Meetings of the WTO TRIPS Council, Developed Country Reports, Communications from Least Developed Countries to WTO TRIPS Council and Decisions of the WTO TRIPS Council may be found at the WTO's Technology Transfer website. <www.wto.org/english/tratop_e/trips_e/techtransfer_e.htm>.
6. WTO, Council for TRIPS, Minutes of Meeting, 2 Mar. 2010, IP/C/M/62. para. 78 addressing views stated in Reflection Paper on Transfer of Technology to Developing and Least-Developed Countries, Communication from the European Communities and their member states, dated 12 Feb. 2003, IP/C/W/398 WT/WGTTT/W/5 para. 39.
7. See WTO, Report on the implementation of Art. 66.2 of the TRIPS Agreement, Switzerland, Addendum dated 18 Oct. 2010, IP/C/W/551/Add.2, 4. Switzerland agreed with the EU and states that its understanding of technology transfer covers the flows of know-how, experience and equipment amongst different stakeholders such as governments, private sector entities, financial institutions, NGOs and research/education institutions. It uses four key modes of technology transfer, according to the UN definition: (i) physical objects or equipment; (ii) skills and human aspects of technology management and learning; (iii) designs and blueprints which constitute the document-embodied knowledge on information and technology; and (iv) production arrangement linkages within which technology is operated, including the enabling environment for such transfer. According to Switzerland foreign direct investment, official development assistance (ODA), commercial lending and equity investment are all important channels through which technology transfer is financed.

efforts to transfer much-needed production technology would be lost among general development aid or technical assistance programmes. The developing world views the technology transfer obligations in international agreements as covering concrete, specific production technology. Further, developed countries must report on their efforts to provide incentives for such transfers.[8]

An agreed definition of technology transfer does accordingly not exist. Generally, the notion is more specific than general development aid, but not as narrow as excluding technical assistance. However, the structure of the TRIPS Agreement implies that a distinction should be maintained between technology transfer and technical assistance – at least from a reporting perspective.[9]

1.2 Sustainable

Finally, does the word 'sustainable' have any bearing on technology transfer? Does the term mean sustainability in the usual environmental sense – reducing the current consumption of the earth's resources to permit long-term environmental viability[10] – or does the notion entail something else within the specific context of technology transfer?

'Sustainable' has become a fashionable expression, easily employed, but perhaps lacking any real meaning. Sustainable technology transfer may include policies that support growth and investment so that businesses may access the critical infrastructure, financial resources and markets they need to succeed. Infrastructure development in the areas of clean environment, renewable energy, biodiversity management, climate change adaptation, waste management, land and ocean resource use, water and sanitation, hazard and disaster management, and transport and water supply all involve elements of technology transfer, because they are long-term projects with lasting effects that may satisfy a sustainability criterion. Such development may also include mentoring and skills training with a view to developing in country expertise. Technology transfer is an important element in achieving sustainability in a variety of environmental regimes, including climate change, independent power production and the maintenance of biological diversity.

Successful sustainable technology transfer does not merely require that the technology in question has been put into operation and resulted in increased productivity in a certain economic activity. It primarily concerns the creation of technological capabilities in the host country that ensure the enduring and efficient use of that technology. Sustainable technology transfer is more than just a single

8. WTO, TRIPS Council Minutes, fn. 5, para. 84.
9. TRIPS Art. 66.2 addresses technology transfer and TRIPS Art. 67 addresses technical assistance.
10. Thus it may conform to the environmental sense of the word by meeting the needs of the present while also securing lasting effects without compromising the ability of future generations. Report of the World Commission on Environment and Development: 'Our Common Future' (2 Aug. 1987), paras 26–28. Published as Annex to General Assembly document A/42/427', Development and International Co-operation: Environment' <www.un-documents.net/wced-ocf.htm>.

transaction. It is rather, one element in a long-term, consistent and complex process, combining technical assistance and technology transfer that results in a verifiable change in the recipient country.

Sustainable technology transfer, as we understand it, oscillates between these two meanings. It refers to the technology transfers between the developed and developing world needed to achieve sustainability in its environmental, health and human resources sense, as well as the need for technology transfer, as an engine for the development of LDCs in particular, to be sustainable.

2 BALANCING FUNDAMENTAL INTERESTS

These definitions and challenges point at underlying uncertainties that have a bearing on how sustainable technology transfer should be addressed. In this section, the intention is to investigate the problem further and define the underlying purposes of technology transfer. Do overriding norms supply a reasoned set of postulates and are there other problems that should be addressed?

2.1 Sustainable Technology Transfer: An Element in a Larger Context

According to a recent report, global wealth increased by 72% in the last decade.[11] Smaller, open economies like Singapore, Switzerland, Norway, Austria and Australia top the list. It is, however, interesting to note that certain developing countries have also experienced a rapid increase, among them India and China, in each of which individual wealth has more than doubled over the last decade. Significantly, 24 million people or 0.5% of the world's population control 35.6% of the wealth, whereas 3 billion of the poorest control only 4.2%.[12] The rich to a large extent live in North America, Europe, Japan and Australia. The poor live in LDCs, primarily in Africa.

2.1.1 UN Millennium Development Goals Require the Eradication of Poverty

The overall UN Millennium Development Goals are to eradicate this uneven distribution pattern and reduce poverty. They aim to create a global partnership for development by addressing the special needs of LDCs, landlocked countries and Small Island Developing States.[13]

11. Credit Suisse, 'Global Wealth Report 2010' (2010 Global Wealth Report), 3. <www.credit-suisse.com/news/doc/credit_suisse_global_wealth_report.pdf>.
12. *Ibid.*, Figure 1.
13. UN, General Assembly, Draft Res. A/65/L.1, 17 Sep. 2010, 'UN Millennium Goals' <www.un.org/millenniumgoals/pdf/MDG_FS_8_EN.pdf.>.

This is not an easy task in an ever more complex trading system with structural differences among the standards, regulations and legal systems of the economies concerned. These differences generate new tensions relating to technology protection, environmental sustainability and resource scarcity.[14]

The *2000 UN Millennium Declaration*[15] promoted globalization as a positive force for all of the world's peoples. While globalization offers great opportunities, its costs and benefits are unevenly shared. Poor countries face special difficulties in responding to this central challenge and any efforts undertaken must include measures which correspond to the needs of developing countries. Based on equity, those who benefit least deserve help from those who benefit most. An environment which is conducive to development and the elimination of poverty must consist of an open, equitable, rule-based, predictable and non-discriminatory multilateral trading and financial system.[16] With the growth of ever more technical, intrusive and binding rules it becomes imperative to assist poor countries in developing productive capacity and participating in international trade.[17]

According to the Millennium Declaration, by 2015 the proportion of the world's people whose income is less than one US dollar a day, who suffer from hunger or are unable to afford safe drinking water shall be halved.[18] By then the spread of HIV/AIDS, malaria and other major diseases that afflict humanity is to be halted.[19]

Both developed and developing countries must construct strong partnerships with the private sector in pursuit of economic growth and of poverty eradication where innovation and new technology are driving forces.[20] Development aid may

14. WTO Director-General Pascal Lamy, *The Doha Round marks a transition from the old governance of the old trade order to the new governance of a new trade order*, Speech at the occasion of the 10th anniversary of the World Trade Institute (1 Oct. 2010), <www.wto.org/english/news_e/sppl_e/sppl173_e.htm>.
15. UN, General Assembly, Res. A/Res/55/2, 18 Sep. 2000. United Nations Millennium Declaration. <www2.ohchr.org/english/law/millennium.htm>.
16. *Ibid.*, Ch. III. Development and poverty eradication, point 12.
17. See Statement by Lamy at the High-Level Plenary Session of the Millennium Development Goals Summit, 20 Sep. 2010, <www.wto.org/english/news_e/sppl_e/sppl170_e.htm>.
18. *Ibid.*, Ch. I Values and principles, point 19.
19. These goals were recently repeated and re-emphasized at the UN General Assembly Summit held 20–22 Sep. 2010 in the General Assembly's Draft Resolution of 17 Sep. 2010 and the Report of the Millennium Development Goal Task Force of 16 Sep. 2010. General Assembly, Draft Res. A/65/L.1, 17 Sep. 2010, 'Keeping the promise: United to Achieve the Millennium Development Goals', Section 23, and 'Millennium Development Goal 8 The Global Partnership for Development at a Critical Juncture.' <www.un.org/en/mdg/summit2010/documents.shtml>.
20. UN, General Assembly, Res. 41/128, Declaration of Right to Development of 4 Dec. 1986, Art. 2. <www.un.org/documents/ga/res/41/a41r128.htm>. See also Diego Comin & Mart Mestieri, 'The Intensive Margin of Technology Adoption', Harvard University and NBER Working Paper 11-026 (May 2010). The authors present a tractable model for analysing the relationship between economic growth and technology adoption. Based on a data set of 15 technologies and 166 countries the model yield four main findings: (i) there are large cross-country differences in adoption; (ii) differences vary substantially across technologies; (iii) the cross-country dispersion of adoption lags has declined over time while the cross-country dispersion in the intensive

provide relief in emergency situations, but 'aid is a catalyst, not a cure', as no country has ever been transformed by aid alone.[21] Transferring innovation and technology from the developed world to LDCs has, for several decades, been an important complement to the general aid schemes provided to poorer nations.

2.1.2 Human Rights Justifications for Technology Transfer

There are different rationales for providing technology to least-developed countries. One important – if not paramount – reason is that the sharing of technology is a principal form of cooperation and assistance required by fundamental international agreements based on social justice and human rights considerations.

The Charter of the United Nations refers in its Preamble to an international mechanism for the promotion of economic and social advancement of all peoples. In its first chapter on Purposes and Principles, members are expected to cooperate in solving international problems of an economic, social, cultural, or humanitarian character.[22] Article 55 further aims at the creation of conditions of stability and well-being necessary for peaceful and friendly relations among nations based on respect for the principle of equal rights and self-determination of peoples by promoting higher standards of living, full employment, and conditions of economic and social progress and development, and also solutions of international economic, social, health, and related problems.

Several other UN instruments develop these commitments though the legal status of these instruments may vary. Conventions are legally binding for those states that ratify or accede to them whereas declarations have no binding legal effect. However, apart from their undeniable moral force, they also serve as guidance to states in the conduct of their policies.

The 1948 Universal Declaration of Human Rights[23] in Article 25.1 proclaims that everyone has the right to a standard of living adequate for the health and well-being of himself and of his family. This principle of the right to health is underpinned by Article 12 of *the 1966 United Nations International Covenant on*

---

margin has not; (iv) the cross-country variation accounts for more than 40% of the variation in income per capita. <www.hbs.edu/research/pdf/11-026.pdf>.

21. Statement of President Barroso at the High-Level Plenary Meeting of the General Assembly on the Millennium Development Goals Millennium Development Goals Plenary Meeting, New York, 20 Sep. 2010. <http://europa.eu/rapid/pressReleasesAction.do?reference=SPEECH/10/465>.
22. Charter of the United Nations, 26 Jun. 1945, 59 Stat. 1031, T.S. 993, 3 Bevans 1153, entered into force 24 Oct. 1945.
23. UN, General Assembly Res. 217 (III), Universal Declaration of Human Rights of 10 Dec. 1948. <www.un.org/en/documents/udhr/index.shtml>. The UDHR is a milestone document in the history of human rights. Drafted by representatives with different legal and cultural backgrounds from all regions of the world, the Declaration was proclaimed by them as a common standard of achievements for all peoples and all nations. It sets out, for the first time, the fundamental human rights which are to be universally protected.

Economic, Social and Cultural Rights which recognizes 'the right of everyone to the enjoyment of the highest attainable standard of physical and mental health'.[24]

More directly, *the 1975 Declaration on the Use of Scientific and Technological Progress*[25] notes in its Preamble that scientific and technological progress has become one of the most important factors in the development of human society and recognizes that scientific and technological progress is both of great importance in accelerating the social and economic development of developing countries and is also one of the principal ways of accelerating their economic development. All States, according to the Declaration, shall cooperate in the establishment, strengthening and development of the scientific and technological capacity of developing countries with a view to accelerating the realization of the social and economic rights of the peoples of those countries.[26] They shall take the necessary measures, including legislative measures, to ensure that the utilization of scientific and technological achievements promotes the fullest realization of human rights and fundamental freedoms without any discrimination whatsoever on grounds of race, sex, language or religious beliefs.[27]

The 1986 Declaration on the Right to Development[28] is of considerable importance in applying these general human rights obligations to the transfer of technology to LDCs and developing countries. It refers to the UN Charter and proclaims, in Article 1, that 'the right to development is an inalienable human right'. States have a duty to collaborate in ensuring development and eliminating obstacles to development so as to promote a new international economic order. Sustained action is required to promote more rapid development of developing countries. As a complement to the efforts of developing countries, effective international cooperation is essential in providing these countries with appropriate means and facilities to foster their comprehensive development.

The 2001 HIV/AIDS Declaration[29] also casts light on the obligations of developed countries to assist developing countries. Under the title 'Global Crisis – Global Action' the UN addressed the HIV/AIDS epidemic, which by that time affected more than 36 million people, 75% of them living in sub-Saharan Africa. Without affordable medicines, supply structures and health care systems there could be no effective response to the HIV/AIDS crisis in poor countries. Promoting domestic industries and granting access to essential drugs was required.[30]

---

24. International Covenant on Economic, Social and Cultural Rights, adopted 16 Dec. 1966, 993 U.N.T.S. 3 (entered into force 3 Jan. 1976), G.A. Res. 2200 (XXI), 21 U.N. GAOR Supp. (No. 16) at 49, U.N. Doc.A/6316 (1966).
25. UN, General Assembly, Res. 3384, Declaration on the Use of Scientific and Technological Progress in the Interests of Peace and for the Benefit of Mankind of 10 Nov. 1975. <www2.ohchr.org/english/law/mankind.htm>.
26. *Ibid.*, point 5.
27. *Ibid.*, point 7.
28. 1986 UN Declaration on the Right to Development, fn. 19.
29. UN, General Assembly Res. S-26/2, Declaration of Commitment on HIV/AIDS, 'Global Crisis – Global Action', of 27 Jun. 2001. <www2.ohchr.org/english/law/hiv.htm>.
30. *Ibid.*, point 26.

The latter point refers to shortcomings in the TRIPS Agreement on compulsory licensing.[31]

In the light of the above penetration of some of the more central human rights declarations, there is, at a minimum, a very strong moral obligation placed on developed countries to support the developing world with hard core technology to secure basic fundamental interests in these countries.

2.1.3 Addressing the Environment: A Question of Survival

Environmental justifications for technology transfer do not seem to be tangible enough to unite the world's leaders into taking serious action.[32] Still, these issues are likely to cause substantial concerns in the future and they are starting to influence the technology transfer agenda. It is of small help if developed countries are successful in reducing present energy consumption if developing countries do not follow suit.[33] The choice of energy technology is one of the few inputs into the global warming agenda that can be influenced by law and policy. It is problematic that countries like China, India and Brazil are not always amenable to proposals from developed countries that they should reduce their energy consumption in parallel. US reluctance to sign 'agreements to agree', such as the Convention on Biological Diversity or the Kyoto Protocol, frustrates the ability to address the problems.

Another important question is what kind of technology should be transferred? Although a great deal of effort has been expended on interpreting and implementing the language relevant to technology transfer contained in the UN Framework Convention on Climate Change[34] disagreement continues to take place over crucial details such as how such technology transfer should take place and the role of the law in facilitating those transfers. Nuclear energy may have certain benefits, but it has major drawbacks as well. Not even wind or solar energy is free from objection. Similar types of balancing are also required with respect to other environmental activities. The fact that these problems must be addressed, however, remains uncontroversial.[35]

---

31. This aspect is further developed by Keith Maskus. In Ch. 6, Professor Maskus reviews arguments for using compulsory licensing as a means of accelerating technology transfer, building industrial capacity and dealing with potential abuses of IPRs. Later, in Ch. 11, Tran Viet Dzung, Le Thi Nam Giang, and Nguyen Nhu Quynh provide a case-law analysis of Vietnam's experience with parallel trade, compulsory licenses and competition law flexibilities.
32. The December 2009 Copenhagen Fifteenth Conference of the Parties of the UNFCCC ended with very little and the Conference of the Parties (COP), 16th Sess., 29 Nov. to 10 Dec. 2010, Cancun, Mexico almost ended on a negative note.
33. Frederick M. Abbott, 'Innovation and Technology Transfer to Address Climate Change: Lessons from the Global Debate on Intellectual Property and Public Health', ICTSD, Issue Paper No. 24 (June 2009) (Abbott). <http://ictsd.org/i/publications/50454/>.
34. Kyoto Protocol to the United Nations Framework Convention on Climate Change, 10 Dec. 1997, 37 I.L.M. 22 (1998) (Kyoto Protocol).
35. See Hari Ososfsky's discussion in Ch. 8.

3 TREATY OBLIGATIONS GIVING EFFECT TO FUNDAMENTAL CONCEPTS

Basic human rights requirements and actual needs are accordingly driving technology transfer activities. There are, of course, a number of other factors that also come into play. Decolonization and its effects have its impact. More importantly, it may simply be a rational behaviour by increasing the basis for world trade. The international agreements also deal with technology transfer and so are several international organizations.

There are more than 80 international agreements and many additional programmes dealing with technology transfer.[36] Today the real question is not which international organizations are involved in technology transfer activities, but rather, which ones are not. Rules are contained in the TRIPS Agreement the CBD and the Framework Convention. Even if the agreements are independent instruments they still have a relationship that must be understood. The fact that so many entities seem to be working at cross-purposes creates a risk of conflict and a need for coordination. Frederick Abbott points out the political dilemma if the discussion may constantly shift forums and there is no one unit in control.[37] It is important to formulate a coherent approach with organizations complementing each other's actions within defined areas. An obvious risk with the multiple fora situation is that states abuse the system by forum shifting and/or forum proliferation. The latest counter development is the creation of an institution based on the Anti-Counterfeiting Trade Agreement[38] (ACTA), which may be seen as a way for developed countries to circumvent the growing influence of developing countries within the established institutions.

Another hurdle is to determine which states have an obligation under a particular international instrument and to which states such obligations are owed. Technical assistance in Article 67 TRIPS is aimed at developing countries (including LDCs) whereas the transfer of production technology in Article 66.2 TRIPS is only directed towards the LDCs. The 1992 Framework Convention on

---

36. Organizations involved in technology transfer include but are not limited to the World Bank and the regional development banks, World Health Organization (WHO), United Nations Conference on Trade and Development (UNCTAD), the International Labour Organization (ILO), the United Nations Industrial Development Organization (UNIDO), the United Nations Development Programme (UNDP), the Food and Agriculture Organization (FAO), the International Fund for Agricultural Development (IFAD), the Consultative Group on International Agricultural Research (CGIAR), the Enhanced Integrated Framework (EIF) and the UN Interagency Cluster on Trade and Productive Capacities.
37. *See* Abbott, fn. 32 *supra*.
38. Anti-Counterfeiting Trade Agreement (ACTA) dated 15 Nov. 2010 among Australia, Austria, Belgium, Bulgaria, Canada, Cyprus, the Czech Republic, Denmark, Estonia, Finland, France, Germany, Greece, Hungary, Ireland, Italy, Japan, the Republic of Korea, Latvia, Lithuania, Luxembourg, Malta, Morocco, the Netherlands, New Zealand, Poland, Portugal, Romania, Singapore, Slovakia, Slovenia, Spain, Sweden, Switzerland, the United Kingdom, the United Mexican States, the United States, and the European Union (ACTA). Final version, Subject to Legal Review. <www.ustr.gov/webfm_send/2379>.

Climate Change has a different approach and identifies in Annex 1, parties subject to specific obligations[39] and, in Annex II, countries with more far reaching obligations.[40] Yet in other provisions, the UNFCCC uses the terms 'developing country' and 'developed country', but does not indicate which parties fall within which categories. Similarly, the Convention on Biological Diversity anticipates a flow of technology transfer benefits aimed at 'developing countries', but does not define which parties qualify for this status.

It is important to understand these notions in order to determine rights and obligations. Perhaps one should not to be categorical. Depending on the situation and the international instrument, a country may actually fall in to different categories.

4 FRAGMENTATION

Accordingly, there are substantial references in international law to mandate forceful activities to support technological transfer from developed to developing countries. The question is whether and to what extent this mandate is used in real life. There are sufficient avenues for addressing technology transfer problems, but the available venues have multiplied and the international law system is now fragmented with a large number of fora.

Against this background, the latest development in the creation of supplementary enforcement provisions to the TRIPS agreement makes sense. Developed States with a clear IPR agenda have agreed that there will be hard core provisions in ACTA, a plurilateral agreement. ACTA has been described as an example of 'TRIPS-plus' law-making and its provisions are intended to be incorporated in future free trade agreements. This method is to be contrasted with the open and transparent process in the CBD and the Climate Change Framework Convention. The plurilateral or bilateral solutions to problems must be scrutinized from a global standpoint. Are they adding to the Millennium Goals or on the contrary counterproductive preventing improvements in the LDCs? In any event they demonstrate the increasing complexities in the multilateral discussions.

---

39. United Nations Framework Convention on Climate Change, FCCC/INFORMAL/84, opened for signature 9 May 1992, entered into force 21 Mar. 1994 (FCCC). Annex 1: Australia, Austria, Belarus, Belgium, Bulgaria, Canada, Croatia , Czech Republic, Denmark, European Economic Community, Estonia, Finland, France, Germany, Greece, Hungary, Iceland, Ireland, Italy, Japan, Latvia, Liechtenstein, Lithuania, Luxembourg, Monaco, Netherlands, New Zealand, Norway, Poland, Portugal, Romania, Russian Federation, Slovakia, Slovenia, Spain, Sweden, Switzerland, Turkey, Ukraine, United Kingdom of Great Britain and Northern Ireland, United States of America. <http://unfccc.int/resource/docs/convkp/conveng.pdf>.
40. *Ibid.*, Annex II (Australia, Austria, Belgium, Canada, Denmark, European Economic Community, Finland, France, Germany, Greece, Iceland, Ireland, Italy, Japan, Luxembourg, Netherlands, New Zealand, Norway, Portugal, Spain, Sweden, Switzerland, United Kingdom of Great Britain and Northern Ireland, United States of America).

Power struggles between developed and developing countries in these new fora challenge the success of the combined efforts to improve conditions for the developing world. The tendency towards forum shifting and creating new international institutions begs the question whether or not the international system is properly equipped to deal with different aspects of technology transfer.

It is therefore imperative to review the true legal status of different obligations of developed countries as well as flexibilities reserved for developing countries and also to investigate what has happened in terms of actual transfer since the time the reporting obligations under TRIPS became mandatory.

Chapter 2

TRIPS Article 66.2: Between Hard Law and Soft Law?

*Thu-Lang Tran-Wasescha & Xavier Groussot**

Article 66.2 of the Agreement on Trade-related Aspects of Intellectual Property Rights (TRIPS Agreement) of the World Trade Organization (WTO) is the subject of increased interest and attention, not only from the organization's Members but also from non-governmental organizations and in academic circles and not without some well-founded reasons. While it would have been interesting to know which parameters and framework conditions have permitted effective technology transfer between companies from developed countries, 'technology transfer' is, in general, understood as an issue of flows of knowledge and technology from developed country private companies or public entities to those in developing countries. The notion of the 'dissemination of technology' knowledge has been the subject of extensive debates in the context of conventions administered by, and work undertaken, in the World Intellectual Property Organization (WIPO), well before the publication of the so-called UNCTAD International Code of Conduct on the

* The views expressed are the personal responsibility of the authors.
The TRIPS Agreement is Annex 1C to the Marrakesh Agreement Establishing the World Trade Organization, done at Marrakesh 15 Apr. 1994, 33 I.L.M. 1125, 869 U.N.T.S. 299.

Hans Henrik Lidgard, Jeffery Atik & Tu Thanh Nguyen, *Sustainable Technology Transfer*, pp. 13–42.

Transfer of Technology.[1] The Code remains, however, an important attempt to deal with technology transfer. The fact that it is a piece of 'soft law', i.e. with no binding effect and of a voluntary character, is of direct interest to this Chapter. It is also testament to the complexity of the issue and the antagonistic approaches that existed. Endeavours have hitherto been multiplied, without achieving any actual operative results. Therefore, the inception of a binding provision in a binding multilateral treaty like the TRIPS Agreement has, understandably, raised great interest. Technology transfer continues to be a topical item on the agenda of many other fora, notably in the context of the Convention on Biological Diversity[2] (CBD) and its subsequent acts, or, more recently, the negotiations on climate change. This article will refer the reader to the analyses made by experts in this publication, notably those dealing with technology transfer in general and more specifically with patents, competition law and the CBD, including the Nagoya Protocol.[3] It will deliberately not address the issue of the relationship between the TRIPS Agreement and CBD.[4]

One issue emerging from the discussions in the International Symposium organized in Ho Chi Minh City in October 2010[5] is the weight to attribute to provisions such as Article 66.2 of the TRIPS Agreement, i.e. one with binding effects on its Members, accompanied by a relatively robust dispute settlement system. This contribution will focus on the legal nature and scope of Article 66.2 and the obligations contained therein, using as points of comparison other provisions of the TRIPS Agreement. Article 66.2 is one among a wide array of negotiated provisions, representing the various interests at stakes at the time of the Uruguay Round negotiations. Since its inception, a number of TRIPS provisions have been scrutinized by WTO Members (either in the Council for TRIPS or in the

1. ECE/UNCTC, Industrialisation and Trade of Developing Countries: Economic and Policy on the Participation of Transnational Corporations, Geneva – ECE/UNCTC Joint Unit Publication Series No. 2, 1984, Working Paper No. 14/Rev. 1. For a more detailed account of the Code of Conduct, see Surendra J. Patel, Pedro Roffe & Abdulqawi A. Yusuf (eds), *International Technology Transfer: The Origins and Aftermath of the United Nations Negotiations on a Draft Code of Conduct* (The Hague: Kluwer Law International, 2000).
2. The Convention on Biological Diversity (CBD), known informally as the Biodiversity Convention, is a legally binding treaty. The Convention was opened for signature at the Earth Summit in Rio de Janeiro on 5 Jun. 1992 and entered into force on 29 Dec. 1993, [1993] ATS 32 / 1760 U.N.T.S. 79 / 31 I.L.M. 818 (1992) <www.cbd.int/>.
3. Nagoya Protocol on Access to Genetic Resources and the Fair and Equitable Sharing of Benefits Arising from their Utilization to the Convention on Biological Diversity, Nagoya, 29 Oct. 2010, Annex I to Depositary Notification, Reference C.N.782.2010.TREATIES-1 ('Nagoya Protocol'), To be open for signature in New York from 2 Feb. 2011 until 1 Feb. 2012, (Nagoya Protocol). <http://treaties.un.org/doc/source/signature/2010/CN782E.pdf>.
4. In particular the question of whether the two treaties are compatible with each other. See <www.wto.org/english/tratop_e/trips_e/art27_3b_e.htm> for a brief overview and other information on relevant discussions and documents.
5. Conference held in Ho Chi Minh City, Vietnam, 19–21 Oct. 2010 'Sustainable Technology Transfer', a joint collaboration between Hanoi, HCMC, Lund, Nagoya & Suffolk Faculties of Law, as part of a twelve year-long Swedish project financed by the Swedish Development Agency (SIDA) under the title 'Strengthening Legal Education in Vietnam.'

Dispute Settlement Body (DSB)), other international governmental organizations, civil society or academicians. Like other international treaties, notably those administered by WIPO, the TRIPS Agreement is a set of mandatory, strongly exhortative and optional provisions. Some of those provisions, e.g. Article 66.2, have been subject to interpretation by ministerial conferences, WTO dispute settlement cases, or scrutiny by Members, NGOs and commentators. The attention paid to Article 66.2 – and other TRIPS provisions as well – is undoubtedly amplified by the fact that the WTO possesses a relatively swift and efficient mechanism for resolving disputes between governments. And, hence, expectations are higher.

At the Ho Chi Minh International Symposium, the debates on the CBD and TRIPS Article 66.2 were immersed in the dichotomy 'hard low-soft law'. This Chapter focuses on Article 66.2, which enshrines a strong obligation for developed country Members with, however, elements that may have been differently interpreted and, in any event, 'lost in translation into implementation'. Such 'loss in translation' may not be due to lack of goodwill from WTO Member governments' desire to implement it but, rather, more to the fundamentals underlying the negotiation and implementation of the provision such as: (1) the concept and purpose of technology transfers linked to intellectual property rights (IPRs), contrasted with the apparent contradiction between inalienable intangible property and free access to technology and knowledge, rights and obligations of IPR holders, competitors and society at large; and (2) the ways and means to implement the provisions of Article 66.2. The views expressed in some quarters are that Article 66.2 is a weak or 'softened' mandatory provision. In terms of results, it could as well – under certain views – be in a 'soft law' piece of text.[6] The ultimate objective of this article is to bring some clarity to the differences between 'soft law' and 'softened hard law', and to give some views on the potential for effective application of Article 66.2 in the future, provided there is a frank and straightforward dialogue between developed and least-developed countries. This article will first address the 'legislative' and 'judicial' roles of the WTO and the 'hierarchy' of texts and acts in those contexts. To better place Article 66.2 in the full picture of the complex 'hierarchy' of TRIPS commitments by Members, this article will examine the following: (1) the various gradations of TRIPS obligations, and their evolution during their implementation at national level, or interpretation by Ministerial Conferences and panels under the WTO Dispute Settlement Understanding (DSU); (2) the issue of 'flexibilities' or 'space for policy options' contained in the TRIPS Agreement, and (3) the Doha Ministerial Declaration on the TRIPS Agreement and Public Health,[7] as the first ministerial confirmation of TRIPS flexibilities leading to a waiver and to an amendment of the TRIPS Agreement which has been diversely commented on with regard to whether or not it is binding. This article will then deal with the concept of 'soft law' in international public law. In examining the relation between soft law and trade

6. See Peter Gottschalk, *Technology Transfer and Benefit Sharing under the Biodiversity Convention*, Ch. 9, *infra*.
7. See WTO Ministerial Conference Fourth Session Doha, 9–14 Nov. 2001WT/MIN (01)/DEC/2 41 I.L.M. (2002), ('Doha Declaration').

law, it will give a brief overview of the use of 'soft law' in the context of EU law and WTO law. In fact, parallels can be useful to the understanding of 'soft law' within the TRIPS Agreement. It finally focuses on Article 66.2, offering some comments on the background of the issue, on the current discussions in the Council for TRIPS, which helps delineate the perimeter of implementation, while keeping in mind the carefully negotiated language. In other words, how could a mandatory provision, falling short of the results expected by LDCs, because of wording reflecting some dilemmas pertaining to technology transfer, be rendered more operational?

1 THE WTO'S 'LEGISLATIVE' AND 'JUDICIAL' ROLES

1.1 Multilateral versus Plurilateral Participation in Treaties

The nomenclature of instruments used by the international legislator varies from binding treaties to softer instruments such as 'recommendations', 'resolutions', 'declarations', 'codes of conduct', 'guidelines', or 'best practices'. The primary source of WTO law is found in the Marrakesh Agreement and its annexes, Annex 1C being the TRIPS Agreement, and Annex 2 the Understanding on Rules and Procedures Governing the Settlement of Disputes (DSU). The WTO Agreement and its annexes form the foundation of WTO Members' specific rights and obligations, enforceable through its dispute settlement system. Other sources have been listed, primarily based on Article 38.1 of the International Court of Justice.[8] As pointed out by Cottier and Oesch:

> [t]his provision is generally regarded to express a complete statement of the sources of international law. Albeit it fails to address the fundamental principles of international law and derived sources such decisions of international organizations, authentic interpretation and the phenomenon of soft law (recommendations and resolutions) in the twilight between law and policy.[9]

Van den Bossche lists the following sources in addition to the fundamental WTO texts: GATT/WTO 'dispute settlement reports, acts of WTO bodies, agreements concluded in the context of the WTO, customary international law, other international agreements, subsequent practice of WTO Members, teaching of the most highly qualified publicists, and finally the negotiating history'.[10]

---

8. The Statute of the International Court of Justice is an annex to the 1945 Charter of the United Nations and forms an integral part thereof.
9. Emphasis added. Thomas Cottier & Matthias Oesch (eds), *International Trade Regulation – Law and Policy in the WTO, the European Union and Switzerland – Cases, Material and Comments* (Bern, Switzerland: Staempfli Publishers Inc., 2005), 109, (citing and quoting an excerpt from David Palmeter & Petros C. Mavroidis, 'The WTO Legal System: Sources of Law', 92 *Am J Int'l L* 398, 398–399 (1998).
10. Peter van den Bossche, *The Law and Policy of the World Trade Organization: Text, Cases and Materials*, 2nd edn (Cambridge, England: Cambridge University Press, 2008), 92.

In terms of the geographical coverage of binding treaties, the GATT and the WTO have so far taken a two-pronged approach: one prong being quite extensive, i.e. a single undertaking with full obligatory adherence to the agreements, and the other prong, much smaller, the so-called 'plurilateral' treaties such the Government Procurement Agreement,[11] participation in which is voluntary. Put in simpler terms, multilateralism in the WTO means full mandatory membership in a set of agreements, with varying degrees of binding provisions, tempered by special and differential treatment for developing countries and least-developed countries, such as exemption from implementation obligations during a determined period of time (transition periods). Multilateralism in WIPO and other UN and non-UN agencies is slightly different to the extent that a country which does not wish to join a multilateral treaty will not have to do so. Treaties in the WIPO are, in general, lawyerly drafted texts, with a relatively high degree of precision and clarity. This does not mean, however, that all provisions are void of ambiguity. In an in-between zone, there is a type of convention which, aiming at a worldwide membership is drafted in such a way as to embrace wide coverage of interests and will therefore have to take into account contradicting concerns. The CBD, negotiated in a relatively short span of time, in view of the ambitious Rio Summit calendar, may fall in this category. As a result some CBD wordings, in granting greater flexibility to contracting parties, may strip the main obligations of the CBD of its initial objectives and expectations.[12] In the WIPO, participation in a multilateral treaty may start with a small number of participants (the flexibility residing in the free decision to participate or not), which could, in time, result in a worldwide membership and hence a universal, solid, and operational platform. This has been the case of the two pillars of the IP system, the Paris Convention for the Protection of Industrial Property ('Paris Convention')[13] and the Berne Convention for the Protection of Literary and Artistic Works ('Berne Convention').[14] The TRIPS Agreement has been built upon the solid and more 'codified' foundations of the Paris and Berne Conventions, reinforced by two elements: it is part of a single undertaking, and it has a relatively swift and effective dispute settlement system.

Resolutions, decisions, declarations and recommendations are often used. Indeed, the exact legal nature of a text must be examined in its context before

11. Agreement on Government Procurement, contained in Proposal for a Council Decision Concerning the Conclusion of the Results of the Uruguay Round of Multilateral Trade Negotiations (1986–1994), COM(94)143 final/2, Brussels, 25 Apr. 1994. Entered into force, 1 Jan. 1994 (GPA).
12. While several provisions of the CBD (1992) have been diluted by escape wording such as 'where necessary', 'as appropriate', Art. 16 (Access to and Transfer of Technology) contains strong commitments, which made certain delegations issue declarations upon signature or ratification.
13. Paris Convention for the Protection of Industrial Property 20 Mar. 1883, as last revised at Stockholm 14 Jul. 1967, 828 U.N.T.S. 305 (Stockholm Act). <www.wipo.int/treaties/en/ip/paris/pdf/trtdocs_wo020.pdf>.
14. Berne Convention for the Protection of Literary and Artistic Works, 9 Sep. 1886, as last revised at Paris 24 Jul. 1971, 1161 U.N.T.S. 30 (Paris Act). <www.wipo.int/treaties/en/ip/berne/trtdocs_wo020.html>.

determining whether it is real 'soft law' or not. It is a well-accepted notion that soft law can build up and become solid pieces of a building, provided they are coherent and not mutually contradicting.

The recourse to instruments such as 'codes' at international level deserves some attention. In the WTO context, the WTO Agreement on Technical Barriers to Trade (TBT Agreement) is an interesting example – while not the only one – of a mix of voluntary practices, strong and softer binding rules.[15] The current TBT Agreement contains, in its Annex 3, a Code of Good Practice for the Preparation, Adoption and Application of Standards, which is 'open to acceptance by any standardizing body within the territory of a WTO Member, . . . '.[16] While acceptance of the code seems voluntary, Article 4 of the TBT Agreement requires Members to 'ensure that their central government standardizing bodies accept and comply with the Code of Good Practice . . . ' whilst they 'shall take such reasonable measures as may be available to them to ensure that local government and non-governmental standardizing bodies within their territories . . . accept and comply with this Code of Good Practice'.

1.2 Application and Interpretation

While not the main point of this Chapter, it is worthwhile to distinguish the concept of the primacy of international treaties obligations, and the notion of 'self-executing? or directly applicable treaties. The latter depends largely on each country's constitutional system, i.e. whether it has adopted a monist or dualist approach. The TRIPS Agreement has incorporated the Paris and Berne Conventions, and complements or supplements provisions of these two WIPO treaties ('Paris plus', 'Berne plus'). These three treaties contain a mix of private and public international law. Bodenhausen[17] extensively addresses the issues of primacy of international treaty over national law and of the 'self-executing' or directly applicable provisions of the Paris Convention. The comments made in the Bodenhausen Guide continue to apply to those provisions of the Paris Convention incorporated in the TRIPS Agreement, subject, however, to how a WTO Member

15. The WTO Agreement on Technical Barriers to Trade is part of the Final Act Embodying the Results of the Uruguay Round of Free Trade Negotiations, 15 Apr. 1994 1867 U.N.T.S. It was first developed in the Tokyo Round as a 'Standards Code'. The Tokyo Round of Multilateral Trade Negotiations begun in 1973 and was concluded in 1979 in Geneva. The resulting multilateral agreement was signed in 1979 by 32 Contracting Parties, setting forth rules for the preparation, adoption and application of technical regulations, standards and conformity assessment procedures. It is interesting to note that he Preamble of the current TBT Agreement refers, in its eighth recital, to the contribution which international standardization can make to the transfer of technology from developed to developing countries.
16. Paragraph B of Annex 3.
17. Georg Hendrik Christiaan Bodenhausen, *Guide to the Application of the Paris Convention for the Protection of Industrial Property* (Geneva: United International Bureaux for the Protection of Intellectual Property (BIRPI) 1968 – WIPO publication 611), 12–16.

has chosen to apply GATT/WTO law.[18] That said, to the extent a WTO Member applies the principle of direct application, there remain two questions: (1) what could, for example a private party as a national of a WTO Member claim as rights under a self-executing treaty in that Member?; (2) what could other WTO Members expect from that Member, including the enjoyment and enforcement of rights under TRIPS for their own nationals?

The two issues of primacy of international treaties and of direct applicability of TRIPS have been addressed in the context of the review of developed countries' legislation in 1996–1997. While some Members have indicated, either voluntarily, or in response to a question posed by another Member, that TRIPS and WIPO conventions were directly applicable, the great majority did not give any indication.[19] Notwithstanding, even in monist countries, direct application is a sensitive issue and depends, *inter alia*, upon the provision. For example, in Switzerland a provision must be very precise, and in particular, one in the IP area. Eventually, since the Swiss legislator will transpose the international rules into domestic law for transparency purposes, the issue may become less pressing but this may not hold true in other countries. Nguyen gives a useful panorama of the issue of effect of the TRIPS Agreement at the domestic law level, including a description of the United States' dualist approach.[20]

2 TRIPS PROVISIONS AND TRIPS FLEXIBILITIES

2.1 THE 'HIERARCHY' OF MEMBERS' OBLIGATIONS

The TRIPS Agreement totals 73 provisions. In fact there are more since the WIPO Conventions are also incorporated. They deal with obligations which vary from strong mandatory (shall) and strong hortatory commitments (should) to optional ones (may) regarding rights and obligations. This is not peculiar to the WTO agreements. Whatever category in which the TRIPS provisions belong, obligations for a Member are subject to the WTO Dispute Settlement System. This is on this point that the WTO provisions differ – in terms of weight – from others multilateral agreements. In the TRIPS/WTO agreements, the non-fulfilment of a mandatory obligation by a Member may trigger a complaint by another Member. We have identified – in terms of obligations or options – the following categories of

18. *Ibid.*
19. See IP/Q/SVN/1 (Slovenia), 2; IP/Q3/CHE/1 (Switzerland), 10–11. It is recalled that the review of legislation in the Council for TRIPS is a 'peer review': questions relating to a Member's legislation are posed by other Members, not by the WTO Secretariat. It was in this context that the issue of direct application was mentioned by these two countries to explain their own legal systems and respond to questions posed by others. For Switzerland, it was a question posed by another Member on a provision in the Swiss patent law on expropriation that triggered that country's response on the primacy of law and the direct applicability of the European Convention on Human Rights.
20. Tu Thanh Nguyen, *Competition Law, Technology Transfer and the TRIPS Agreement – Implications for Developing Countries* (Cheltenham: Edward Elgar Publishing Limited, 2010), 7–11.

provisions, although the frontiers between certain categories may not be clear in all instances:

1. The strict mandatory provisions, drafted as 'shall', which represent an important part of the agreement, obligating a WTO Member to provide protection or certain measures ('positive obligation') or to refrain from certain policies or acts ('negative obligation'). The provision may be worded in general terms – leaving Members some discretion to formulate provisions – or in a very precise wording. Failure to fulfil the obligation, e.g. through its inclusion in the national law, may result in the possibility of other Members to ask for a dispute settlement procedure.
2. The mandatory provisions, either positive or negative, can be worded in a direct way to address the matter in question, e.g. Article 33 on the duration of protection for a patent. This category is typically a technique used in private international law. These provisions necessarily imply that a Member provides such mandatory provisions in its legislation. Category 1 coincides with 2 in terms of substance. Comments on dispute settlement under 1 above also apply.
3. The mandatory 'protective' provision, i.e. a provision which prohibits a Member to require or force another Member to accord certain type or level of protection or to take certain measures. This is the case of Article 2.2, first sentence,[21] or of Article 24.6 on the continued use of generic terms in the area of geographical indications. This kind of protective provision acts as a 'shield' against any temptation from a Member to force, for example in the case of Article 24.6, another Member to roll-back the use of a geographical indication as a generic. This kind of protective 'shall provision' is also, typically, Article 6 on exhaustion.[22]
4. Mandatory provisions, the wording of which is ambiguous, or at least understood differently by Members. Typically this is the case of Article 23.4 which establishes, at first sight, a clear built-in agenda, but without any deadline, and is open to diverging views about whether the system of notification and registration of geographical for wines (and spirits) would be mandatory or voluntary, with legal binding effect or not.
5. Mandatory provisions providing for positive or negative obligations with some flexibility in terms of concrete implementation. In the area of enforcement, certain provisions impose strong obligations to Members to foresee certain measures, with however the freedom for competent authorities to have the authority to apply the measure or not. A panel has confirmed the discretion that national authorities have under Article 59.[23]

---

21. TRIPS Art. 2.2: 'Members may, but shall not be obliged to, implement in their law more extensive protection than is required by this Agreement. . . . '
22. See also examination of Exhaustion under Art. 6 below.
23. China – Measures Affecting the Protection and Enforcement of Intellectual Property Rights, WT/DS362/R (2009).

6. Provision of 'hortatory' character, drafted as 'should', namely Article 7.[24]
7. Permissive or optional provisions granting Members the freedom to Members to adopt or not certain levels of protection, or to exclude from protection certain areas. The permissive or optional provision is important. For example, Article 27.3(b) protects the Member opting not to provide patent protection to plants and animals inventions from being attacked before a panel. In the same category, Members have certain discretion in providing exceptions to exclusive rights, but the adoption of such exceptions is subject to certain strict criteria. This is the case of Article 30, which gives some freedom of manoeuvre to countries. However, measures taken under a 'may provision' should be compatible with the TRIPS Agreement: this is the case, for example, of Article 8(1), which allows Members to take measures in the public health area, 'provided that such measures are consistent with [the TRIPS] Agreement'.[24]

The interpretation made by Ministers in the case of the Doha Declaration or made by panels may reinforce a 'may provision', or a 'hortat provision', or soften a mandatory provision.

Some provisions do not address all elements as they should have because it was impossible for negotiators to compromise on those aspects. This is the case of Article 30 on exceptions to exclusive rights. However, the most eloquent example is found in Article 31, which does not actually prescribe grounds for 'use without authorization', but only the conditions for their application. The few references to certain grounds for granting compulsory licensing mainly resulted from the fact that certain types of licenses require specific conditions of use, and have, therefore, to be explicitly mentioned in Article 31. It may well be that this was not clear enough for some Members and quarters, provoking the crisis of IP-related issues in the context of public health, which necessitated a clarification by Ministers in Doha in 2001.[25]

---

24. See also examination of certain TRIPS provisions and *infra*.
25. See recital 5 of the Doha Declaration:

> 5. Accordingly and in the light of paragraph 4 above, while maintaining our commitments in the TRIPS Agreement, we recognize that these flexibilities include the following:
>
> (a) In applying the customary rules of interpretation of public international law, each provision of the TRIPS Agreement shall be read in the light of the object and purpose of the Agreement as expressed, in particular, in its objectives and principles.
> (b) Each Member has the right to grant compulsory licences and the freedom to determine the grounds upon which such licences are granted.
> (c) Each Member has the right to determine what constitutes a national emergency or other circumstances of extreme urgency, it being understood that public health crises, including those relating to HIV/AIDS, tuberculosis, malaria and other epidemics, can represent a national emergency or other circumstances of extreme urgency.
> (d) The effect of the provisions in the TRIPS Agreement that are relevant to the exhaustion of intellectual property rights is to leave each Member free to establish its own regime for such exhaustion without challenge, subject to the MFN and national treatment provisions of Articles 3 and 4.

The TRIPS provisions, when interpreted, may be softened or hardened by external soft law.[26] Provisions may also be softened by contradicting or mutually nullifying elements which compose them. While not falling in the category of soft law, Article 66.2 – and several mandatory provisions of the CBD as well – has raised concerns because of its little operability, making it be compared to some de facto 'soft law'. The example of the CBD provisions on technology transfer is interesting. Developing countries, disappointed by the 'soft law' emanating from the Nagoya negotiations, are trying to transpose the debate and negotiations into the WTO/TRIPS framework.[27]

2.2 Flexibilities and Constructive Ambiguity

Before addressing the interpretation of hard law contained in a few dispute settlement cases in the area of TRIPS, it seems indispensable to make a distinction between 'constructive ambiguity' and 'flexibilities'. They are not synonyms. 'Constructive ambiguity' is positive as long as it permits Members to accept a proposed wording and interpret it as going in the direction that fits their interests. It remains positive as long as the interpretation is not challenged or has not given rise to differences at domestic level and between Members. Flexibility would stand at a higher level: it should in general be worded in a clear way. It implies a *marge de manoeuvre* to achieve certain policy options. In any event, constructive ambiguity should ideally not be used by a Member to avail itself of flexibility, which in principle, clearly derives from the text. In case of disputes, the constructive ambiguity may lead to an interpretation unfavourable to the complaining Member.

The wide gamut of categories of provisions, with sometimes constructively ambiguous wording or seemingly contradicting wording within the same corpus, do offer flexibility to countries to decide on one policy option or another. Article 6 on exhaustion of rights, which seems quite clear, in spite of the fears of some developing countries about its exact meaning and the necessity of a Ministerial Declaration, is a good example. A Member free to opt for international, regional or national exhaustion,[28] depending on how it envisions its short, medium and long-term strategy, on its domestic framework – size; low, middle or high income – and on the neighbouing environment – competitive neighbours and size of their markets.

---

26. See s. 5 *infra*.
27. See WTO TN/C/W/59, of 19 Apr. 2011, Draft Decision to Enhance Mutual Supportiveness between the TRIPS Agreement and the Convention on Biological Diversity, Communication from Brazil, China, Colombia, Ecuador, India, Indonesia, Peru, Thailand, the ACP Group, and the African Group.
28. While 'regional' exhaustion is a type of 'international' exhaustion, permitting free circulation of IP-related goods, it is limited to the countries of the region concerned; IP-related goods coming from countries not members of that region are generally governed by the regime of 'national' exhaustion.

2.3 BALANCING INTERESTS

The essence of IP law lies in the successful balancing of the interests involved. This is particularly the case with patents for inventions, which concentrate all the elements representing the interests at stake. If adequately taken into account, we could have a perfect triangle: right holders-competitors-public at large. If the right holder obtains excessive rights which unbalance competition, the pendulum will bounce back strongly. We have been watching this pendulum effect, culminating with the Doha Declaration. If the pendulum now swings in the other direction, there would be a possible deterring effect on innovation and transparent flows of information on new technologies.

2.3.1 The case of TRIPS-public health

The example of TRIPS-public health is interesting in many respects. It reflects the issues addressed above, flexibilities, constructive ambiguity and the balance between interests. It is also a good example of the hierarchy of texts contained in, and resulting from, the Doha Declaration. The Declaration is a mix of interpretative recitals and instructions. The decisions resulting from the negotiations under the Doha Declaration, (i.e. the waiver[29] and the TRIPS amendment[30]), which are hard law, are accompanied by Chairman's statements on 'key shared understandings of Members', e.g. on diversion prevention, including a list of best practices to avoid such diversion, on manufacturing capacities, on review by the TRIPS Council, and on a list of Members opting out of the so-called 'Paragraph 6 system'. What is the legal nature of such a statement? Views diverge. The footnote giving some indication about the statement has been amended in light of the differing views on the status of the statements. In any event, the statements – from some insiders' views – have helped in forming the consensus towards the conclusion of the negotiations, while other insiders tend to consider these statements as having, in some way, altered the waiver/amendment negotiated.

3 OVERVIEW OF PROVISIONS RELATED TO PATENTS AND DSU

Article 6

For the purposes of dispute settlement under this Agreement, subject to the provisions of Articles 3 and 4 nothing in this Agreement shall be used to address the issue of the exhaustion of intellectual property rights.

29. For the 'waiver', see WT/L/540 and Corr.1 ('Implementation of Paragraph 6 of the Doha Declaration on the TRIPS Agreement and Public Health').
30. For the TRIPS amendment (or 'Protocol'), see WT/L/641 ('Amendment of the TRIPS Agreement').

This clear 'protective' mandatory provision has provoked heated debates, which are albeit far from finished, on the complete freedom of Members to choose the regime it deems appropriate to its public policy and strategy. While the issue of exhaustion has not been challenged under the DSU, the Doha Declaration on TRIPS and Public Health confirms Members' freedom, and in doing so, has considerably eased the political tension preceding the Doha Conference.

> ***Article 7***
> The protection and enforcement of intellectual property rights should contribute to the promotion of technological innovation and to the transfer and dissemination of technology, to the mutual advantage of producers and users of technological knowledge and in a manner conducive to social and economic welfare, and to a balance of rights and obligations.

This hortatory provision – the only one in TRIPS – while not fully operative as the various specific provisions that follow in each section deal specifically with the substantive norms, is an objective traditionally directed at one of the most solid category of intellectual property right – patents for inventions. It has extensively served as support material for the parties to the disputes to make their cases and for the panel to examine the claims. This was, in particular, the case for Article 30 in *Canada – Patent Protection of Pharmaceutical Products*.[31] This role was later confirmed by the Doha Declaration. Placed at a higher level than the preamble for purposes of interpretation, it has gained 'weight' over time, without indeed becoming hard law.[32] This is not a negligible achievement in terms of the development of international law.

> ***Article 8***
> 1. Members may, in formulating or amending their laws and regulations, adopt measures necessary to protect public health and nutrition, and to promote the public interest in sectors of vital importance to their socio-economic and technological development, provided that such measures are consistent with the provisions of this Agreement.
> 2. Appropriate measures, provided that they are consistent with the provisions of this Agreement, may be needed to prevent the abuse of intellectual property rights by right holders or the resort to practices which unreasonably restrain trade or adversely affect the international transfer of technology.

These 'may' provisions of TRIPS were heavily negotiated, in particular paragraph 1. It serves the purpose of avoiding a situation where, a Member taking a measure in public health may be attacked under the Dispute Settlement system. The proviso is important, if the measure is not in conformity with the pertinent TRIPS provisions, such measure could be challenged. Paragraph 2, which is linked to Article 40, concerns voluntary licenses. However, its potential for further development and use – again as a support for claims and interpretation – may attract greater attention

31. WT/ DS114/R (WTO Panel Adoption 7 Apr. 2000).
32. See Doha Declaration, recital 5(a).

in the future as developing countries are getting aware of the possibility of using compulsory licensing for violation of anti-trust law. Article 8 was also extensively examined in the *Canada Patent Protection of Pharmaceutical Products* Case. The remarks made above on Article 7 and the Doha Declaration also apply.[33]

> ***Article 30***
> Members may provide limited exceptions to the exclusive rights conferred by a patent, provided that such exceptions do not unreasonably conflict with a normal exploitation of the patent and do not unreasonably prejudice the legitimate interests of the patent owner, taking account of the legitimate interests of third parties.

This 'may' provision gives Members some freedom to determine the exceptions, such as private use or use for teaching purposes. However, as negotiators were unable to agree on an exhaustive list of exceptions, the freedom was accompanied by some safeguards, namely the three-step test, presaged in the Berne Convention, and examined by a panel in a landmark case between the EU and the US in the area of copyright.[34] Article 30 was subjected to detailed analysis in the *Canada – Patent Protection of Pharmaceutical Products* case.[35] The panel's conclusion – a landmark one too – may be considered as a major development in the sensitive debate between the research-based pharmaceutical industry and the generic one. While in some quarters the view may be that the panel decision weakened patentee's exclusive rights, it seems to have eased the tension and political sensitivity regarding TRIPS and access to medicines and to correspond to a trend in many countries, including developed ones, towards the use of generics on grounds alien to IP protection alone, e.g. alleviation of public health burdens and costs.

> ***Article 59***
> Without prejudice to other rights of action open to the right holder and subject to the right of the defendant to seek review by a judicial authority, competent authorities shall *have the authority* to order the destruction or disposal of infringing goods in accordance with the principles set out in Article 46. In regard to counterfeit trademark goods, the authorities *shall not allow* the re-exportation of the infringing goods in an unaltered state or subject them to a different customs procedure, other than in exceptional circumstances.[36]

The mandatory or prescriptive provision, softened by the authority granted to courts and administrations to decide to adopt or not the measures was closely examined by a panel in *China – Measures Affecting the Protection and enforcement of Intellectual Property Rights.*[37] It confirms the flexibilities afforded at the domestic authorities' level.

---

33. Recital 5(a).
34. WT/DS160/R (WTO Panel Adoption 27 Jul. 2000).
35. WT/DS/114/R (WTO Panel Adoption on 7 Apr. 2000).
36. Emphasis added.
37. WT/ DS362/R (WTO Panel Adoption 20 Mar. 2009).

4 'SOFT LAW' AND TRADE LAW

An understanding of 'soft law' within the general context of trade law is essential if one wishes to cope with its potential effects in the area of the TRIPS Agreement. This section first considers the definition of 'soft law' in international/European trade law. It then assesses the impact of 'soft law' in the WTO context. In that respect, two WTO cases are analysed in light of 'soft law' and their possible repercussions on the interpretation of the TRIPS Agreement.

4.1 Defining Soft Law in the Context of Trade Law

According to Senden, soft law is formed by 'rules of conduct that are laid down in instruments which have not been attributed a legally binding force as such, but nevertheless may have certain indirect legal effects that are carried out and may produce practical effects'.[38] The idiom 'soft law' relates to quasi-legal instruments which do not have any legally binding force, or whose binding force is somewhat 'weaker' than the binding force of traditional law, often differentiated with soft law by being named to as 'hard law'.

Article 66.2 TRIPS may be viewed by some as an example of soft law. It is found within TRIPS – and while it creates some obligation, it is unlikely to be enforceable in dispute settlement. Besides this type of (internal) soft law, one may identify another type of soft law that fits a wider definition. This is the situation for instance when the DSU reaches outside of the WTO Agreements to borrow concepts developed in other international law regimes as we will develop in this section (external soft law). Frequently, the term 'soft law' (both internal and external) is connected with international law, even if more recently it has been relocated to other branches of domestic law as well. More particularly, in the area of international law, soft law is particularly associated with trade law as it is, in our view, clearly exemplified by the WTO and EU legal orders.

In EU law, it appears clear that soft law instruments are heavily relied on and even institutionalized by Article 288 of the Treaty on the Functioning of the European Union (TFEU). Codes of conduct, guidelines, resolutions and programmes like the State Aids Action Program (SAAP) are not binding per se but may be useful for the interpretation of binding acts and will sometimes bind the institution that produced the document. As explained by Senden, a code of conduct may be assessed as a sort of pledge on the part of an EU political institution to adopt a certain approach and departure from that commitment. Departure from it may infringe a general principle of EU law, e.g. that of legitimate expectations. Similarly, guidelines issued by the Commission in the State aids and competition law fields are considered to be an expression of the interpretation which the

38. Linda Senden, *Soft Law in European Community Law* (Oxford, England: Hart Publishing, 2004), 112. See also, from the same author, 'Soft Law and Its Implications for Institutional Balance in the EC', *Utrecht Law Review* 1, no. 2 (December 2005), 79, 81.

Commission will apply to particular circumstances and which binds the Commission in its assessment of alleged aid. Soft law instruments may also be seen as a way of implementing the duty of cooperation incumbent on the institutions and Member States by virtue of Article 4(3) of the Treaty on the European Union (TEU).[39] Furthermore, the European Convention on the Protection of Human Rights and Fundamental Freedoms,[40] which became binding on the EU and its institutions December 2009, offers an interesting example of a soft law instrument used by the Court of Justice before the Treaty of Lisbon's entry into force.[41]

The relationship between the EU and WTO legal orders is also of interest for understanding the concept of soft law. For instance, in certain cases, the Court of Justice of the European Union (CJEU) has referred to decisions of WTO dispute settlement bodies as tools for the interpretation of relevant agreements and thus also for Union legal acts.[42] So it may be said that WTO rules are, to a certain extent, used as soft law instruments in the CJEU case law. Generally speaking, the WTO rules do not boast direct effect – i.e. cannot be relied on by individuals to challenge the legality of measures adopted by Union institutions – in the EU legal order unless the Union has intended to implement a particular obligation in the context of the WTO. It is in the *Portugal v. Council* case that the CJEU articulated most properly its reasoning on their direct effect.[43] The operative part of the ruling is worth quoting (paragraphs 42–47):

> As regards, more particularly, the application of the WTO agreements in the Community legal order, it must be noted that, according to its preamble, the agreement establishing the WTO, including the annexes, is still founded, like GATT 1947, on the principle of negotiations with a view to 'entering into reciprocal and mutually advantageous arrangements' and is thus distinguished, from the viewpoint of the Community, from the agreements concluded between the Community and non-member countries which introduce a certain asymmetry of obligations, or create special relations of integration with the Community, such as the agreement which the Court was required to interpret in *Kupferberg*.[44]

39. *Ibid.*, 352–356 and 442–446.
40. OJ 2000, C 364/1, Proclaimed by the Member States at the Nice European Council Summit, 18 Dec. 2000, Art. 6(2) TFEU (ex Art. 6(2) TEU), and Protocol No. 8 Relating to Art. 6(2) of the Treaty on the Accession of the Union to the European Convention on the Protection of Human Rights and Fundamental Freedoms. See, e.g. Case C-540/03, *European Parliament v. Commission*, 27 Jun. 2006, [2006] ECR I-5769 (Grand Chamber) para. 38.
41. See Case C-144/04, *Mangold v. Helm*, 22 Nov. 2005 [2005] ECR I-9981(Grand Chamber), and Case C-555/07, *Kücükdeveci v. Swedex GmbH & Co. KG*, 19 Jan. 2010, [2010] ECR, (Grand Chamber) n.y.p.
42. Case C-260/08, *Bundesfinanzdirektion West v. HEKO Industrieerzeugnisse GmbH*, 10 Dec. 2009, [2009] ECR I-11571, para. 22.
43. Case C-149/96, *Portugal v. Council*, 23 Nov. 1999 [1999] ECR I-8395.
44. Reference to Case 104/81, *Hauptzollamt Mainz v. Kupferberg & Cie KG A.A.*, 26 Oct. 1982 [1982] ECR 3641.

> It is common ground, moreover, that some of the contracting parties, which are among the most important commercial partners of the Community, have concluded from the subject-matter and purpose of the WTO agreements that they are not among the rules applicable by their judicial organs when reviewing the legality of their rules of domestic law. Admittedly, the fact that the courts of one of the parties consider that some of the provisions of the agreement concluded by the Community are of direct application whereas the courts of the other party do not recognise such direct application is not in itself such as to constitute a lack of reciprocity in the implementation of the agreement (*Kupferberg,* paragraph 18).
>
> However, the lack of reciprocity in that regard on the part of the Community's trading partners, in relation to the WTO agreements which are based on reciprocal and mutually advantageous arrangements and which must *ipso facto* be distinguished from agreements concluded by the Community, referred to in paragraph 42 of the present judgment, may lead to disuniform application of the WTO rules. To accept that the role of ensuring that those rules comply with Community law devolves directly on the Community judicature would deprive the legislative or executive organs of the Community of the scope for manoeuvre enjoyed by their counterparts in the Community's trading partners. It follows from all those considerations that, having regard to their nature and structure, the WTO agreements are not in principle among the rules in the light of which the Court is to review the legality of measures adopted by the Community institutions.

The CJEU's reasoning is based primarily on the so-called 'reciprocity argument' that focuses on determining the nature and broad logic of the agreement. The condition that a provision be unconditional and sufficiently precise is simply a standard to assess whether a judge is able to identify the content of the rule he is being asked to apply. However, there is another condition, the nature and broad logic of the agreement.[45] If direct effect had been granted by the Court of Justice, it would have restricted the scope of manoeuvre enjoyed by EU institutions in the WTO agreement vis-à-vis its trade partners. Therefore it may also be said that it is the bearing of direct effect on EU political institutions which is of importance in *Portugal v. Council.*[46]

The CJEU has constantly ruled that the WTO agreements are not, as a matter of principle, among the rules in which it is to review the legality of measure adopted by Union institutions. As lucidly explained by Rosas and Armati, the WTO case law of the CJEU appears to be based almost entirely on the particular nature of the WTO dispute settlement system which provides for three optional

45. See Case C-308/06, *The Queen, ex rel. International Association of Independent Tanker Owners and others v. Secretary of State for Transport,* 3 Jun. 2008, [2008] ECR I-4057 (Grand Chamber). In that case, the nature and logic of the UN Convention on the Law of the Sea prevented the Court from assessing the validity of a Union measure in light of that Convention (para. 65).
46. Piet Eeckhout, *External Relations of the European Union – Legal and Constitutional Foundations* (Oxford: Oxford University Press 2004), 305–306.

means of implementation: full implementation of a DSB's recommendation (which, according to DSU, should be preferred), compensation (temporary) and tolerance of trade sanctions (suspension of concessions).[47] As a result, allowing Member States or private parties to invoke the substantive rules of WTO directly before national courts would imply direct implementation on the EU side, while the Union political institutions as well as the trade partners might prefer to choose an alternative method, e.g. compensation.[48]

As posited by Rosas and Armati:

> The case law is therefore not based on any hostile attitude towards direct effect but is concerned rather with recognizing that international agreements are concluded outside the constraints of the EU legal order, with third states or other international organizations . . . [the judge] must ask whether the nature and broad logic lends itself to being applied in the situation before him. This will often depend on whether the agreement provides for individual rights, but also may be influenced by other considerations, for example, of reciprocity. On that basis, non-discrimination provisions in a number of bilateral association agreements and other cooperation agreements have been successfully invoked by individuals.[49]

Therefore, it is not surprising that, under CJEU case law, it is only where the Union has intended to implement a particular obligation assumed in the context of the WTO, or where the Union measure refers expressly to the precise provisions of the WTO agreements, that it is left to the Court to review the legality of the EU measure in question in light of WTO rules.[50]

4.2 Soft Law and Theory in the WTO Context

It has often been maintained that soft law has a tendency, over time, to harden into international legal obligations.[51] In that sense, soft law may also serve as a foundation for the evolution of binding international custom, which is accepted as a rule of international law. In the end, soft law norms act as precursor to hard law.[52] This effect of soft law is clearly visible in the WTO law context. Interestingly, the

47. Alan Rosas & Lorna Armati, *EU Constitutional Law: An Introduction* (London: Hart Publishing, 2010), 70–71. See e.g., Art. 22(1) DSU.
48. *Ibid.*
49. *Ibid.*
50. See Case C-93/02 P, *Biret International SA v. Council*, 30 Sep. 2003, [2003] ECR I-10497 (Full Court), para. 53.
51. Kenneth W. Abbott & Duncan Snidal, 'Hard and Soft Law in International Governance' 54 *Int'l Organ.*, 421, 421–422 (2000); see also Christine M. Chinkin, 'The Challenge of Soft Law: Development and Change in International Law', 38 *Int'l & Comp L Q.* 850, 856–859 (1989).
52. See Mary E. Footer 'The Role of Soft Law Norms in Reconciling the Antinomies of WTO Law', *Society of International Economic Law (SIEL) Inaugural Conference 2008 paper.* <http://ssrn.com/abstract=1159929>.

disputes under the DSU offer exemplifications of the complex interaction between soft law and hard law. In that sense, three types of relationship may be identified in the WTO case law:[53]

- Soft Law to interact with Hard Law
- Hard Law to interact with Soft Law
- Soft Law to interact with Soft Law.

According to Guzman and Meyer, there are four explanations for soft law in the international context: Soft law as coordinating device, loss avoidance theory, delegation theory and international common law theory.[54] These two authors, taking the trade rule in WTO as an example, notably introduce the concept of international common law (ICL). They define the ICL as a:

> nonbinding gloss that international institutions, such as international tribunals, put on binding legal rules. The theory of ICL is based on the observation that, except occasionally with respect to the facts and parties to the dispute before it, the decisions of international tribunals are non-binding interpretations of binding legal rules. States grant institutions the authority to make ICL as a way around the requirement that states must consent in order to be bound by legal rules. ICL affects all states subject to the underlying rule, regardless of whether they have consented to the creation of the ICL. As such, ICL provides cooperation-minded states with the opportunity to deepen cooperation in exchange for surrendering some measure of control over legal rules.[55]

This theory is of relevance when looking more closely at the WTO case law which has developed under the DSU.

In that respect, it is well-known that the main objective of the WTO is to promote trade and economic development. WTO disputes are governed by the DSU, which establishes a multi-tier system of review.[56] That review consists, most notably, of a panel of legal and technical experts that review the challenged conduct, and the WTO Appellate Body, which is empowered to review the panel's legal conclusions.[57] Panel reports and Appellate Body decisions are adopted by the

53. See Gregory Shaffer &Mark A. Pollack, 'Hard vs. Soft Law: Alternatives, Complements, and Antagonists in International Governance', 94 *Minn L Rev.* 706 (2010). According to the authors, interaction of hard law and soft law depends on the respective powers of the key players. In addition, hard law and soft law sometimes interact in a complementary and mutually reinforcing evolutionary fashion but only under certain circumstances. For them, in the presence of distributive conflicts among states, states and other actors may strategically use different hard and soft law instruments to advance their respective claims in the international arena.
54. Andrew T. Guzman & Timothy Meyer, 'International Soft Law', 2 *J. Legal Analysis* 171 (2010).
55. *Ibid.*, 171–172.
56. *Ibid.*, 215–216.
57. *Ibid.* The effect was a dramatic increase in the number of disputes adjudicated under the DSU as measured against the GATT's dispute resolution provisions. While 452 complaints were filed during the forty-six year tenure of the GATT, 279 complaints were filed in the first eight years of the WTO.

DSB. The WTO Appellate Body has played an important role in defining and thus clarifying WTO obligations. For instance, in the US *Shrimp Turtle* case,[58] the Appellate Body was asked to build up a framework for when a State may not rely on derogations under Article XX of the General Agreement on Tariffs and Trade[59] (GATT). This case, as we shall examine soon in more detail, is essential to understand the impact of soft law in the WTO legal order.

Also, it is important to keep in mind that the DSU may constitute a notable move toward hard law within the multilateral trading system. As put by Abbott and Snidal:

> [T]he virtues of hard law include the ability to reduce transaction costs, strengthen the credibility of commitments, and resolve problems associated with incomplete contracts. Indeed, it appears now clear that non-compliance with the treaty can be challenged through the WTO's effective dispute settlement system, in which rulings by WTO panels and Appellate Body are backed up by the threat of trade sanctions. The enforcement mechanism of WTO enshrines a difficult dilemma. . . . If enforcement is too harsh, states will comply with trade rules even in the face of high economic and political costs, and general support for liberalization is likely to decline. On the other hand, if enforcement is too lax, states will cheat, leading to a different dynamic that could similarly undermine the system.[60]

It is worth noting that the effect of moving intellectual property protection rules into the WTO is the creation of a capability to enforce them through a strong dispute settlement system. The panels and Appellate Body have always emphasized that their main duty is to interpret the WTO agreements. Yet, this interpretive task is not realized in 'clinical isolation' from other fields of international law.[61] This raises the legitimate query of how adjudicating bodies will operate in response to a claim that, for example, a TRIPS Agreement's obligation is incompatible with the treaty obligations and soft law norms of other international law regimes. Indeed, the TRIPS Agreement unequivocally enshrines certain provisions established in other international treaties regarding intellectual property rights, e.g. the Berne Convention and the Paris Convention. These 'external' rules of public

58. WTO Appellate Body Report on United States – Import Prohibition of Certain Shrimp and Shrimp Products, WTO Doc. WT/DS58/AB/R (8 Oct. 1998). Appellate Body Report, United States – Import Prohibition of Certain Shrimp and Shrimp Products, paras 79–91, WT/DS58/AB/R (12 Oct. 1998). <http://docsonline.wto.org/DDFDocuments/t/WT/DS/58ABR.doc>.
59. General Agreement on Tariffs and Trade, 30 Oct. 1947, 55 U.N.T.S. 194.
60. Abbot & Snidal, fn. 51, *supra*, the authors pinpoint the need for flexibility in international trade law as well as the danger that the new DSU system.
61. See Gabriel Marceau, 'A Call for Coherence in International Law: Praises for the Prohibition against "Clinical Isolation" in WTO Dispute Settlement', 33 *J World Trade* 87 (1999); and Joost Pauwelyn, 'The Role of Public International Law in the WTO: How Far Can We Go?', 95 *Am J Int'l L* 535, 577 (2001). See WTO Appellate Body Report on United States – Standards for Reformulated Gasoline, at 17, WTO Doc. WT/DS2/AB/R (29 Apr. 1996) (stating that WTO Agreements 'are not to be read in clinical isolation from public international law').

international law thereby become an integral part of the WTO treaty and claims under these rules can, thus, be enforced before a WTO panel.[62] It is recalled that the 'compatibility' between the incorporated WIPO conventions and the TRIPS Agreement was a highly contentious negotiating issue in the Uruguay Round negotiations. It was therefore not without reason that negotiators from both developed and developing countries wanted to ensure that in the event of unclear, ambiguous or seemingly contradictory readings, the WIPO treaties rights and obligations would not be endangered by the TRIPs additions and complements. Article 2.2 reads:

> Nothing in Parts I to IV of this Agreement shall derogate from existing obligations that Members may have to each other under the Paris Convention, the Berne Convention, the Rome Convention and the Treaty on Intellectual Property in Respect of Integrated Circuits.

This clearly aims at preserving the integrity and autonomy of WIPO Conventions, not only in light of the current TRIPS provisions as such, but also in the event of future interpretation by the WTO DSB of WIPO conventions as incorporated.

WIPO incorporated conventions are systematically examined by panels in the cases involving TRIPS. For example, Article 9.2 (on the three-step test) of the Berne Convention was key to the examination of TRIPS Article 13 on limited exceptions to exclusive rights in the *US – Copyright ACT* case.[63] It was quite noteworthy that the panel did not limit itself to an examination of the Berne Convention but further examined other WIPO treaties, the WCT and WPTT, (which were not incorporated into the TRIPS Agreement), as well as WIPO's Tunis Model Law on Copyright.[64] This is testament to the point made that for DSU purposes international treaties outside the purview of the WTO as well as soft law like model provisions are to be considered, should their consideration by the panel be pertinent to the merits of a case. However, such examination would not be possible if one of the parties disagrees.

The TRIPS Agreement mentions, *inter alia*, international technology transfer promotion. The TRIPS Agreement relies on the argument that increased technological capacity fosters development.[65] In that respect, the TRIPS Agreement codified the rights and obligations of Members in light of a balance

62. Non-WTO rules as reference material to interpret WTO provisions Arts 31 and 32 of the Vienna Convention on the Law of Treaties (explicitly incorporated in DSU Art. 3.2), in turn, refer to non-WTO rules for purposes of interpreting WTO provisions.
63. WT/DS/160R. See also, respectively, for the implementation of the recommendation, the award of the Arbitrators WT/DS160/ARB25/1, and the Mutually Satisfactory Temporary Arrangement WT/DS160/2.
64. WT/DS/160R, 133 and 132.
65. The fifth recital of the TRIPS Agreement preamble recognizes the underlying public policy objectives of national systems for the protection of intellectual property, including development of technologies objectives. Sixth recital emphasizes the need for 'maximum flexibility' for LDC Members.

between the protection of private IPRs and socio-economic and technological development.[66] It should also be noted that the use of soft law human rights norms in relation to WTO law has increased in the recent years. Already, in August 2000, the Sub-Commission on the Promotion and Protection of Human Rights adopted Resolution 2000/7 on Intellectual Property Rights and Human Rights. This resolution took a negative approach to TRIPS as it highlighted that 'actual or potential conflicts exist between the implementation of the TRIPS Agreement and the realization of economic, social and cultural rights'.[67] Generally speaking, it appears true that post-TRIPs intellectual property law making in the biodiversity, public health, and human rights regimes has mostly been of the soft law variety. Although soft law instruments are non-binding, they can be used as parameters for voluntary government conduct or templates for national legislation. However, soft law human rights norms may have an effect only with regard to principles, norms, and rules that do not conflict with TRIPS. In contrast, TRIPs-inconsistent soft law, voluntary government action or implementation into national laws would create a risk of exposure to WTO dispute settlement proceedings and WTO trade sanctions. WTO dispute settlement panels will probably not accept argumentation based on the use of soft law as derogation to States' intellectual property obligations.[68] As expressed by Marceau, '[F]or a conflict to exist between a WTO provision and a provision of a human rights treaty, evidence must be put forward that the WTO mandates or prohibits an action that a human rights treaty conversely prohibits or mandates. Such situations would be rare'.[69] It is unlikely that claims alleging an infringement of TRIPS, on the basis of soft law human rights instruments are successful.

4.2.1 *Shrimp/Turtle* Case

The possibility that WTO jurists might use soft law to interpret TRIPS and other WTO agreements was raised by the Appellate Body in the *Shrimp/Turtle* case, though not per se, a TRIPS case.[70] An interpretation in good faith in light of the 'ordinary meaning' of a term may permit for a reference to non-WTO rules. This was indeed realized by the Appellate Body in *US – Shrimp* when interpreting the term 'exhaustible natural resources' in GATT Article XX(g). To construe these words, reference was made to a dictionary, but also to certain Multilateral

66. Jerome H. Reichmann, 'The TRIPS Agreement Comes of Age: Conflict or Cooperation with the Developing Countries?', 32 *Case Wt Res J Int'l L* 441 (2000).
67. See, for more details, Laurence R. Helfer, 'Constitutional Analogies in the International Legal System', 37 *Loy LA Int'l & Comp L Rev* 193 (2004).
68. Laurence R. Helfer, 'Regime Shifting: The TRIPS Agreement and New Dynamics of International Intellectual Property Lawmaking', 29 *Yale J Int'l L* 1, 61 (2004).
69. Gabrielle Marceau, 'WTO Dispute Settlement and Human Rights', 13 *Eur J Int'l L* 753, 792 (2002).
70. WTO Appellate Body Report on United States – Import Prohibition of Certain Shrimp and Shrimp Products, fn. 58 *supra*, paras 79–91.

Environmental Agreements (MEAs) not binding on all WTO Members. These MEAs were, in fact, considered to express 'common concerns of nations', leading the Appellate Body to read the term 'exhaustible natural resources' so as to include also living things.[71]

The Appellate Body considered that the United States' ban was temporarily justified under Article XX(g) as a measure required for the 'conservation of exhaustible natural resources'. In so doing, the Appellate Body rejected a previous understanding of Article XX(g), which required that the resource in question be located within the jurisdiction of the State invoking the derogation. Yet the Appellate Body went on to assess whether the ban was also compatible with the chapeau of Article XX which states Article XX's exception may be invoked '[s]ubject to the requirement that such measures are not applied in a manner which would constitute a means of arbitrary or unjustifiable discrimination between countries where the same conditions prevail, or a disguised restriction on international trade'. Goldstein and Steinberg have viewed this type of law making by the Appellate Body in the same way than scholars conceptually analyse common law courts, i.e. gap-filling and clarification of ambiguities.[72] As emphasized by Guzman and Meyer:

> In the case of the former, courts create obligations for which there is no textual hook; in the case of the latter, courts define text left ambiguous, often deliberately, by its drafters. In reality, the distinction between these two modes of lawmaking is a difference in degree rather than a difference in kind. Regardless of whether an ambiguous textual basis for a tribunal's ruling exists, in both cases a court is crafting an obligation not explicitly defined by the text of an agreement.[73]

The Appellate Body in *Shrimp/Turtle* had to interpret the meaning of the phrase 'exhaustible natural resources' in Article XX(g) of the GATT – a provision that allows Members to impose derogations to GATT's free trade rules. Relying upon a reference in the WTO Agreement's Preamble to 'sustainable development', the Appellate Body considered that 'exhaustible natural resources' was an evolutionary concept to 'be read by a treaty interpreter in the light of contemporary concerns of the community of nations about the protection and conservation of the environment'. Those concerns, in turn, were reflected not only in treaties regulating natural resources but also in non-binding declarations addressing that topic.[74]

This type of 'external' situation when the DSU reaches outside of the WTO agreements to borrow concepts developed in other international law regimes may be viewed as a kind of soft law (external soft law) but differs from a more narrow definition that can be used in relation to non-justiciable or hortatory language that can be found in the WTO agreements.

---

71. Guzman & Meyer, fn. 54 *supra*, 171 et seq.
72. Quoted in *Ibid.*, 214–215.
73. *Ibid.*
74. *Ibid.*

4.2.2 *The Canada – Pharmaceutical Patents* Case[75]

In the *Canada – Patent Protection of Pharmaceutical Products*, a WTO dispute settlement panel considered a complaint brought by the EC with respect to two provisions of Canada's patent law. The first provision permitted uses of patented pharmaceuticals without the patent owner's authorization for the purpose of obtaining approval of a generic product before the patent term expired. The second provision permitted production and stockpiling of generic drugs during the patent term in order that they could be immediately released after the term's expiration. Both provisions were intended to promote the prompt marketing of generic versions of pharmaceuticals after patent term expiration. The panel decision held that only the first of these exceptions was consistent with TRIPS. Neither the EC nor Canada appealed the panel's ruling to the Appellate Body, a ruling that the WTO DSB later adopted. The precise scope of Article 30's authority will depend on the specific meaning given to its limiting conditions. The words of those conditions must be examined with particular care on this point. Both the goals and the limitation stated in Articles 7 and 8.1 must obviously be taken into consideration as well as those of other provisions of the TRIPS Agreement which indicate its object and purposes.

Howse encourages precisely this approach in his critique of the panel's decision in the GMO case.[76] He argues that in construing the patent derogations clause in TRIPS Article 30, the panel should have considered Article 8.1 TRIPS that identifies Members' right to, *inter alia*, 'adopt measures necessary to protect public health' provided that they are consistent with TRIPS. Subsequently, according to Article 31.3(c) of the Vienna Convention on the Law of Treaties,[77] to establish the meaning of 'necessary to protection public health', one would have recourse to relevant international health law', a body of rules that includes soft law sources, such as resolutions and authoritative reports and policy statements of relevant international organizations. If the panel had used these soft law sources, thus following Howse's interpretation, it would have given 'clear priority to the legitimate health interests in question over any competing interests of the rights holder', and interpreted TRIPS 'in a manner consistent with what is required for the protection of public health, as defined by world health policy'. In his article, Howse clearly emphasizes the need for using soft law to construe TRIPS. It is, however, not so clear whether the Appellate Body's analysis in the *Shrimp/Turtle* case can be expanded so far. As an initial matter, Article 31.3(c) of the Vienna Convention requires panels to 'take into account . . . any relevant *rules of international law* applicable in the relations between the parties'. Therefore, non-binding resolutions and policy statements do not fall within this definition, although they may provide

75. WT/DS/114R fn 31 *supra*.
76. Robert Howse, 'The Canadian Generic Medicines Panel: A Dangerous Precedent in Dangerous Times', (2000) 3 *J World Int Prop* 493 (2000).
77. Vienna Convention on the Law of Treaties (Done at Vienna, 23 May 1969), 1155 U.N.T.S. 331, 8 I.L.M. 679, entered into force 27 Jan. 1980.

evidence of a political agreement among States to which WTO jurists may feel inclined to defer.[78]

Furthermore, both treaty law and non-binding declarations reinforced the Appellate Body's interpretations in the *Shrimp/Turtle* case. As stated by Helfer:

> Whether WTO jurists would give similar interpretive weight to soft law alone is uncertain. Moreover, even were the Appellate Body is willing to rely exclusively on non-binding principles to add interpretive color to TRIPs, it would first need to determine which of the treaty's provisions embody 'evolutionary' principles that justify a reference to post-TRIPs legal developments. Although the important if open-ended reference in Article 8(1) to 'measures necessary . . . to promote the public interest in sectors of vital importance to [members'] socio-economic . . . development' would seem to qualify, the more detailed intellectual property protection rules in Part II of TRIPs might not. At a minimum, however, *Shrimp/Turtle* is likely to invite competing arguments from states as to how WTO panels should (or should not) take soft law norms generated outside the trade regime into account when interpreting the TRIPS Agreement. It follows from this restrictive approach to conflicts that soft law is not on par with legally binding treaty obligations and thus can never excuse compliance with those obligations.[79]

5 TECHNOLOGY TRANSFER AND ARTICLE 66.2

Yamane points out – quite correctly – that 'there is no well-established definition of "transfer of technology", either in law or in economics, although it can generally be characterized as a process of absorption, by enterprises of knowledge and technology through various channels of learning'.[80] In a more specific context, a WIPO Secretariat's paper said that '[t]he term "transfer of technology" may be understood in a narrow or broad sense when used in the context of intellectual property, in particular, patents'. Broadly stated, transfer of technology is a series of processes for sharing ideas, knowledge, technology and skills with another individual or institution (e.g., a company, a university or a governmental body) and of acquisition by the other of such ideas, knowledge, technologies and skills. In the context of transferring technologies from the public sector and universities to the private sector, the term 'transfer of technology' is sometimes used in a narrower sense: as a synonym of 'technology commercialization' whereby basic scientific research outcomes from universities and public research institutions are applied to practical, commercial products for the market by private companies.'[81] These attempts show the complexity of defining two concepts, 'technology' and 'technology transfer'.

---

78. Robert Howse, fn. 76 *supra*.
79. Helfer, fn. 68 *supra*, 79.
80. Hiroko Yamane, *Interpreting TRIPS: Globalisation of Intellectual Property Rights and Access to Medicines* (Oxford: Hart Publishing Ltd, 2011), 52.
81. See WIPO document SCP/14/4 Rev. of 13 Apr. 2011. <www.wipo.int/edocs/mdocs/scp/en/scp_16/scp_14_4_rev.pdf>.

There have been references to a 'UN definition' in the context of Article 66.2 reporting mechanism.[82] It was mentioned in the report made by New Zealand in 2007 first,[83] followed by other Members such as Switzerland[84] as well as by some authors.[85] This wide definition,[86] which has the merit of covering a wide array of measures and acts, might not have fully assisted in a better understanding of the

---

82. See Ch. 34 of Agenda 21 (United Nations Conference on Environment and Development (UNCED)), notably the following:

> 34.3. Environmentally sound technologies are not just individual technologies, but total systems which include know-how, procedures, goods and services, and equipment as well as organizational and managerial procedures. This implies that when discussing transfer of technologies, the human resource development and local capacity-building aspects of technology choices, including gender-relevant aspects, should also be addressed.

See UNCTAD's approach in 'Transfer of Technology', *UNCATD Series on Issues in International Agreements* (New York and Geneva: United Nations, 2001). In its introduction, the paper recognizes that 'technology' can be defined in various ways. It further says that:

> [t]he UNCTAD draft international code on the Transfer of Technology (the draft TOT Code), in its definition of 'technology transfer' . . . describes 'technology' as 'systematic knowledge for the manufacture of a product, for the application of a process or for the rendering of a service' which does not extend to the transactions involving the mere sale or mere lease of goods. . . . 'Technology transfer' is the process by which commercial technology is disseminated . . . (pp. 4–6 and footnote 2).

It should be noted that the TOT Code is focused on transnational corporations (TNCs) and its context is prior to and different from the TRIPS Art. 66.2 and CBD negotiations. TRIPS Art. 66.2 deals with how to foster technology transfer by companies, small and medium enterprises and public institutions included, from developed countries to LDCs. From that perspective, the Agenda 21 'definition' may, in some quarters' views, offer wider avenues for further reflection.

83. See New Zealand's reports in WTO documents IP/C/W/551/Add. 3 of 25 Oct. 2010 and IP/497/Add.3 of 3 Dec. 2007:

> 3. Technology transfer is interpreted in this report broadly to include training, education and 'know-how', along with any capital component. Using a United Nations definition, we see four key modes of technology transfer: (i) physical objects or equipment; (ii) skills and human aspects of technology management and learning; (iii) designs and blueprints which constitute the document-embodied knowledge on information and technology; and (iv) production arrangement linkages within which technology is operated.

(The report in IP/497/add.3 of 2007 refers to 'the' United Nations definition.)

84. See Switzerland's report in WTO document IP/C/W/551/Add.4 of 18 Oct. 2010.
85. Suerie Moon, 'Meaningful Technology Transfer to the LDCs: A proposal for a Monitoring Mechanism for TRIPS Article 66.2', *ICTSD Programme on Innovation, Technology and Intellectual Property Project on IPRs and Sustainable Development Policy Brief Number 9*, April 2011, t, at 4, citing para. 3 of New Zealand's report.
86. See Ch. 34 of Agenda 21 (United Nations Conference on Environment and Development (UNCED)), notably the following:

> 34.3. Environmentally sound technologies are not just individual technologies, but total systems which include know-how, procedures, goods and services, and equipment as well as organizational and managerial procedures. This implies that when discussing transfer of technologies, the human resource development and local capacity-building aspects of technology choices, including gender-relevant aspects, should also be addressed.

operation of Article 66.2, which would have informed developed country Members in reporting activities possibly falling under the UN definition but might not immediately, or more visibly, reflect the fostering of technology transfer.

This chapter will not address the work done in the WTO Working Group on Transfer of Technology established by the Ministers in Doha which aims at examining the relationship between trade and the transfer of technology from developed to developing countries, and ways to increase the flow of technology to developing countries. For further details, see Chapter 3. There is also a particularly rich and inspiring literature on this issue.[87]

5.1 Article 66.2: A Paper Tiger?

> Developed country Members *shall* provide *incentives* to *enterprises and institutions in their territories* for the purpose of *promoting and encouraging technology transfer* to least-developed country Members in order to enable them to create a *sound and viable technological base*.[88]

The wording, in its apparent simplicity and concision, reflects in the condensed phrase the difficulties and dilemmas faced by negotiators of developed countries in responding to a legitimate aspiration of least-developed countries, namely the desire for access to technology as one among other necessary steps to break the vicious circle of lack of development and lack of attractiveness for foreign direct investment. The provision establishes a 'positive legal obligation'[89] for developed countries. It enjoins Members to take measures to incentivize companies in their own territory to transfer technology to LDCs.

Some background information is necessary to understand the *ratio legis* of this provision. IPRs are 'private rights', as defined in the preamble of the TRIPS Agreement, whether owned by private physical or moral persons or by public institutions. Apart from certain limitations, such as government use and compulsory licensing under Article 31 TRIPS, allowing some limited forms of technology transfer, the notion of technology transfer is, in general understood as voluntary. Technology transfer under compulsory licenses may work in large, emerging developing countries possessing some manufacturing capacity or great direct investment attractiveness due to their size or other factors. It does not work for all countries.

IPRs are, as a general fundamental rule, of a territorial nature. Technology transfer of IP-protected subject-matter presupposes that in the recipient country some protection exists. Taking patents as an example, if there is no patent protection, and assuming that the foreign enterprise is prepared to transfer technology through

87. See General documents of the WTO Working Group on Transfer of Technology. <www.wto.org/english/tratop_e/devel_e/dev_wkgp_trade_transfer_technology_e.htm>.
88. Emphasis added.
89. Suerie Moon, 'Does TRIPS Art. 66.2 Encourage Technology Transfer to LDCs?: An Analysis of Country Submissions to the TRIPS Council (1999–2007)', *Policy Brief Number 2, December 8, UNCTAD-ICTSD project on IPRs and Sustainable Development*, 2.

undisclosed information (know-how), it would help if there were a framework of operative rules to protect against any breach of contract or unfair business practices. May authorities 'force' or encourage private right holders to transfer technology to a least-developed country on the ground, for example, that it is part of a national policy to assist those countries? The 'solution' to this conundrum would be to give incentives to enterprises and institutions within the territory of the developed country.

Maskus offers the view that:

> Article 66.2 is not likely on its own merits to achieve significant increases in ITT. There are two essential difficulties. For one, ITT largely relies on private market incentives and this article does little to redress the basic problems mentioned above. Second, if governments in developed countries were willing to offer substantial incentives they would face domestic political opposition in doing so.[90]

The current wording of paragraph 2 of Article 66 came to the table relatively late in 1990 a year where negotiations entered in its intensive phase, in the Uruguay Round. For example, it did not appear in July 1990 in the Chairman's report in document MTN.GNG/NG11/W/76. The issue surfaced later and was reflected – in its current wording – in a room document dated 23 November 1990 in article 69, paragraph 2. Not all negotiating delegations agreed on the scope of the obligation. Neither was there a sufficient debate on what 'incentives' could be.

It would perhaps help to understand the aspirations of LDCs on one side, and the concerns of some negotiators on the other, by looking back at the negotiations of the CBD in Nairobi before the convention was transmitted to the Rio Summit. It should be recalled that in December 1990 the Uruguay Round came to an impasse at the Ministerial Conference in Brussels, and the work undertaken went into 'hibernation', with great doubts from many quarters about its resumption and success. The issue of Technology Transfer was, on the contrary, on the table in the CBD negotiations in Nairobi, and was heavily negotiated. The declarative interpretations read out by the delegations of the European Community, France (upon adoption and upon signature), Ireland, and Switzerland (upon signature and ratification), are testament to the concerns regarding the operability of transfer technology, even if it was circumscribed to the CBD.[91] The Swiss delegation, for example, felt it was feasible to take some engagement regarding technology developed by public-funded institutions but was reluctant to commit for technology detained by private firms. The last declaration adopted at ratification insists, *inter alia*, on voluntary cooperation for technology transfer.[92] Contrary to the

90. Keith Maskus, 'Encouraging International Technology Transfer' (Geneva: International Centre for Trade and Sustainable Development, Project on IPRs and Sustainable Development, Issue Paper 7, 2004), 3.
91. See *Secretariat of the Convention on Biological Diversity, Handbook of the Convention on Biological Diversity Including its Cartagena Protocol on Biosafety*, 3rd edn (Montreal Canada, 2005), s. VIII (Declarations), 385–395. <www.cbd.int/convention/refrhandbook.shtml>.
92. 'Switzerland wishes to reaffirm the importance it attaches to transfers of technology and to biotechnology in order to ensure the conservation and sustainable use of biological diversity.

TRIPS Agreement and the WTO, the CBD has a financial mechanism, and its scope – biodiversity – is more limited, and might have facilitated a relatively 'positive' declaration by Switzerland. One may point out that TRIPS Article 66.2 could be an easier matter since it is limited to the needs of LDCs. Others may object that the scope of technologies involved is greater, and it would therefore be more difficult to delineate the frontiers. There is still a long way to go to find a solution. As the most recent attempt at the international level, the Nagoya Protocol does not seem to have made noticeable progress regarding technology transfer.[93] This shows the difficulty of going beyond voluntary transfer and any financial funding mechanism.

The situation in TRIPS slightly differs. The crux of the matter resides in 'incentives'. What kind of incentive may be envisaged without having them turn into some disguised measures prohibited by national constitutional law, WTO law, other international treaties, and last, but not least, without such measure being diverted from their initial noble objective? Direct financing, tax exemptions, protection against export risks, and so on, which one is most suited and 'safe' in terms of national treatment, most-favoured nation (MFN), subsidies, to cite a few points? Perhaps this debate and some clarifications are indispensable to circumscribe what is doable from what is not.[94] Another concern, not spelt out, but in many delegation's minds, is the fact that sound and viable technology transfer requires some form of stable local infrastructure and minimum legal assurances. Another one, (not spelt out either), but which has some importance, is the fear that, once technology transferred – especially know-how – the proprietary information owner would be deprived of the use of its own technology by the local authority, in favour of domestic firms. Having said that, this seems to be a concern more related to experience of developed countries in some developing and emerging countries' markets.

The compliance with intellectual property rights constitutes an essential element for the implementation of policies for y transfer and co-investment. For Switzerland, transfers of technology and access to biotechnology, as defined in the text of the Convention on Biological Diversity, will be carried out in accordance with article 16 of the said Convention and in compliance with the principles and rules of protection of intellectual property, in particular multilateral and bilateral agreements signed or negotiated by the Contracting Parties to this Convention.' Section VIII (Declarations), Handbook, n. 82, 393–394.'

Switzerland will encourage the use of the financial mechanism established by the Convention to promote the voluntary transfer of intellectual property rights held by Swiss operators, in particular as regards the granting of licences, through normal commercial mechanisms and decisions, while ensuring adequate and effective protection of property rights.' Section VIII (Declarations), Handbook, fn. 85 *supra*, 393–394.

93. See Nagoya Protocol, fn. 3 *supra*, Art. 22 (capacity), para. 5.
94. This is reminiscent of the long-debated issue of whether patent applicants from developing countries should be exempted from patent fees, including issues such as the determination of eligibility, discrimination, diversion/circumvention, etc. Over the years, a solution has been found within the framework of WIPO's Patent Cooperation Treaty (PCT). But the parallel stops there as the issue of incentives is more complex.

The LDC Group took the issue of the implementation of Article 66.2 seriously. Upon its repeated queries and initiative, the TRIPS Council adopted in February 2003 a decision on the 'Implementation of Article Council for 66.2 of the TRIPS Agreement'.[95] This Chapter will not deal with the details, as they can be found in Chapter 3 as well as the minutes of the TRIPS Council.

The concerns voiced by LDCs on the implementation of Article 66.2 have recently been better heard thanks to a more concerted collective action from their side. In some quarters, the possibility of bringing the case of Article 66.2 under the DSU for non-compliance with a 'strong' obligation has been recently contemplated. What could be noted is that the LDCs have triggered a process of more structured discussions. Workshops organized for delegations from LDCs and developed countries most interested in the reporting have contributed more to this progress in three years than in the entire history of the reporting process started in 2003.Whilst for many observers this might not be sufficient, it is not presumptuous to say that in light of complex machinery like the WTO, and of the limited human resources of LDC delegations, this 'building up' process is worth being underscored. It shows how a 'skeleton' of a mandatory provision may be fleshed out to form a set of elements that could give greater weight to the provision.

In terms of substance the LDC Group has 'identified two critical areas: the definition of technology transfer and an illustration of 'types of incentives'. On technology of transfer, there is 'no common understanding of what constituted technology transfer, hence the inclusion in some reports of technical cooperation activities'.[96] On incentives, in addition to a list of incentives, the LDC asked for specific examples of incentives that have effectively led to technology transfer.

At the Ho Chi Minh International Symposium on Sustainable Technology, a few points were made by participants, which may serve as a platform for further reflection and concrete action. First, there is a wide range of technologies at the disposal of all countries at no cost, namely technology in the public domain. It should not be a problem to implement such technology and there are avenues to help Small and Medium Sized Enterprises (SMEs) in LDCs to use it. Second, the size of a market and the infrastructural conditions are determinant. If a country is too small, it might not be efficient to persist in taking solitary action but rather to take concerted action along with countries of the region to share the burden of creating a sound and viable basis for development, while avoiding too great dependence and ensuring a fair division of 'labour'. Third, a 'good' technology transfer implies full involvement of the recipients (an 'active' rather than 'passive' technology transfer approach). Fourth, identification of 'cost-effective technology' is indispensable. Ironically, looking at developed countries facing the difficult choice of adequate renewable energies to replace nuclear energy and still trying to fully measure the impacts, one point for reflection is that assistance in identification of useful technology should be one of the priorities in the cooperation between

95. WTO IP/C/28 of 20 Feb. 2003 (decision of 19 February) – Implementation of Art. 66.2 of the TRIPS Agreement.
96. IP/C/M/64, para. 341 (Representative of Angola's comments on Developed Country reports).

developed countries and LDCs. Fifth, 'sustainability' of technology transfer should be ensured in terms of 'time'- projects which permit better absorption by local recipients and which should not be launched in haste without establishing medium and long-term objectives – as well as in terms of 'space' (see the second point mentioned above). Sixth, sustainable technology transfer should aim at 'quality'– quality of the technology chosen and quality of contract and relationship between the transferor and the transferees. There are indeed many other avenues which have been envisaged by international organizations, NGOs and universities, to cite a few. Finally, the 'incentives' issue is not insuperable. There are incentives that do not create constitutional or legal problems in the developed countries, and that are not incompatible with WTO law. This is also an area requiring more attention and examination. (See for further reflection the contribution of Christoffer Wong et al., in Chapter 12).

The road is winding, because the legal and practical problems implied by Article 66.2 are real. But at least there is a road. After all, it is in the interest of LDCs, and above all, of developed countries to ensure a sound and viable technological basis. Apart from a few cases of extremely vulnerable economies, many LDCs have more cards in their hands, e.g. natural resources or manpower, which could serve as negotiating leverage. The implications of sound and viable technology development are not only found in trade,[97] there are also implications for the security and stability of countries and regions.

It should be noted that the Doha Declaration on the TRIPS Agreement and Public health has reaffirmed the commitment of developed country Members to provide incentives to their enterprises and institutions to promote and encourage technology transfer to LDCs pursuant to Article 66.2. This Declaration, or at least its recitals, soft law or 'softer law', may provide a push for creative pragmatic incentives. The pharmaceutical sector might be a good way to start with a model voluntary project, combining incentives and real and effective technology transfer.

6 CONCLUSIONS

This examination of the TRIPS provisions within the framework of the WTO law and public international law, the various categories of texts or acts to be applied or used for interpretation at national level, and in the WTO context, shows that soft law and hard law interact. Soft law reflecting modern needs and developments is taken into consideration by adjudicating bodies at the domestic and at the WTO levels when it is pertinent. Article 66.2 is hard law, albeit weakened by elements not yet well defined. While some may call this 'soft law', the conclusion arrived at in this article is different: it is 'softened' hard law and, hence the softer elements could be reinforced. It is not a paper tiger; just a young one.

97. See also the Fourth United Nations Conference on Least Developed Countries (LDC-IV), Istanbul 9–11 May 2011. Both WTO Director-General Lamy and UN Secretary-General Ban Ki-moon referred to 'untapped potential' of LDCs and envisioned a win-win situation.

Chapter 3

Assessing Reporting Obligations under TRIPS Article 66.2

Hans Henrik Lidgard

In the preceding chapters the legal issues have been framed as well as consideration of the significance of the Trade-related Aspects of Intellectual Property Rights (TRIPS) stipulations in creating some form of obligation to transfer technology as a quid pro quo for undertakings assumed by the developing countries. Are these provisions mere window dressing or do they lead to real and significant transfers? As the obligations lie somewhere between hard law and soft law it is interesting to investigate the extent to which they have had any real impact.

The reporting obligations under the TRIPS Agreement should provide a basis for discovering how the developed countries provide incentives to enterprises and institutions in their territories for the purpose of promoting and encouraging technology transfer and if these activities lead to concrete and helpful best industry practices. The obligations must be seen in their context. Other international activities also address technology transfer from developed to developing countries. The United Nations Conference on Trade and Development (UNCTAD) has such activities on its agenda and the Convention on Biodiversity (CBD) and the UN Framework Convention on Climate Change contain fairly far-reaching language indicating that technology transfer is also a priority in fields unconnected to trade in goods. The CBD is still, more than 15 years after its execution, at a formation stage with some way to go before it can generate substantial technology transfer. The situation is similar for the environmental development under the Framework

Hans Henrik Lidgard, Jeffery Atik & Tu Thanh Nguyen, *Sustainable Technology Transfer*, pp. 43–64.

Convention. In any event, TRIPS is a good measuring tool as reporting has not been limited to intellectual property rights activities. Member countries in their Article 66.2 TRIPS reporting cover exchanges in any area or sector including health, biodiversity, climate and energy alike.

This chapter summarizes the concrete results achieved in relation to technology transfer in the narrow sense, searching for best practices of how a developed country can provide incentives to its private industry to actively participate in transferring catch up, state-of-the-art technology. It is a descriptive empirical account, evaluating if reported technology transfer activities are substantial and sustainable or if they are merely reporting on technical assistance or even general development aid.

1 THE UNCTAD DISCUSSION

UNCTAD, with its focus on the least developed countries' (LDCs') situation, is a natural forum to discuss and follow sustainable-technology-transfer developments and the organization has provided a multitude of studies and policy documents over the years regarding these issues.

1.1 2002 Policy Document

In 2002, an UNCTAD policy agreement addressed best practices for technology transfer to LDCs.[1] The commission underlined that both governments and the private sector have an important role to play in the implementation of TRIPS commitments, *inter alia* through public and private partnerships. It was agreed that UNCTAD should provide assistance to developing countries in strengthening their capacities to discuss and examine transfer of technology issues, to negotiate international instruments, and to explore ways and means for effective implementation of international commitments in the area of transfer of technology and capacity building. UNCTAD should also assist interested countries with regard to the interface between commitments in the TRIPS Agreement and national implementation requirements. Furthermore, the organization should disseminate information concerning existing home country measures that encourage transfer of technology in various modes to developing countries, in particular LDCs.

Also of interest is the 2003 UNCTAD report on Transfer of technology[2] which, based on three case studies, concluded that countries rely for their

1. TRADE AND DEVELOPMENT BOARD, Commission on Investment, Technology and Related Financial Issues, Sixth session, Geneva, 21–25 Jan. 2002, Agenda item 4, <www.unctad.org/en/docs/c2l16_en.pdf based on secretariat reviews>.
2. Three studies were carried out under the UNCTAD/UNDP Global Programme on Globalization, Liberalization and Sustainable Human Development: Best Practices in Transfer of Technology under the title Transfer of Technology for Successful Integration into the Global Economy, (United Nations, New York and Geneva 2003), 6, <www.unctad.org/en/docs/iteipc20036_en.pdf>.

advancement on influxes of ideas. The poorer economies had to learn from others to catch up and experience of emerging economies demonstrated the importance of foreign technology in the form of licensing arrangements, collaboration agreements and foreign direct investment.

1.2 UNCTAD Conferences

In reviewing past UNCTAD Conferences on the LDCs and the extent to which they addressed sustainable technology transfer, the conclusion must be that the Conferences are proceeding on a fairly abstract level with general recommendations, which do not easily translate into hard-core technology transfer activities. Rarely does a Conference refer to specific technology transfer or private industry in more than general statements underlining the importance of the involvement and participation of the private sector in both the developed and developing countries.

The 1990 Paris Conference discussed the role of the private-enterprise sector and the need to stimulate entrepreneurship by policies creating a more favourable economic environment for local initiatives of the private sector. The LDCs had to create an appropriate legal and institutional framework for the conduct of activities by the private entrepreneurs (fiscal, financial and credit policies as well as protecting private investment). Such measures were (and still are) important in order to attract foreign companies to invest to any significant degree in the LDCs' economies. Programmes for private-enterprise development should address the promotion of domestic and foreign direct investment, management training, non-traditional exports, promotion of small-scale ventures and micro-enterprises, including those within the informal sector, and the development of entrepreneurial skills.[3]

Ten years later in Brussels the Conference adopted the 'Brussels Declaration' in line with the overall Millennium Goals. Again the resolution and the general discussion appear to have moved on a principles level, perhaps even less concrete than the prior Conference as to how sustainable technology transfer should be achieved, and even less so as to the concrete results that had actually been reached.[4]

Nor did technology transfer as such feature as an independent important topic in the Fourth Conference on LDCs in Istanbul, Turkey in May 2011.[5] Improvements had been achieved during the last decade and the combined LDC economy had grown 7% during the prior decade,[6] but many problems remained and had

3. Report of the Second United Nations Conference on the Least Developed Countries, Paris, 3–14 Sep. 1990, Doc. A/CONF.147/18. <www.un.org/wcm/webdav/site/ldc/shared/Outcome_%20document_Second_UN_Conference_%20LDCs.%20pdf.pdf>.
4. Report of the Third United Nations Conference on the Least Developed Countries, Brussels, Belgium, 14–20 May 2001, Doc. A/CONF.191/13. <www.unctad.org/en/docs/aconf191d13.en.pdf>.
5. The Fourth United Nations Conference on the Least Developed Countries (LDC-IV), 9–13 May 2011, Istanbul, Turkey. <www.un.org/wcm/content/site/ldc/home>.
6. WTO Director-General Pascal Lamy, Global trading system must deliver more to the least-developed, speech 9 May 2011, <www.wto.org/english/news_e/sppl_e/sppl191_e.htm>.

actually worsened due to a.o. the financial crises, fast urbanization, weak institutions, political instability, inadequate health and education infrastructures, environmental degradation, and the impact of climate change.[7] The national reports suggested that a new programme of action should be more specific and make use of the wealth of natural resources available in many LDCs. International initiatives to promote transparency in the extraction of natural resources, investing in education, technical skills and the building up of technological capabilities, must be strengthened, to speed up the technology catch-up process. Increased external resources would not result in sustained progress without continued financial support complemented with trade and long-term development of productive capacities.

The initiatives in Istanbul resulted in a Programme of Action for the Least Developed Countries for the decade 2011–2020[8] summarizing policy measures and mechanisms on Science and Technology.[9] LDCs should set up and strengthen institutions to support local research and development, science and technology, whereas the development partners should, for example:

(a) allocate at least 5% of the agreed 1% of GNI to LDCs for the development of science and technology;
(b) create a Science and Technology Bank for LDCs to facilitate LDCs' access to technologies and technological know-how;
(c) provide start-up finance for LDC firms which venture into new technologies by setting up technology sharing consortia by LDCs;
(d) improve access to technology and strengthen research in Science and Technology by creating necessary enabling provisions under World Intellectual Property Organization (WIPO), World Trade Organization (WTO) and other relevant frameworks;
(e) provide incentives for firms as called for in Article 66.2 of the TRIPS Agreement in expanding LDC access to technology and knowledge across all relevant sectors;
(f) extend LDC waiver under TRIPS until they graduate from LDC status

2 THE TRIPS DEVELOPMENT

2.1 No Reporting Obligations in the Agreement

The intent expressed in TRIPS Article 7 is that protection and enforcement of intellectual property rights contribute to the promotion of technological innovation and the transfer and dissemination of technology, in a manner conducive to social

7. National Reports on the Implementation of the Brussels Programme of Action for the Decade 2001–2010. <www.un.org/wcm/webdav/site/ldc/shared/documents/UN_Synthesis_Report_032711_w.pdf>.
8. <www.un.org/wcm/webdav/site/ldc/shared/Compilation%20of%20Inputs.pdf>.
9. *Ibid.*, point 16.

and economic welfare, while balancing rights and obligations to the mutual advantage of both the producers and the users of technological knowledge.

Article 66 of the TRIPS Agreement addresses the LDCs' Members' special needs by referring to these special needs and requirements, their economic, financial and administrative constraints, and the need for flexibility to create a viable technological base.[10] Paragraph 1 provides a temporary exemption from the TRIPS obligations for the least-developed country Members, which have now been extended on two occasions. In the same spirit Article 66.2 stipulates that:

> **Article 66.2**: Developed country Members shall provide incentives to enterprises and institutions in their territories for the purpose of promoting and encouraging technology transfer to least-developed country Members in order to enable them to create a sound and viable technological base.

The stipulation does not contain any reporting requirement.

Similarly Article 67, which deals with technical cooperation, and is addressed to all developing country members, not only LDCs. Article 67 does not require developed countries to provide information about their activities to meet these obligations:

> **Article 67**: In order to facilitate the implementation of this Agreement, developed country Members shall provide, on request and on mutually agreed terms and conditions, technical and financial cooperation in favour of developing and least-developed country Members. Such cooperation shall include assistance in the preparation of laws and regulations on the protection and enforcement of intellectual property rights as well as on the prevention of their abuse, and shall include support regarding the establishment or reinforcement of domestic offices and agencies relevant to these matters, including the training of personnel.

As a matter of fact, the TRIPS Agreement only contains minimal reporting obligations. The TRIPS Council shall under Article 68 monitor the operation of the Agreement and compliance with its obligations. In carrying out its functions, the Council may consult with, and seek, information from any source it deems appropriate. Furthermore under Article 63 Members shall notify amended legislation to the Council in order to assist the Council in its review of the operation of the Agreement. The remainder of the TRIPS Agreement is silent on such obligations and that reporting was initiated at all, was rather the result of voluntary undertakings. Initial compliance with the Article 66.2 requirements was also spotty and reporting unclear. The 2000 TRIPS annual report contained five categories, only two of which referenced technology transfer.[11]

---

10. This argument is built on the Preamble, which uses the same formula:

> *Recognizing* also the special needs of the least-developed country Members in respect of maximum flexibility in the domestic implementation of laws and regulations in order to enable them to create a sound and viable technological base.

11. See WTO, TRIPS Council, 'Implementation of Article 66.2: Information from Developed Country Members – Types of Incentive Measure Notified', Note 3 May 2000 by the Secretariat,

2.2 THE 2001 DOHA DECLARATION

The developed countries reaffirmed, in paragraph 11.2 of the 2001 Doha Declaration, their commitment to provide enterprises and institutions with incentives to promote and encourage technology transfer to LDCs pursuant to Article 66.2.[12]

These undertakings were discussed in the November 2002 Meetings of the TRIPS Council where LDC representatives stated that, in their view, it was important that developed countries show precisely what laws, identifiable policy or regulations they had enacted rather than merely providing general statements.[13]

In its Decision of 19 February 2003, the TRIPS Council agreed that developed countries should submit annual reports on actions taken or planned in pursuance of their commitments under Article 66.2.[14] The reports were required to identify the specific legislation involved, the type of incentive, the entity making it available, the eligible enterprises and how the system functioned in practice. Reports must contain statistics on the use of the incentives, the terms of transfer, the recipient LDCs and the information needed for the assessment of the effects of these measures.[15] It was further agreed that new detailed reports should be delivered every third year and that updates be provided in the intervening years.

IP/C/W/169: (i) Export and investment promotion for developed countries' enterprises, (ii) Export finance for developed countries' enterprises, (iii) Tax credits and other financial incentives for developed countries' enterprises, (iv) Loans and other support for enterprises in developing countries, and (v) Infrastructure projects in developing countries.) <www.wtocenter.org.tw/SmartKMS/fileviewer?id=5256>.

12. See WTO, Ministerial Conference Fourth Session Doha, 9–14 Nov. 2001, Decision of 14 Nov. 2001, Implementation-Related Issues and Concerns, WT/MIN (01)/17, DEC/1, 41 I.L.M. (2002) (WTO 2001 Doha Decision), Art. 11.2.
13. WTO, TRIPS Council, Communication from Least-Developed Countries, 5 Jul. 2002, IP/C/W/357 and Minutes of Meeting on 25–27 and 29 Nov. and 20 Dec. 2002, IP/C/M/38 para. 262.
14. WTO, TRIPS Council on Report on the implementation of Art. 66.2 of the TRIPS Agreement of 19 Feb. 2003, IP/C/28 (2003 Decision).
15. *Ibid.*, para. 3 mandates the inclusion in Developed Country Reports of certain information. Para. 3(d) requires:

    Any information available on the functioning in practice of these incentives, such as:

    - statistical and/or other information on the use of the incentives in question by the eligible enterprises and institutions;
    - the type of technology that has been transferred by these enterprises and institutions and the terms on which it has been transferred;
    - the mode of technology transfer;
    - least-developed countries to which these enterprises and institutions have transferred technology and the extent to which the incentives are specific to least developed countries; and
    - any additional information available that would help assess the effects of the measures in promoting and encouraging technology transfer to least developed country Members in order to enable them to create a sound and viable technological base.

2.3 LDC, Developing and Developed

The Article 66.2 reference to least developed country Members is clear, as the WTO defines LDCs as those countries specified as such by the United Nations.[16] Graduating from LDC to developing country status is also a fairly well defined process.[17]

If the LDCs are the recipient countries less clear is precisely who should be the technology suppliers. In spite of the reference in Article 66.2 to 'developed countries', there is no clear definition of who is a 'developed' or 'developing' country.

The current trend in a larger developmental context is to introduce new and shifting categories into the general language, which are aimed at further differentiating between statuses. 'LDCs', 'landlocked countries', and 'small island developing states' have long been terms in official use to cover a lower degree of economic development. Newly industrialized countries are sometimes referred to as 'NICs'. 'Economies in transition' refers to former eastern block countries. They and the BRICs cluster (Brazil, Russia, India and China), commonly referred to as 'emerging economies', are more advanced, but still not quite developed. To the extent any of these countries have become members of the Organisation for Economic Co-operation and Development (OECD) family, they have taken another step towards developed status. However, it remains unsettled when a country has moved from the developing to a developed country position.

In the end it depends more on how countries act than on a precise definition. This creates uncertainties, as obligations in international agreements often require developed countries to assist developing countries. From this perspective it becomes convenient to simply identify one's own status as that of a developing country and thus evade responsibility. It also raises a question whether technology transfer rights and obligations should be more differentiated in international law.

With respect to the TRIPS reporting obligations, one radical interpretation could be that Article 66.2 of the TRIPS is a mandatory obligation for any country other than an LDC. In reality, however, only the most developed OECD countries

16. Of these 32 are WTO members, 10 are in the accession process, 1 is an observer, and 7 are not WTO members. <www.wto.org/english/thewto_e/whatis_e/tif_e/org7_e.htm>.
17. See, United Nations Office of the High Representative for the Least Developed Countries, Landlocked Developing Countries and the Small Island Developing States (UN-OHRLL-S)',Criteria for Identification of LDCs. <www.unohrlls.org/en/ldc/related/59/>. A low-income criterion is established by the UN-OHRLLS based on a three-year average estimate of the gross national income (GNI) per capita (under USD 905for inclusion, above USD 1,086 for graduation). The UN-OHRLL 2009 review recommended that Equatorial Guinea be graduated from the list of LDCs. Others were considered eligible but not recommended for graduation due to doubts about the sustainability of their progress. In 2004 Cape Verde and Maldives graduated from the list of LDCs. See A/61/173, Midterm Comprehensive Global Review of the Implementation of the Programme of Action for the Least Developed Countries for the Decade 2001–2010, Report of the Secretary-General (21 Jul. 2006) A/61/173. p.3. <http://www.unohrlls.org/en/ldc/166/>.

have submitted Article 66.2 reports of their technology transfer programmes in accordance with the Doha 2001 obligation.[18] No reports from emerging economies, including the BRICs cluster or countries with a relatively mature economy such as Turkey and Mexico, have been forthcoming. It remains unclear if they were understood to be subjects of Article 66.2s obligations in the early 1990s when the TRIPS Agreement was negotiated. Even if they are not specifically excluded from these reporting obligations, and even if much has happened since the early 1990s, these countries have, in fact, asserted a third world status in the TRIPS Council by either active self-identification or by simply not assuming developed country obligations.

2.4 Unspecific Initial Reporting

Following the 2003 Council Decision, approximately eighty Article 66.2 reports and amendments have been filed by developed countries in the period 2004–2010.

In 2008 Suerie Moon analysed the developed countries' reporting during the preceding period in a critical assessment to UNCTAD.[19] She described their shortcomings and the lack of clear definitions. She included training, education and know-how, along with any capital component in the technology transfer definition. Her report specifically excluded an assessment of the volume or nature of the technology transfer reported. Also excluded were market-based technology transfers that largely occur through private channels because, 'measuring private technology transfer is very difficult in the absence of a unified reporting mechanism'.[20]

Many high-income OECD countries had never submitted a report, and submissions had been irregular among those that had. A majority of the programmes and policies reported did not specifically target LDCs or technology transfer.[21] If they did benefit LDCs, they did not provide hard data useful to determine whether or not Article 66.2 provided incentives beyond usual development aid.[22] There was also a need for understanding what actually comprised an acceptable level of compliance.

Moon concluded that IP remains one of the most contentious policy arenas within the WTO, and the institution's credibility could suffer if developed countries were perceived to be falling short in their commitments. Both developed and LDC members could benefit from an effective monitoring system promoting

18. See fn. 12 *infra*.
19. Suerie Moon, 'Does TRIPS Art. 66.2 Encourage Technology Transfer to LDCs? An Analysis of Country Submissions to the TRIPS Council (1999–2007)', Policy Brief No. 2, December 2008. <http://ictsd.org/i/publications/37159/>. Moon has updated her report 'Meaningful Technology Transfer to LDCs: A Proposal for a Monitoring Mechanism for TRIPS Article 66.2', Policy Brief No. 9, May 2011, <http://ictsd.org/i/publications/106434/>, and she now provides a proposal for a monitoring mechanism.
20. *Ibid.*, 1.
21. *Ibid.*, 5.
22. *Ibid.*, 6.

accountability, as it would recognize those developed members that have taken bona fide measures to comply, focus attention on those members that have not, and provide guidance on effective modes of technology transfer. Alternatively, members should consider revising and strengthening Article 66. Without the technology transfer component, it was difficult to see why LDC members should implement other parts of the TRIPS Agreement.

3 RECENT ARTICLE 66.2 REPORTING

The Moon study is indicative of developments after 2008 as well. New reports are often long and non-specific, covering a variety of issues, many of them falling under general development aid. After a 2008 Workshop on Article 66.2 of the TRIPS Agreement, some 2010 reports showed improvement.[23]

In what follows, the presentation focuses on true production transfer arrangements, i.e. 'incentives to enterprises and institutions in developed countries' to transfer technical information to LDCs, which allow the production of modern commodities, taking the form of blueprints, instruments, machinery, manuals, quality control programmes as well as access to intellectual property rights (IPR) and know-how.

3.1 United States Provides Tax Incentives and Forges Public-Private Sector Alliances

United States considers that the TRIPS Article 66.2 reporting obligations refer to technical assistance in the areas of trade and investment policy, institutional analysis and reform, trade capacity building, technical training, and the promotion of private sector development, including helping in expanding LDC trade and investment with the United States.

In October 2006, United States provided the required triennial report. An update[24] was submitted in October 2007 which focused on LDCs and technology transfer in the field of public health and US tax incentives for the private sector.[25]

23. WTO, TRIPS Council Meeting, 26–27 Oct. 2010, Eighth Annual Review under para. 2 of the Decision on the Implementation of Art. 66.2 of the TRIPS Agreement, IP/C/M/64 (17 Feb. 2011), paras 340–342. On behalf of the LDCs the representative of Angola acknowledged the improvements, but more could be done. Reports had to ensure LDC specificity, results achieved, distinction to technical assistance, funding provided to incentives in a harmonized format. The Council had to develop parameters on what constitutes 'technology transfer', 'types of incentives', 'a uniform reporting mechanism' and 'a toolkit for assessing the impact of technology transfer'. *Ibid.*, at para. 342.
24. WTO, TRIPS Council, United States 2006 Addendum Report on the Implementation of Art. 66.2, 20 Oct. 2006, IP/C/W/480/Add.6.
25. WTO TRIPS Council United States 2007 Addendum Report on the Implementation of Art. 66.2, 12 Oct. 2007, IP/C/W/497/Add.5 (3 Dec. 2007). US tax law provides incentives to private institutions engaged in activities relevant to the transfer of technology. Charities, private

In October 2008, the US included information on bilateral agreements providing technology transfer incentives to the private sector.[26] The US 2009 addendum [27] complemented the earlier reports and provided as an example the Global Development Alliance (GDA), which mobilizes public-private alliances. The GDA business model links US foreign assistance to governments, business, and civil society. During the fiscal year 2009, US Agency for International Development (USAID) cultivated more than 900 alliances with 1,700 partners. For example, USAID and Bridgestone/Firestone, Inc. planned to reinvigorate Liberia's rubber industry and support the development of a clean energy power plant in Monrovia. The partnership is expected to increase export-driven trade and growth while boosting Liberia's clean energy initiatives.

The US October 2010 report expands on earlier reports by giving a systematic presentation of on-going activities, with special focus on the LDCs.[28] The report is in line with a broader understanding of technology transfer and includes technical assistance information. On the other hand, the US initiatives are so all-embracing that a report has to remain on the macro level, rather than identify individual initiatives. The report is only 20 pages and presents a number of different US agencies and how they promote technical dissemination in all sectors around the globe. It notes that public-private alliances continue to stimulate economic growth, develop businesses and workforces, addresses health and environmental issues, and expanded access to education and technology. USAID has now cultivated more than 1000 alliances with 3000 partners.

More concrete examples are presented in passing: The report refers to the Firestone venture and the clean energy power plant in Monrovia. Other alliances have been created to help climate-friendly technologies find financing. The United States supports the availability and affordability of access allowing mobile and broadband services to reach out to rural areas and provide low cost, environmentally friendly electricity. Support has been provided to African electrical utilities' construction of gas pipelines via a consortium of US and West African companies. Additional financial resources are established for new technologies, intellectual capital, and technical and managerial expertise that can meet increasingly complex challenges.

---

foundations, religious institutions, and certain non-governmental organizations) qualify for exemptions from federal income taxation by virtue of being organized and operating so as to carry out international charitable work. Individuals who make personal contributions generally benefit from a reduction in income tax liability. US Non-governmental organizations delivered USD 6.3 billion in grants to low-income developing countries in 2003. The existence of such tax incentives has a strong impact on decision-making, by both individual donors and institutions.

26. WTO TRIPS Council, United States 2008 Addendum Report on the Implementation of Art. 66.2, 9 Oct. 2008, IP/C/W/519/Add.3.
27. WTO, TRIPS Council, United States 2009 Addendum Report on the Implementation of Art. 66.2, 2 Oct. 2009, IP/C/W/536/Add.1.
28. WTO, TRIPS Council, United States 2010 Addendum Report on the Implementation of Art. 66.2, 25 Oct. 2010, IP/C/W/551/Add.5.

It is interesting to note that whereas most other countries make note of their efforts to train LDC government officials in IPR protection and enforcement, the United States focuses on training its own Small and Medium Sized Enterprises (SMEs) in the same field in order to encourage them to do business abroad – also in LDCs. It thereby clearly makes the point that acting against piracy is actually in the interest of the poorer economies too: 'Strong IPR protection systems, in particular, can help assure foreign companies that they can transfer technology without fear that it will be stolen, and can provide incentives for domestic innovation'.[29]

3.2 European Union Emphasizes R&D Collaboration[30]

In its 2003 Report the European Union suggested that Foreign Direct Investment (FDI) in the private sector was the most important channel for technology transfer. The reported actions were based on a broad perspective, covering joint research programmes, the provision of advice and expert assistance, investment assistance programmes, and the establishment of contacts between partners from the North and the South.[31]

In 2009 the EU issued a 103 page-long report, which also covered some of the Member States' activities.[32] The first section deals primarily with Research and Development (R&D) collaboration and the fact that LDCs and developing countries are encouraged to participate in the 6th and 7th Framework programme and other scientifically oriented activities in the EU.[33] There is not a single line in the EU report about creating incentives for European companies to engage in technology transfer. The report is more a report on technical assistance and R&D collaboration under TRIPS Article 67 as an element of general development aid. Furthermore it does not single out LDCs, rather including them along with its dealings with all developing countries. If anything, the report demonstrates that the EU tends to favour those countries with which Member States have had relations in the past.[34]

3.3 Individual Country Reports

Ten EU member states included their reports in the overall EU report. The remainders have not provided comments. The European reports are here presented with

29. *Ibid.*, para. 38.
30. WTO, European Communities Addendum Report on the Implementation of Art. 66.2, 13 Nov. 2009, IP/C/W/536/Add.7, 27 Apr. 2010.
31. *Ibid.*, paras 150–152.
32. *Ibid.* The reporting Member States consist of Austria, Belgium, the Czech Republic, Denmark, Finland, France, Germany, Spain, Sweden and the United Kingdom.
33. Under the 6th Framework Programme for the period 2002–2006 a total of 263 such programmes involved LDCs for a total expenditure of EUR 33,185,824. See European Commission, Research Directorate General, FP6 Final Review: Subscription, Implementation, Participation (June 2008). <ec.europa.eu/research/reports/2008/pdf/fp6-final-review.pdf>.
34. The 2010 EU report was not available at the time of this writing.

reports provided by other OECD countries. The level of detail varies and LDCs are included in a generic way under the notion of developing countries. The reports do not provide enough information to allow an evaluation of the effects of these efforts.

The main impression is that individual countries want to report on development aid in order to show an overall commitment to such important items as eradication of poverty, health care activities, environmental improvements, implementation of social standards and eco-standards, good governance, higher education, sustainable tourism, sustainable land and forest management, food quality and safety system development and training and capacity building for SMEs. The list of submitted activities contains a number of 'buzz words' in line with the political agenda, such as the promotion of employee and human rights, measures for equal rights and the improvement of living conditions for women, but rarely connect to TRIPS Article 66.2. Thus almost anything seems to fall under the notion of technology transfer.

Several country reports, however, contain elements of what could indeed be termed 'incentives provided to their enterprises or institutions for the purpose of promoting and encouraging technology transfer to least developed country Members' though these are sometimes combined with efforts directed to developing countries as a whole.

3.3.1 Spain – A Bottom-Up Approach[35]

Spain reports on its Iberoeka projects, which support technological business cooperation in Latin America. These projects promote direct investment, licensing, franchising, sub-contracting, support joint research projects between private partners and encourage trade in technological goods in all fields except the military.

Projects are based on the fundamental premise of being bottom-up. Participants are free to choose their own criteria with which to formulate, develop and finance Research & Development & Investment (R&D&I) projects in accordance with their needs. Iberoeka promotes the participation of Spanish companies by advising on the presentation of new proposals, the search for partners, and on access to sources of financing. The incentives are both non-financial and relate to identifying potential partners (advice, contacts and the like) but also include up to 100% financing of or insurance for prospective activities. Loans are provided as zero interest credits up to 75% of the budget of Spanish participation with an amortization period of up to 10 years. There is also a possibility of a non-reimbursable tranche of 25% of the credit granted.

In 2008, 36 projects with Spanish participation were approved, with a total budget of EUR 42.9 million.

35. *Ibid.*, s. XI paras 332–337.

3.3.2 United Kingdom – Africa Enterprise Challenge Fund[36]

The UK's Department of International Development co-funds start-ups primarily in sub-Saharan Africa, whether LDCs or not, through yearly competitions with a view to promoting direct investment, licensing, franchising and sub-contracting in agricultural technology.

Applicants for Africa Enterprise Challenge Fund (AFEC) grants must normally be a for-profit company with at least 2 years of audited accounts. The minimum grant is USD 250,000 and the maximum is USD 1.5 million. The applicant must at least match (in cash or kind) the contribution. The business idea must be innovative (a new product, service or business model in an existing market or an existing product, service or business model in a new market). The funds requested must be additional to the applicant's investment and be required for the launch of the project. The business idea must have the potential to be commercially viable and the objective is to leave behind a market system that works better for a large number of rural poor rather than helping a single company to do better business.

Over a six-year period, AFEC is expected to stimulate the raising of over USD 200 million in additional private sector investment for financial and agricultural markets across Africa.

3.3.3 Germany – Public-Private Partnership[37]

PPP are alliances between German Development Cooperation (GDC) and private industry. GDC supplies know-how and contacts, supports concept-development and financing, coordinates the various measures and networks. Private companies contribute technology, capital and expertise and are usually responsible for carrying out the projects on site. The partners share costs and risks equally. The projects are not specific to LDCs and support is provided to European firms as well as to private companies in African partner countries.

Companies submit brief project proposals to GDC which are selected in accordance with five political parameters: (i) Compatibility with German development policy principles; (ii) Contributions of the partners must complement one another, so that all sides achieve benefits and commercial success more economically, effectively and rapidly than they otherwise would; (iii) The private partner must bear 50% of the project cost; (iv) Project scope must clearly go beyond the limits of normal commercial activity; (v) Competitive neutrality should be a guiding principle.

Public Private Partnership (PPP) projects should improve access to available techniques and industrial processes, support the introduction of new technologies in local firms, assist research and educational institutions; provide training in the

36. *Ibid.*, s. XIII paras 349–358.
37. *Ibid.*, s. X paras 317–331.

state-of-the-art and in technology management and production methods. PPP programmes in Nepal and Cambodia are given as examples:

> **Organic Production of Tea in Nepal**: A German company aims to satisfy the current demand for organic and fair trade tea among its customers with organically grown tea from Nepal. The objective of the project is to train tea farmers and processors in the eastern part of the country to produce and market tea which meets the international standards of organic tea production and fair trade. Local advisors are trained who will be able to provide advice to the farmers after the project has ended.
>
> **Good Governance in Cambodia**: A German company allocates software which is adapted for the electronic provision of services to the public administration in Cambodia. This software application has been implemented in two provincial capitals as a showcase. Users are citizens who can register their businesses online and buy licences. The public administration may thus possess a useful instrument in the fight against corruption and the shadow economy.[38]

3.3.4 Denmark: The Business-to-Business Programme[39]

The Danish B2B programme aims at developing the private sector in a range of developing countries by supporting the establishment of long-term and mutually committing partnerships between Danish companies and companies in those countries.

Over 900 partnerships have been supported in their efforts to attract Danish technology and investment. A project should improve access to available techniques, industrial processes, technological information and on-the-job-training. Quality control provided by the Danish partner should also lead to trade in technological goods. The concept appears to be that the B2B programme acts as a facilitator for 'new-born' partnerships that become commercially viable after a period of time and which the partners intend to continue once the support has ceased. Direct Danish investments, licensing, franchising, and sub-contracting arrangements can receive non-financial support in identifying potential partners and financial incentives (100% or co-financing) of prospective activities. The Danish companies must be registered according to Danish law, comply with certain financial requirements and have operated a business in the field covered by the cooperation for at least five years:

> **Vagn Plenge ApS and Sub-Saharan Publishers in Ghana**. Since 1993, the publishing company Sub-Saharan Publishers (SSP) has specialized in printing, distributing and promoting the books of Ghanaian authors. SSP initiated a cooperation with the Danish publishing house Vagn Plenge ApS to upgrade its skills in the fields of publishing, promoting, typography, layout and computer

38. *Ibid.*, para. 322.
39. EU 2009 Report, fn. 30 *supra*, s. VII paras 199–219.

software. A solid understanding of international publishing and the tools required to operate in this line of business has resulted from the partnership. Their improved skills have made it possible for SSP to compete with international publishers and publish a range of books written by Ghanaian authors, which are today incorporated in the curricula of a variety of schools. Additionally, the partnership has made it possible for rights for several Ghanaian works to be sold abroad.

Bording Data A/S in Denmark and TechnoVista Ltd. in Bangladesh have established a Joint Venture – Bording Vista. The overall aim of the partnership is to set up a new software development company and take advantage of the joint capabilities of the partners to develop a competitive position in both their respective home markets and new markets. Bording Vista's goal is to employ IT Bangladeshi professionals, and train them in international standards, technologies and practices related to the Information and Communication Technology (ICT) industry and to work with the newest and most promising software products so as to be trendsetters for the Bangladeshi ICT industry.[40]

3.3.5 Norway: Clean Technology Support[41]

The 2010 Norwegian report[42] only contains 7 pages on LDC focused activities. Norway provides support to countries with a GDP per capita of less than USD 5,295 in accordance with the World Bank's threshold. Norway's focus is primarily on LDCs in Southern Africa, Eastern Africa, Central America and the Mekong area. Technical assistance in areas where Norway has a special competence is the main mode of technology transfer. The schemes include investments in infrastructure (road, water/sanitation, electricity and telecommunication) and environmental technology (clean and renewable energy). Technology and competence, both regarding operational and industrial management and health, safety and environmental issues, are required.

Norway promotes private sector development in developing countries by providing expertise and risk capital (50% support with a higher percentage for LDCs). Commercial projects shall be viable with economic, social and environmental considerations balancing. Support is given to technology transfer cooperation, feasibility studies for establishing joint ventures or foreign subsidiaries as well as public-private partnership arrangements. The arrangements must have an important technology transfer component.

An example of Norway's investment policy is the partnership with the international banana company Chiquita in Matanuska, Northern Mozambique,

40. *Ibid.*, paras 205–206.
41. WTO, Norway 2009 Addendum Report on the Implementation of Art. 66.2 of the TRIPS Agreement, 16 Oct. 2009 IP/C/W/536/Add.4.
42. WTO, Norway 2010 Addendum Report on the Implementation of Art. 66.2 of the TRIPS Agreement, 11 Oct. 2010, IP/C/W/551.

> in a banana plantation motivated by new technology and skills and leading to an upgrading of export-oriented farming. In Uganda, Norfund and the Norwegian hydropower company TronderEnergi have built the 13 MW Bugoye hydropower plant covering 7% of Uganda's electricity needs and reducing dependency on imported oil. TronderEnergi provides the training of the technical staff in Norway as well as back-up technical support.[43]

3.3.6 Switzerland: A Broad and Committed Engagement[44]

The 2010 Switzerland report is in the new format, which takes into account the comments made by LDCs at the 2008 WTO workshop. Switzerland reaffirms its commitment to actively engage in the provision of incentives for enhanced technology transfer to LDCs and is working on continuously improving its activities and reporting in this regard. Emphasis is put on measures which are mainly targeted at LDCs and on a new category of projects with a minimum budget value.

Swiss SMEs are encouraged to invest in commercially viable projects, which meet recognized environmental and social standards. Such investments involve increased business risks and the Start-up Fund co-finances the initial investment phase in the form of a loan that must be repaid within five years. Switzerland operates in domains where it has special expertise such as the manufacturing of precision instruments and chemical products. It is also promoting public health, water supply and sanitation, agriculture, food industry, textiles, clean energy and eco-technology.

The Swiss activities include tax exemptions for Swiss charitable institutions and the provision of credit facilities which support private infrastructure projects by taking on the high costs and risks of early stage project development in situations where the private sector would not otherwise be willing or able to invest. LDC domiciled companies are encouraged to enter European markets and they receive support in the form of marketing consultancy, business partner search, training in export marketing and for participation in trade fairs. A special focus is given to the establishment of public-private partnerships for infrastructure services, environmentally sound technologies and in the field of energy efficiency. In addition, Switzerland supports the transfer of know-how, training and infrastructure in the field of environment monitoring and chemicals management.

> Switzerland has also engaged in capacity building and technology transfer to developing and LDCs in accordance with the CBD and its provisions on access and benefit-sharing. It also seeks new ways of promoting, incentivizing and testing models for technology transfer by way of public-private partnerships. It is in search of vehicles that help to better match the supply of and demand for technology in developed countries and LDCs respectively and to bridge the

43. *Ibid.*, paras 13–14.
44. WTO, Switzerland 2010 Addendum Report on the Implementation of Art. 66.2 of the TRIPS Agreement, 18 Oct. 2010, IP/C/W/551/Add.2, which updates the 2009 detailed report, IP/C/W/536/Add.3.

gap between the pre-commercial phase and the point when the project becomes commercially viable.

Even if Switzerland enumerates a series of projects in line with the LDC countries request, the identification of the projects is often too summary to provide information for best practices. If this improvement were made, the report could serve as a guiding document for other countries.

3.3.7 Canada: Confirms a New Reporting Trend[45]

Canada's 2010 report follows the new trend in national reporting by accommodating itself to the criticisms advanced by the LDCs. With reference to the 2008 Workshop, Canada is committed to enhancing its reporting and addressing the concerns of the LDCs. It focuses on reporting activities that are targeted at LDCs rather than developing countries more generally. In addition, the intention is to report on technology transfer specifically, as distinct from technical assistance (TRIPS Article 67). The report thus has a much narrower focus than those in previous years.

The report explains how different Canadian agencies are providing support to enhance the environment in LDCs and much is being done by Canada. The annex lists a number of more specific projects, which Canada would like to characterize as technology transfer under the TRIPS Article 66.2. All these projects, however, appear to be directed to non-profit seeking activities. The national entity is often an agency, a fund or a trust and the objective is to train and educate rather than to build technology based commercial activities which may move a LDC into a more advanced stage. Two examples are summarized as follows:

> **Water Harvesting and Institutional Strengthening in Amhara.** The purpose of the project is to increase the food security of poor male and female farmers in Ethiopia through improved water management and it is managed by a Canadian consortium consisting of Hydrosult Inc., Agrodev Canada Inc. and Oxfam Canada. It will provide expertise and the know-how required to plan, design and implement sustainable water management through capacity building: training and demonstrations.
>
> **Prefabricated Bamboo Housing in Africa.** The overall goal of the project is to assess the value of bamboo as an eco-friendly, low-cost building material which can be used to meet urgent housing needs for the poor in Africa. The project supports research, technology transfer, capacity building, networking and innovative design strategies to set up pre-processing and production centres. It builds local production capacities in Uganda and Kenya. Again this is not a project involving Canadian firms. Technology, technical training and skilled individuals are provided by China and India.[46]

---

45. WTO, Canada 2010 Addendum Report on the Implementation of Art. 66.2 of the TRIPS Agreement, 25 Oct. 2010, IP/C/W/551/Add.6.
46. *Ibid.*, paras 20, 33, Annex I and III.

There are several other projects mentioned in the report. They all provide support to LDCs, but often through non-commercial entities and only one case contains a solid reference to intellectual property rights. The impression the catalogue gives is that Canada is still hesitant to provide incentives to commercial entities interested in starting commercial ventures. However, the report is a substantial improvement over earlier ones.

3.3.8 Australia Broadly Supports Education[47]

The Australian 2009 filing is another example of a technical assistance oriented report. Australia provides training, education and knowledge including governance capacity building and technical assistance to LDCs. Australia reports on a large number of government funded projects – primarily in the Pacific Rim and in certain Association of Southeast Asian Nations (ASEAN) countries. Technology transfer occurs through exports and outward or direct investment by Australian firms in other countries. The education of LDC nationals at Australian universities comprises the most effective means of transferring technology and knowledge. Nothing in the Australian report suggests that it is providing incentives to private companies in Australia or elsewhere to disseminate technology to LDCs.

The Australian 2010 report is no different and again enumerates general aid and technical assistance projects without any reference to a single production technology project. Australia is focusing its resources on South East Asia and the Pacific Rim – especially small island nations. The 2010 report again emphasizes the education of LDC nationals at Australian universities. Its training programmes are to a large extent oriented towards the environment, energy and climate change.[48]

3.3.9 New Zealand Promotes Water and Waste Projects[49]

Both the 2009 and the 2010 New Zealand reports emphasize the technology transfer aspects of its development assistance. They mention a number of bilateral aid programmes with Pacific LDCs for which it provides funding. These include funding government research into how New Zealand R&D may promote Pacific development, water quality and waste treatment projects and a variety of training and capacity building projects aiming to assisting LDCs and developing countries. Specific projects in named LDCs are mentioned as well as the sums expended upon each, but the reports do not indicate anything specifically linked to the fulfilment of Article 66.2 obligations.

---

47. WTO, Australia 2009 Addendum Report on the Implementation of Art. 66.2 of the TRIPS Agreement, 12 Oct. 2009, IP/C/W/536/Add.2.
48. WTO, Australia 2010 Addendum Report on the Implementation of Art. 66.2 of the TRIPS Agreement, Addendum, 14 Oct. 2010, Addendum, IP/C/W/551/Add.1.
49. WTO, New Zealand 2009 Addendum Report on the Implementation of Art. 66.2 of the TRIPS Agreement, 22 Oct. 2009, IP/C/W/536/Add.6.

3.3.10 Japan Supports Training[50]

Japan's 2009 report contains 7 pages, which do not deal with Article 66.2 issues. The report merely highlights different development initiatives, which take the simple form of training organized by Japanese authorities. Unfortunately the October 2010 Japanese report is very much in line with the 2009 report and shows no sign of the new reporting requirements. Japan lists a number of technical assistance projects primarily aiming at training governmental officials in South East Asia. It also emphasizes aid for trade projects, but does not report on a single activity with any transfer of actual production technology.[51]

4 DEVELOPING COUNTRIES REQUIRE ADHERENCE TO THE OBLIGATIONS ESTABLISHED IN THE CONVENTIONS

The developed countries generally aim at a broader interpretation of the technology transfer notion as encompassing training, education and know-how, along with any capital component.[52] For example, the EU has access to more information than it discloses in its reports regarding its partnerships with many African LDCs, under the Cotonou Agreement.[53] A strategic study on the Enforcement of Intellectual Property Rights in Third Countries[54] is also likely to examine the results of technology transfer to LDCs (amongst many other issues). Such detailed information is, however, reduced to very general statements in the reports to the TRIPS Council which also do not address 'best practices' in transferring production technology.

Developing and least developed countries appear to favour a more narrow definition of technology transfer covering physical objects or equipment and skills and the human aspects of technology management and learning; hands on information; and production arrangement linkages within which technology is operated. They have registered frustration with the current level of reporting and the consequent difficulties in measuring the result of such technology transfer as has actually occurred.

50. WTO, Japan 2009 Report on the implementation of Art. 66.2 of the TRIPS Agreement, dated 2 Oct. 2009, IP/C/W/536.
51. WTO, Japan 2010 Addendum Report on the implementation of Art. 66.2 of the TRIPS Agreement, 20 Oct. 2010, IP/C/W/551/Add.3.
52. Foray, Dominique, 'Technology Transfer in the TRIPS Age: The Need for New Types of Partnerships, between the Least Developed and Most Advanced Economies', ICTSD Issue Paper No.23 at 10 (May 2009), <http://ictsd.org/downloads/2009/07/foray_may2009.pdf>.
53. See Partnership Agreement between the Members of the African, Caribbean and Pacific Group of States on the One Part and the European Community and its Member States on the Other Part, Signed in Cotonou Benin on 23 Jun. 2000, entered into force 1 Apr. 2003 (as amended).
54. See 2005 Strategy for the enforcement of intellectual property rights in third countries, OJ 2005 C 129/5 (26 May 2005).

The complaints continue.[55] This issue has consistently been addressed by LDCs at TRIPS Council Meetings during the period 2005–2010.[56]

5 CONCLUSION: ENIGMATIC PROVISIONS AND UNCREATIVE REPORTING

The technology transfer provisions in TRIPS establish in Article 7 a principle, which is then followed by an operative provision in Article 66.2. If the latter provision identifies the addressee and the obligation is clear it would, in EU terminology, be regarded as a provision with direct effect, creating binding obligations. Yet experience suggests that technology transfer provisions in TRIPS and other international conventions[57] end up as soft law promises only.[58]

55. WTO, TRIPS Council meetings 25–26 and 28 Oct., 29 Nov., and 6 Dec. 2005 IP/C/M/49 para. 301 The Representative of Rwanda suggests establishing committee of experts to evaluate results of programmes; 25–26 Oct. 2006 IP/C/M/52 para. 159 and 5 Jun. 2007IP/C/M/54 para. 106 The Representative of Bangladesh, on behalf of LDC Group requesting that developed countries clarify how reported assistance led to technology transfer; 13 Mar. 2008 IP/C/M/56 para. 73 The Representative of Uganda suggested establishing a monitoring mechanism to determine whether technology transfer had occurred; 27, 28 Oct. and 6 Nov. 2009 IP/C/M/61 para. 165, The Representative of Tanzania, on behalf of the LDC's, reminded the developed countries of the LDC's October request to establish a monitoring mechanism; 2 Mar. 2010 IP/C/M/62. The Representative of Bangladesh admonished the reporting countries to 'avoiding mixing their contents with usual technical assistance.'
56. WTO, TRIPS Council, Communication from Bangladesh, Priority Needs for Technical and Financial Cooperation, dated 1 Mar. 2010 IP/C/W/546, (23 Mar. 2010), para. 9: 'Although Article 66.2 of the TRIPS Agreement requests developed country Members to provide incentives for technology transfer to build the capacity of LDCs, there remain questions about the effectiveness of the provision in the context of Bangladesh'. See also WTO, TRIPS Council, Communication from Rwanda, Priority Needs for Technical and Financial Cooperation, 7 Jun. 2010 IP/C/W/548, Annex I, Report of Needs Assessment Diagnostic, para. 33: 'Most of the technology used in Rwanda originates from abroad. This technology is transferred into the Rwandan economy through various means including FDI, trade, and migration as well as licensing . . . In the context of TRIPS Article 66.2, there was, however, no evidence that Rwanda is benefiting from specific programmes from developed countries.'
57. The technology transfer provisions in the CBD and Climate Change Framework are worded similarly to the language of the TRIPS Agreement and should also be seen as mandatory. However, what has developed is very different. These conventions have been the subject of a protracted discussion in organized fora on how to interpret the 'mandatory' stipulations. This discussion has been on-going for more than a decade and could perhaps be best described as a joint learning process on all sides where the positions are gradually approaching each other as a result of which enforceable provisions may eventually appear. This is, however, at least as slow and complex as changing a provision in the TRIPS Agreement.
58. See WTO, Ministerial Conference Fourth Session Doha, 9–14 Nov. 2001, Decision of 14 Nov. 2001, Implementation-Related Issues and Concerns, WT/MIN (01)/17, DEC/1, 41 I.L.M. (2002) (WTO 2001 Doha Decision), Art. 11.2:

    Reaffirming that the provisions of Article 66.2 of the TRIPS Agreement are mandatory, it is agreed that the WTO TRIPS Council shall put in place a mechanism for ensuring the

Article 66.2 is both a far-reaching obligation and one which is extremely vague and enigmatic. There is no definition of the notions, 'technology transfer' and 'technology assistance'. Nor is it clear who carries the relevant obligations. The notion 'LDC' is precise, but it is unclear which countries are 'developed'.[59] There is no statistical reporting mechanism on actual transfers. Nor is it clear that Article 66.2 contains a requirement that transfer activities be separated from technical assistance or development aid actions directed to developing countries in general. Article 66.2 does not require that incentives have any effect and it does not provide that such effects be monitored.

It is sometimes suggested that it is time to amend the international conventions in order to provide more substance to the technology transfer provisions and how they are reported. However tempting this may sound, it is not an easy process. In spite of political agreement in 2003 to amend TRIPS Article 31f to allow more liberal compulsory licensing activities in the pharmaceutical field, formal steps have still not been finalized.[60] Making a seemingly uncontroversial change to any international agreement has proved to be an extremely burdensome process. Conflicting opinions between developed and developing countries and different regions add to the difficulties and may explain the new trend to prefer bilateral agreements such as the TRIPS Plus agreements or plurilateral forum shifting such as the Anti-Counterfeiting Trade Agreement (ACTA agreement), discussed by Atik in Chapter 6 below.

It is also frequently suggested that new institutions or new mechanisms should be instituted. There are, however, reasons to move cautiously. Today the problem rather seems to be that there are too many institutions and too much forum shifting for measures to be really efficient. At present it appears realistic to coordinate the international activities and search for solutions within existing international agreements.

---

monitoring and full implementation of the obligations in question. To this end, developed country Members shall submit prior to the end of 2002 detailed reports on the functioning in practice of the incentives provided to their enterprises for the transfer of technology in pursuance of their commitments under Article 66.2. These submissions shall be subject to a review in the WTO TRIPS Council and the information shall be updated by Members annually. The receipt of a strong mandate from the Council of Ministers in 2001 is not a clear indication that there will, one day, be a hard law mandatory provision.

59. There are no WTO definitions of 'developed' or 'developing' countries and each Member announces its status, subject to challenge. See 'Who are the developing countries in the WTO?' <http://www.wto.org/english/tratop_e/devel_e/d1who_e.htm>.

60. As called for in para. 11 of the 2003 Decision, the TRIPS Council adopted a Protocol Amending the TRIPS Agreement by a Decision of 6 Dec. 2005 (WT/L/641). The Protocol is open for acceptance until 31 Dec. 2011 or such later date as may be decided by the Ministerial Conference (WT/L/785). In accordance with Art. X para. 3 of the WTO Agreement, the Protocol will enter into force upon acceptance by two-thirds of the WTO Members. As of 18 Dec. 2010 some three delegations including the EU had notified their acceptance. See WTO, WTO TRIPS Council, 10 Dec. 2010, Annual Review of the Decision on the implementation of para. 6 of the Doha Declaration on the TRIPS Agreement and Public Health, IP/C/57 n.y.p.

The fact that there are divergent views on how to interpret the obligations in TRIPS Article 66.2 is no major surprise. Most country reports are vague on the actual transfer of production technology. None of them provides any measurement of the technology transfer resulting from any incentives taken. It is not possible to establish what results the international agreements and the ensuing activities have led to and whether there is indeed a flow of technology that has benefited the developing world, the LDCs especially.

On the other hand it appears as if the reports are becoming more sophisticated when it comes to addressing the needs of LDCs. Several countries say in their 2010 reporting that they have adapted to the 2008 WTO workshop discussion and are now more specific. The United States believes that the effective functioning of Article 66.2 of the TRIPS Agreement requires a robust dialogue between developed and LDC Members in order to target incentives in a way that is most responsive to the self-identified technology transfer interests and needs of LDC Members.[61] In its 2010 report[62] Switzerland states that Members have come to a better understanding of what the terms 'technology transfer' and 'incentive' should cover in the context of the implementation of Article 66.2 of the TRIPS Agreement. This understanding and the continued dialogue between developed countries and LDCs are promising and may contribute to further improving the developed countries' annual reports, both in terms of format and content.

In conclusion, the overall impression after a penetration of Article 66.2 TRIPS reporting is that only a few countries actually report and that the reporting countries overload their reports with matters which are not dealt with under the provision. The reports do not focus on LDC needs nor do they primarily deal with incentives to their own companies to foster technology transfer. The sense is that reporting countries are anxious to show a commitment to the overall goal of reducing poverty and that there is a general lack of imagination and creative thinking how to approach the obligation in Article 66.2. The reports do not provide a basis for extracting best practices recommendations. Thereby a crucial reason for the reporting is not satisfied.

However, on a positive note, the national reports appear to be improving after the 2008 workshop. Some countries have started to provide focused production technology reports, but too much still focuses on technical assistance or even development aid in general. A stronger focus on the specific needs of the LDCs can also be detected in the reports. A clear benefit is that these reports are briefer and more to the point. Still, more can be done to describe successful private industry initiatives.

61. US 2010 Report, fn. 28 *supra*, para. 4.
62. Switzerland 2009 Report, fn. 44 *supra*, para. 36.

Chapter 4

Technology Transfer and Competition Law: Options for Developing Countries

*Tu Thanh Nguyen**

Technology is vital to competitiveness, economic growth and sustainable development for both developed and developing countries. It provides 'a solid potential for improving the well-being of all peoples';[1] but a large technological gap between the developed and developing worlds remains. This gap is widening due to rapid technological advances in developed countries in comparison to relatively slow advances in most developing countries.[2] As a result, most developing countries continue to be net importers of technology. However, it is almost impossible for

* This chapter draws on the author's previous works, including T.T. Nguyen, *Competition Law, Technology Transfer and the TRIPS Agreement: Implications for Developing Countries* (Cheltenham: Edward Elgar, 2010); and T.T. Nguyen, 'Competition Law and Access to Pharmaceutical Technology in the Developing World', 28 *Biotech L Rep 693* (2009).

1. United Nations Declaration on the Establishment of a New International Economic Order, Resolution 3201(S-VI) of 1 May 1974, A/RES/S-6/3201, para. 1.
2. UNCTAD, *Intellectual Property in the World Trade Organization: Turning It into Developing Countries' Real Property* (New York and Geneva: United Nations, 2010), 1; World Bank, *Global Economic Prospects 2008: Technology Diffusion in the Developing World* (Washington DC: World Bank, 2008), 92; UNCTAD, *The Least Developed Countries Report 2007: Knowledge, Technological Learning and Innovation for Development* (Geneva: UNCTAD, 2007), 3.

Hans Henrik Lidgard, Jeffery Atik & Tu Thanh Nguyen, *Sustainable Technology Transfer*, pp. 65–88.

them to undertake the extensive reconstruction of their domestic intellectual property laws in response, without infringing well-established globalized minimum standards under the Agreement on Trade Related Aspects of Intellectual Property Rights (TRIPS)[3] of the World Trade Organization (WTO) or TRIPS-plus standards.

In practice, most of the intellectual property rights (IPRs) comprising this technology are multinational companies' private property of which can only be transferred through negotiated commercial agreements. On the one hand, these companies offer good opportunities for developing countries to obtain access to technology. On the other hand, if not effectively controlled by appropriate legislation and its enforcement in host countries, this often leads to maximizing the profits of the companies from technology transfer by, among other things, (i) refusing to use and/or license, (ii) charging excessive prices for transferred technology or technology-embodied products, or (iii) incorporating anti-competitive restrictions into technology transfer agreements.[4] These practices hinder the efforts of local entrepreneurs seeking to acquire technology on open markets at reasonable prices and under reasonable terms and conditions. It also prevents consumers' access to technology-embodied products at affordable prices. In addition, such practices may frustrate the socio-economic policies of developing country governments. The benefits of developing countries and consumer welfare may thus be adversely affected due to both the market power of the multinational companies and asymmetric information in technology transfer.

Consequently, investigating and arriving at a re-definition of the interface between IPRs and competition law in a manner conducive to promoting access to technology has attracted a great deal of attention at both national and international levels.[5] Theoretically, developing countries may invoke competition flexibilities in the TRIPS Agreement in order to adopt and apply an appropriate measure of domestic competition law to address anti-competitive practices with respect to technology transfer. However, the adoption and/or application of these flexibilities have so far proved difficult for most developing countries.

Sections 1 and 2 of this chapter review respectively the interface between IPRs and competition law, and the reaction of various international forums to this issue. Section 3 then examines whether and to which extent developing countries may enact and enforce domestic competition law in order to promote access to technology. Finally, Section 4 arrives at some conclusions.

---

3. WTO, *The Legal Texts: The Results of the Uruguay Round of Multinational Trade Negotiations* (Cambridge: Cambridge University Press, 2010), 320–353.
4. See C.P. Jefferies, 'A Preliminary Evaluation of the Proposed Text', in *International Technology Transfer: The Origins and Aftermath of the United Nations Negotiations on a Draft Code of Conduct*, ed. S. Patel et al. (The Hague: Kluwer Law International, 2001), 22–25.
5. J.H. Reichman, 'Intellectual Property in The Twenty-First Century: Will the Developing Countries Lead Or Follow?', 46 *Hous L Rev* 1160 (2009).

1 INTERFACE BETWEEN IPRS AND COMPETITION LAW

The interface between IPRs and competition law has been controversial. Exclusive rights granted by intellectual property law seek to protect IPRs. In doing so, the intellectual property system seems to permit right holders to 'raise price above marginal cost, creating deadweight losses by raising the price to consumers' and limit competition.[6] Viewed in this light, intellectual property protection has been criticized for creating monopoly rights that reduce consumer welfare. In contrast, competition law generally reflects the premise that consumer welfare is best served by removing impediments to competition. From the intellectual property perspective, competition law may be considered as an interventionist instrument, which infringes right holders' entitlements, thereby affecting the very foundation of intellectual property law. Consequently, intellectual property law may be considered to conflict with competition law and vice versa.

However, the relationship between competition law and intellectual property law may be reflected by the relationship between static efficiency and dynamic (or innovation) efficiency.[7] Both measure the efficiency of the market respectively at a particular point of time and over time. Static efficiency refers to the reduction of costs and prices, and the increase of quantity at given prices. It consists of productive efficiency and allocative efficiency.[8] On the other hand, dynamic efficiency involves the increase of novelty and improvement through new products, processes, services, or even new markets.

Competition law protects static efficiency by eliminating artificial restraints, i.e. collusion and abuse of a dominant position, and promoting the entry of new competitors in order to leave more benefits and surplus in the hands of consumers. Nevertheless, the pursuit of static efficiency alone will not suffice. The major impetus to growth and welfare comes from dynamic efficiency, which refers to gains from entirely new ways of doing business through investment in innovation, aiming at creating new or improved products and/or reducing the costs of existing ones.[9] Innovation furthers economic efficiency more than any policy that reduces production costs and/or drives prices to marginal cost. However, that does not come without an investment of time, effort, and human and financial resources. If competitors can in the short-term, free ride on a new innovation and drive production costs and prices to their lowest levels, it adversely affects innovators' and investors' chances to recoup capital expenditures and benefit from such investment. That, in turn, impairs innovation and growth. Dynamic efficiency,

---

6. M.A. Lemley, 'Property, Intellectual Property, and Free-riding', 83 *Tex Law Rev* 1059 (2005).
7. J.T. Rosch, 'The Role of Static and Dynamic Analysis in Pharmaceutical Antitrust', (18 Feb. 2010). <www.ftc.gov/speeches/rosch/100218pharmaantitrust.pdf>.
8. Productive efficiency is achieved when products are produced cost-effectively under existing technology, i.e. at the lowest possible cost-level. Allocative efficiency occurs when the prices of products are driven down to a level close to production costs, i.e. marginal cost.
9. G.F. Masoudi, 'Intellectual Property and Competition: Four Principles for Encouraging Innovation', (11 Apr. 2006) <www.usdoj.gov/atr/public/speeches/215645.pdf>; Rosch, fn. 7 *supra*.

therefore, cannot be gained without some limited form of exclusivity, which may, nevertheless, lead to static efficiency losses.

It should be noted that the meaning of 'monopoly' conferred on a right holder under intellectual property law is merely the temporary right to exclude others from commercialization efforts, i.e. a negative right. It is analogous to a purely legal monopoly; but it is not equivalent to an economic monopoly, i.e. a dominant position, under competition law, where a firm gains power enabling it to 'behave, to an appreciable extent, independently on the market'.[10] Accordingly, a short-term approach is replaced by a longer-term one acknowledging some restrictions of competition today in order to promote competition in new products and processes tomorrow. It is now generally agreed that the goals of intellectual property law and competition law, rather than being in conflict, are complementary and mutually reinforcing. Both of them share the common purpose of promoting competition, innovation and commercialization of such innovation, and enhancing and benefiting consumer welfare as well as efficiently allocating economic resources. In other words, the 'intellectual property system further advances follow-on innovation through public disclosure of the invention'; while competition law, by preventing anti-competitive conduct threatening free markets, 'is to appropriately foster and protect the competitive environment [which] pushes companies to constantly innovate and allows them to profit when they do'.[11] Both law systems 'work in tandem to bring new and better technologies, products, and services to consumers at lower prices'.[12]

Although intellectual property and competition laws aim to promote innovation, both will discourage it if they are pursued either too vigorously or too weakly. Intellectual property law often promotes innovation, but may also impede it if IPR-transaction costs 'are high or rights are fragmented in a way that makes them hard to access'.[13] Too much intellectual property protection 'can produce costly monopolies or exclusive rights that others must either license or innovate around'.[14] In practice, many companies, especially those in high technology sectors, maintain and use patents to 'block the development of a new, competing product rather than

10. See, e.g., Case T-321/05, *AstraZeneca v. Commission*, Judgment of 1 Jul. 2010, not yet reported, para. 265, appeal pending, Case C-457/10 P.
11. C. Varney, 'Promoting Innovation through Patent and Antitrust Law and Policy', (26 May 2010). <www.justice.gov/atr/public/speeches/260101.pdf>.
12. U.S. Department of Justice and Federal Trade Commission, 'Antitrust Enforcement and Intellectual Property Rights: Promoting Innovation and Competition' (April 2007), <www.justice.gov/atr/public/hearings/ip/222655.pdf>, 1; U.S. Department of Justice and Federal Trade Commission, 'Antitrust Guidelines for the Licensing of Intellectual Property' (6 Apr. 1995). <www.usdoj.gov/atr/public/guidelines/0558.pdf>; European Commission Guidelines on the Application of Art. 81 of the EC Treaty to Technology Transfer Agreements, OJ 2004 C 101/2, para. 7.
13. Ian Hargreaves, 'Digital Opportunity: A Review of Intellectual Property and Growth' (May 2011), <www.ipo.gov.uk/ipreview-finalreport.pdf>, 10.
14. H. Hovenkamp, *The Antitrust Enterprise: Principle and Execution* (Cambridge, MA: Harvard University Press, 2005), 249.

for protecting' their own products.[15] On the other hand, competition law may, but not always, promote innovation. If its enforcement is pursued so aggressively that competitors are permitted easy access to use their competitor's innovation, few incentives will remain to invest in and commercialize such innovation.[16]

In any case, the fact that intellectual property law grants exclusive rights for commercial exploitation does not imply that IPRs are immune from competition law's intervention.[17] IPRs do not confer some privileges to violate competition law. The existence of such legal monopoly rights does not, in itself, infringe competition law; yet such rights may be exercised and exploited in such a way, and to such an extent, as to infringe competition law. Competition law should be applied to ensure that consumers can benefit from the best quality products at the lowest prices. That is why, despite the fact that the exercise of IPRs is already extensively regulated under intellectual property law by rules of scope and duration as well as various exceptions,[18] an additional filter of regulation is provided by competition law. This second filter aims at ensuring that a grant of exclusivity under intellectual property law is not abused (or misused) by anti-competitive licensing agreements and monopolistic conduct that denies parties access to the technology as well as technology-embodied product markets and harms consumer and social welfare. Competition law, nevertheless, cannot serve as a substitute for bad intellectual property law which cannot establish an internal balance within itself, i.e. as an initial filter.

Briefly, IPRs should be regarded as being in themselves pro-competitive unless unilateral or concerted practices transform them into an anti-competitive tool. From the competition law perspective, IPRs are viewed as private property rights which may be abused. Both intellectual property law and competition law, respectively the first and the second filters, should work in tandem to control and prevent IPR-related abuses. They should both be seen as part of a competition and innovation policy in which intellectual property law creates a market for innovation and commercialization of innovation, while competition law controls and corrects malfunctions of that market caused by anti-competitive practices.

---

15. European Commission, 'Pharmaceutical Sector Inquiry: Final Report' (Sector Inquiry) (8 Jul. 2009), <http://ec.europa.eu/competition/sectors/pharmaceuticals/inquiry/staff_working_paper_part1.pdf>, paras 1117–1118.
16. OECD, 'Intellectual Property Rights', DAF/COMP(2004)24 (21 Jan. 2005), <www.oecd.org/dataoecd/61/48/34306055.pdf>, 17–18.
17. The U.S. Federal Circuit in *US v. Microsoft Corp.*, 253 F.3d 34, 63 (DC Cir. 2001) held that the proposition that the exercise of IPRs lawfully acquired cannot give rise to antitrust liability 'is no more correct than the [one] that use of one's personal property, such as a baseball bat, cannot give rise to tort liability'.
18. Under the TRIPS Agreement, an invention is patentable if it is new, non-obvious, and useful (Art. 27.1). Furthermore, there are conditions concerning sufficiently clear and complete disclosure of invention on patent applicants (Art. 29). Even if the invention meets those criteria, it may be excluded from patentability if it is contrary to *ordre public*, or morality, or other specific circumstances (Art. 27.2 and 27.3). The monopoly rights of the patent holder are limited to a specific period of time, but at least 20 years counted from the filing date (Art. 33), with some exceptions and other use not needing authorization of the patent holder (Arts 30 and 31).

2 INTERNATIONAL REACTIONS

In order to promote, set forth principles on, and eliminate restrictive business practices in technology transfer, negotiations on a draft of the International Code of Conduct on the Transfer of Technology (ToT Code) took place under the aegis of the United Nations Conference on Trade and Development (UNCTAD) from the 1970s to 1985. The 1985 version of the ToT Code (Chapter 4) listed fourteen restrictive practices, which were forbidden to incorporate into a technology transfer agreement.[19] Due to dissension between developed and developing countries and changes in the world economic and political situation,[20] the ToT Code negotiations never resulted in the adoption of an international legal document.

After the unsuccessful negotiations regarding the ToT Code, there remained serious concerns in developing countries regarding the adverse effects of IPR-related anti-competitive practices. These came to the fore during the negotiations regarding the TRIPS Agreement from 1986 to 1994.[21] As a result of mutual concessions, even though the TRIPS Agreement is an international convention dealing with IPRs, it also contains a modicum of competition rules, such as in Articles 8.2, 31(k), and 40.[22] Under Article 8.2, a WTO Member is permitted to apply 'appropriate measures' to prevent the abuse of IPRs, or practices that 'unreasonably restrain trade or adversely affect international transfer of technology'. Article 31(k) permits the use of a patent's subject matter without a patent holder's authorization in order to remedy an anti-competitive practice conducted by the holder. Article 40 acknowledges that some terms and conditions in a technology transfer agreement may be anti-competitive; therefore, a WTO Member may appropriately prevent or control them. These provisions were intensively discussed in the WTO Working Group on the Interaction between Trade and Competition Policy from 1996 to 2003.[23] In addition, issues concerning technical cooperation and assistance in the area of IPR-related competition law under Article 67 of the TRIPS Agreement were raised by Brazil in the TRIPS Council meeting in June 2008;[24] and they continue to be under discussion in this Council.

There is no definition of the term 'anti-competitive practice' in the TRIPS Agreement as well as in other WTO agreements. It is up to national law to

19. UNCTAD, *Compendium of International Arrangements on Technology Transfer: Selected Instruments* (Geneva: United Nations, 2001), 266–269.
20. S.K. Sell, *Power and Ideas: North South Politics of Intellectual Property and Antitrust* (New York: State University of New York Press, 1998).
21. UNCTAD-ICTSD, *Resource Book on TRIPS and Development* (Cambridge University Press, Cambridge, 2005), 543–546.
22. Articles 6, 31(c) & 37.2 of the TRIPS Agreement, to some extent, may be also regarded as competition rules.
23. R.D. Anderson, 'Competition Policy and Intellectual Property in the WTO: More Guidance Needed?', in *Research Handbook on Intellectual Property and Competition Law*, ed. J. Drexl (Cheltenham: Edward Elgar, 2008), 451–473.
24. WTO, 'Minutes of Meeting of the TRIPS Council on 17 June 2008', IP/C/M/57.

determine what an 'anti-competitive practice' is, as long as consistency and appropriateness requirements[25] under the TRIPS Agreement are satisfied (but such determination may be subject to the review of a panel and the Appellate Body under the WTO dispute settlement mechanism). This is, to some extent, confirmed by the Court of First Instance (now the General Court) of the European Union (EU) – a WTO Member – in *Microsoft v. Commission*.[26] In this case, the Court held that the TRIPS Agreement does not prevent WTO Members' competition authorities from imposing remedies limiting or regulating the exploitation of IPRs 'held by an undertaking in a dominant position where that undertaking exercises those rights in an anti-competitive manner', and WTO Members 'are entitled to regulate the abusive use of such rights in order to avoid effects which harm competition'.[27]

Competition flexibilities under the TRIPS Agreement provide broad discretion for WTO Members to enact and apply domestic competition law in addressing both anti-competitive unilateral abuses of right holders and anti-competitive contractual restraints in technology transfer agreements. However, developing countries rarely apply domestic competition law to promote access to technology. This is due to many reasons, especially the under-development of competition law, lack of capacity, and the complexity of the interface between technology-related IPRs and competition law in these countries. That is likely not to remain the case in the near future because strong competition policy in general and technology-related competition law in particular are 'not just a luxury to be enjoyed by rich countries, but a real necessity for those striving to create democratic market economies'.[28] Competition policy/law has been increasingly recognized as indispensable to growth and development in the interdependent knowledge-based world; although there are 'different perspectives on the right balance between markets and states, reflecting different backgrounds and historical experiences'.[29]

Consequently, issues concerning the use of competition law to promote access to technology have, once again, recently received much attention in developing countries and in international forums. There is growing awareness among the nations of the developing world of the adverse effects of IPR-related anti-competitive practices on their technological levels and economic development.

---

25. The consistency requires that domestic competition laws and regulations and their enforcement must: (i) comply with the fundamental principles of the TRIPS Agreement; and (ii) not deprive other Members' nationals of the minimum standards of IP protection established by that Agreement. The appropriateness is basically scrutinized under the so-called 'necessity test' or even 'proportionality test'. See T.T. Nguyen, *Competition Law, Technology Transfer and the TRIPS Agreement: Implications for Developing Countries* (Cheltenham: Edward Elgar, 2010), 48–52.
26. Case T-201/04, *Microsoft v. Commission*, [2007] ECR II-3601.
27. *Ibid.*, para. 1192.
28. J.E. Stiglitz, 'Competing over Competition Policy', (2001), <www.project-syndicate.org/commentary/stiglitz5>.
29. P. Supachai, 'Opening Address for Sixth UN Conference to Review all Aspects of the Multilaterally Agreed Equitable Principles and Rules for the Control of Restrictive Business', (8 Nov. 2010). <www.unctad.info/en/6th-UN-Conference-on-Competition-Policy/Conference/Opening-Address-by-Secretary-General-of-UNCTAD/>.

They increasingly recognize the potential benefits derived from IPR-related competition law enforcement. According to the UNCTAD Model Law on Competition, the competition law of a country is recommended to be applied to IPR-related transactions; technology transfer should be subject to scrutiny for restrictions or abuses, but the legal exclusivity granted by intellectual property law must be taken into account in that process. UNCTAD also contends that it is 'necessary to draw up specific regulations' for the application of competition law in the field of IPRs/technology transfer.[30]

IPR-related competition issues were addressed in four out of forty-five recommendations under the Development Agenda adopted in 2007 by the World Intellectual Property Organization (WIPO).[31] On that basis, a project was initiated in January 2010 by the WIPO Committee on Development and Intellectual Property (CDIP) in order to promote a better understanding of the interface between intellectual property and competition law/policy.[32] In the context of IPRs and public health, the World Health Organization (WHO) adopted its global strategy and plan of action on public health, innovation and intellectual property in May 2008 (revised in 2009), which encourages and promotes the use of competition law and policy in order to improve availability and affordability of pharmaceutical technology. In particular, one of the actions recommended to developing countries by the WHO is taking appropriate measures consistent with the TRIPS Agreement to prevent IPR abuse or practices which adversely affect the transfer of pharmaceutical technology.[33] Moreover, from the perspective of human rights, the United Nations Commission on Human Rights has encouraged 'States to consider the elaboration of competition laws that prevent abuses of IPRs that lead to violations of the right to health – in particular restrictive licensing practices or the setting of high prices for essential drugs'.[34]

In practice, it appears that developing countries have been attempting to use exceptions and limitations as permitted in the TRIPS Agreement and regulated in domestic intellectual property law, to increase access to technology, particularly that relating to public health. However, such limited flexibilities alone cannot solve all of the issues relating to access to technology. This is illustrated by the fact that IPR-related anti-competitive practices, particularly in the pharmaceutical sector, are now subject to intensive competition/antitrust law scrutiny in both the EU and

30. UNCTAD, *Model Law on Competition: Substantive Possible Elements for a Competition Law, Commentaries and Alternative Approaches in Existing Legislation* (New York and Geneva: United Nations, 2010) (definitions and scope of application (Art. 2) and Commentaries 19 and 20).
31. Recommendations 7, 22, 23, and 32 under the WIPO Development Agenda, (2007). <www.wipo.int/export/sites/www/ip-development/en/agenda/recommendations.pdf>.
32. WIPO Committee on Development and Intellectual Property, 'Progress Reports on Development Agenda Projects', WIPO Doc. CDIP/6/2, (1 Oct. 2010).
33. WHO, 'Global Strategy and Plan of Action on Public Health, Innovation and Intellectual Property' WHA61.21 (24 May 2008), revised in WHA62.16 (22 May 2009).
34. United Nations Commission on Human Rights, 'The Impact of the Agreement on Trade-Related Aspects of Intellectual Property Rights on Human Rights – Report of the High Commissioner', E/CN.4/Sub.2/2001/13, (27 Jun. 2001), para. 64.

the United States (US). In July 2009, the European Commission released a final report on pharmaceutical sector inquiry dealing with the alleged obstacles to market entry for pharmaceutical products, including generic and innovative ones. The main conclusion of the report is that competition law enforcement, among other things, can be utilized in order to improve the competitive functioning of the pharmaceutical market.[35] This is also reflected in the *AstraZeneca* judgment delivered by the EU General Court in July 2010,[36] under which the Court confirmed that AstraZeneca had abused its dominant position in breach of Article 102 of the Treaty on the Functioning of the European Union (TFEU) by: (i) providing patent authorities misleading launch date information when applying for a special protection certificate for Losec, and (ii) withdrawing its marketing registration for an older version of Losec in certain Member States to impede generic producers and parallel traders. In the US, antitrust authorities have been very active in antitrust law enforcement in the field of health care services and products.[37] One of the most prominent issues in US antitrust law today is how to assess settlements of patent disputes involving payments from innovative to generic drug manufacturers, the so-called pay-for-delay settlements or reverse payments.[38] The goal of such antitrust enforcement is to ensure access to innovative, safe and affordable pharmaceutical technology. In 2010, the US Supreme Court in *American Needle v. NFL*[39] held that antitrust allegations relating to technology transfer should be scrutinized under the 'rule of reason'.

3 TECHNOLOGY TRANSFER-RELATED COMPETITION LAW ENFORCEMENT IN DEVELOPING COUNTRIES

Developing countries have lagged behind in the area of competition law in general and that relating to technology transfer in particular.[40] However, some countries

35. European Commission, 'Pharmaceutical Sector Inquiry', fn. 15 *supra*, paras 1563–1577.
36. *AstraZeneca*, fn. 10 *supra*.
37. M.H. Meier, B.S. Albert & S.C. Brau, 'Overview of the FTC Antitrust Actions in Health Care Services and Products', (June 2010). <www.ftc.gov/bc/0610hcupdate.pdf>.
38. See *Arkansas Carpenters Health and Welfare Fund v. Bayer AG*, 604 F.3d 98 (2d Cir. 2010), *reh en banc denied*, 2010 WL 3464382. On 6 Dec. 2010, the unsuccessful antitrust challengers at the Second Circuit petitioned the U.S. Supreme Court to determine the antitrust status of reverse payment settlements. See also Amici Curiae of 86 Intellectual Property Law, Antitrust Law, Economics, Business and Public Health Professors in *Louisiana Wholesale Drug v. Bayer, cert. denied*, 2011 WL 767662.
39. *American Needle v. National Football League*, 130 S. Ct. 2201, 2217 (2010) (Held: an arrangement among the 32 separately owned member teams of the National Football League to license their IPRs collectively through their jointly owned licensing affiliate (National Football League Properties) constituted concerted activity under s. 1 of the Sherman Act; the challenged agreement, on remand, would be reviewed under a rule of reason analysis).
40. In 1980 only 15 countries had a competition law; but in 2010 there are now 112 countries, and a further 14 countries are in the process of preparing competition laws. See Supachai 'Opening Address', fn. 29 *supra*.

have enacted competition law and/or competition policy (intellectual property law for instance) in order to control abuses of IPRs. In 2007 China adopted its Anti-monopoly Law confirming that, in principle, the normal exercise of IPRs should not violate competition law, but conduct which eliminates or restricts competition by abusing IPRs will be subject to scrutiny under competition law.[41] In its Patent Law as revised in 2008, compulsory licensing may be granted for failure to work, anti-competitive enforcement of patent rights, or even refusal to license on reasonable terms and conditions after a reasonable period of negotiation time.[42] The Patent Act of India (as the patent acts of some other developing countries) goes even further by stating that certain practices of patent holders are subject to compulsory licensing, including excessive prices, refusal to license affecting downstream markets, tying, exclusive grant-back, non-challenge, package licensing for instance.[43]

In practice, there have been only a handful of cases relating the application of competition law to technology transfer in developing countries.

3.1 Refusal to License Pharmaceutical Patents in South Africa

As noted, IPRs give holders rights to exclude others from exploiting their intellectual property. Refusal to license cannot, by itself, be in breach of either competition or intellectual property law. However, if refusal to license by a company holding a dominant position constitutes an IPR abuse under a country's competition law, the TRIPS Agreement permits that country to grant compulsory licensing to remedy such anti-competitive conduct. It is worth reiterating that refusal to license technology and excessive pricing of technology-embodied products may be considered as the two sides of anti-competitive conduct.

3.1.1 *Hazel Tau v. GSK & BI*

In September 2002, a group of individuals and organizations led by the Treatment Action Campaign (TAC) initiated a complaint against GlaxoSmithKline (GSK) and Boehringer Ingelheim (BI) with the South African Competition Commission (SACC). The complainants alleged that these two pharmaceutical companies, who had dominant positions in the anti-retroviral medicine (ARV) markets for

41. See Anti-monopoly Law of the People's Republic of China (Art. 55) 2007, <www.china.org.cn/government/laws/2009-02/10/content_17254169.htm>.
42. See the Patent Law of the People's Republic of China (Arts 48, 51 and 54) 1984, as amended in 2008, <www.ipr2.org/ipsearch/file.php?id=188#> (compulsory licensing may be granted 'where it has been legally determined that the enforcement of the patent right by the patentee is an act of monopoly, to avoid or to eliminate the adverse effects caused to competition').
43. See ss 84–90 of the Patent Act of India 1970, as amended in 2005, <http://ipindia.nic.in/ipr/patent/patents.htm>. See also WIPO, 'Patent Related Flexibilities in the Multilateral Legal Framework and Their Legislative Implementation at the National and Regional Levels', WIPO Doc. CDIP/5/4 REV, (18 Aug. 2010).

HIV/AIDS in South Africa, engaged in excessive pricing of ARV to the detriment of consumers in violation of Section 8(a) of the Competition Act.[44]

This competition case provided the basis for a larger public campaign attracting the attention of both local and international communities. SACC, after conducting its investigation, agreed with the complainants' arguments. It even went further by concluding that GSK and BI had both abused their dominance and contravened not only Section 8(a) of the Competition Act, due to the excessive pricing complained of, but also Section 8(b), relating to refusal to supply an essential facility, and Section 8(c), regarding exclusionary conduct. Faced with the three alleged violations, GSK and BI still refused to admit them but initiated settlement negotiations. Finally, in December 2003 the parties entered into settlement agreements. Under these agreements, GSK and BI agreed to grant voluntary licenses of their ARV-related patents to generic manufacturers with the royalty rate of 5%.[45]

3.1.2 *TAC v. Merck/MSD*

In the latter part of 2007, TAC filed a competition complaint with SACC against Merck, a pharmaceutical company with headquarters in the US, and its South African subsidiary, MSD. TAC alleged that Merck/MSD had abused their dominant position in the market for the drug efavirenz (EFV), an ARV marketed under the brand name Stocrin, by refusing to license generic producers to manufacture and/or import and sell not only (i) a generic version of this medicine but also (ii) co-formulated and/or co-packaged generic products containing EFV and at least one other ARV on reasonable and non-discriminatory terms in the South African market. Instead of focusing on excessive pricing as the consequence of refusal to license as in *Hazel Tau v. GSK & BI*, TAC considered the two types of the refusal to license by Merck/MSD as exclusionary conduct whose anti-competitive effects outweighed their pro-competitive gains as prohibited by Section 8(c) of the Competition Act.[46]

To prove an infringement under Section 8(c), a complainant must prove two factors: first, that the alleged infringer has a dominant position; and second, that its conduct (i) is exclusionary, (ii) has anti-competitive effects, and (iii) such effects outweigh any pro-competitive justifications. Regarding the dominant position, TAC contended that, as in *Hazel Tau v. GSK & BI*, since ARVs are generally not substitutable for one another,[47] EFV constitutes its own relevant market in

---

44. T.T. Nguyen, 'Competition Law and Access to Pharmaceutical Technology in the Developing World', 28 *Biotech L Rep 693* (2009); S.M. Flynn, 'Using Competition Law to Promote Access to Knowledge', in *Access to Knowledge in the Age of Intellectual Property*, ed. G. Krikorian & A. Kapczynski (New York: Zone Books, 2010), 451–474.
45. International Competition Network, 'Unilateral Conduct in the Pharmaceutical Sector' (2 Nov. 2010), <http://internationalcompetitionnetwork.org/uploads/library/doc671.pdf>.
46. Nguyen, fn. 44 *supra*, 698–705.
47. According to standard regimens for HIV/AIDS treatment, each HIV/AIDS patient must take at least three ARVs simultaneously and must also be able to change to other ARVs, i.e. other regimens, in case of unmanageable side effects or treatment failure.

respect of manufacturers, marketers, and consumers. In the alternative, TAC alleged that MSD's market share in South Africa with respect to the relevant therapeutic class (non-nucleoside reverse transcriptase inhibitors) accounted for over 45% in terms of both turnover and units sold.[48]

The second factor involves three sub-conditions. With respect to the exclusionary conduct, TAC argued that due to Merck/MSD's refusal, no firm (except Aspen and Adcock, two licensees of Merck/MSD) could import into, manufacture, use, offer to dispose of and/or dispose of, in South Africa, generic EFV products; and no firm (including Aspen and Adcock) could import into, manufacture, use, offer to dispose of and/or dispose of, in South Africa, co-formulated and/or co-packaged generic products containing EFV and at least one other ARV. Therefore, the refusal impeded and prevented a generic producer from entering not only the generic EFV market but also the co-formulated and/or co-packaged product market, i.e. the fixed dose combination ARV market, in South Africa.

Regarding anti-competitive effects, TAC argued that Merck/MSD's refusal to license additional generic producers unreasonably harmed consumer welfare and prevented new products for which there was a potential customer demand. It contended that consumers in South Africa suffered unreasonable harm not only in terms of prices but also in terms of access to new products. In addition, consumer welfare was harmed by shortages in supply of EFV as well as other ARVs. Therefore, Merck/MSD's refusal to license was asserted to be anti-competitive.

In order to weigh the anti-competitive effects and the pro-competitive justifications for such refusal, TAC analysed case law in the US and the EU, especially with regard to the doctrine of 'refusal to supply', and argued that a right holder could not justify exclusionary conduct solely on the basis of its IPRs. TAC also asserted the importance of fundamental issues relating to the rights to property, life, dignity, and access to medicines under the Constitution of South Africa and in respect of international law. Based on all of the reasoning, TAC contended that Merck/MSD's refusal to license violated Section 8(c) of the Competition Act.

During the pendency of SACC's investigation, Merck/MSD: (i) licensed four generic drug companies, two local producers and two locally-based importers, to bring stand-alone patented EFV medicines to the South African market; (ii) agreed that all four licensees are entitled to bring co-packaged or co-formulated products containing EFV to the market; and (iii) waived any right to royalties. These Merck/MSD's responses were responsive to the main demands in TAC's complaint. On 30 May 2008, TAC therefore announced that Merck/MSD was no longer acting in an anti-competitive way; and SACC did not refer the case to the Competition Tribunal for adjudication.[49]

48. Nguyen, fn. 44 *supra*, 698–705.
49. *Ibid.*

3.1.3 Remarks

The above-described cases demonstrate that competition law may be used to achieve significant effects in reducing the prices of patented medicines and/or gaining access to relevant patents by compulsory or voluntary licenses. However, they may also give rise to concerns relating to the relationship between intellectual property and competition laws in a developing economy.

First, regarding the effects of applying IPR-related competition law, application of domestic competition law in a developing country may threaten the existence of IPRs and reduce innovation. That country may no longer attract investment in research and development (R&D) and foreign investment (and even be vulnerable to a WTO dispute). However, IPRs protected under domestic intellectual property law in a developing country have, for the most part, been created in developed countries.[50] In general, companies expect to make most of their potential profits in the markets of developed countries when making initial investment decisions about R&D.[51] The pharmaceutical sector, an area requiring very large investments and costs and having risks in terms of innovation illustrates this point. Out of approximately 160 billion USD spent annually worldwide on medicinal R&D, only 3% of that money 'is directed at diseases that disproportionately affect developing countries'.[52] As a result, the world sales of medicines are very highly directed towards developed country markets. North America, Europe and Japan account for 85.6% of global sales. Meanwhile, developing countries, accounting for more than 80% of the world's population, are responsible for only about 10% of the global sales; and Africa only 1.1%.[53] Abbott and Reichman, after carefully addressing the need to preserve incentives for R&D of new medicines, conclude the following:

> [O]riginator pharmaceutical companies based in the OECD recover the great part of their R&D expenditures in the more affluent OECD markets, and invest a small part of their R&D budgets on diseases of special relevance to developing countries. Consequently, the use by developing countries of compulsory licensing to ensure public access to affordable medicines is unlikely to have a material effect on the level of research currently undertaken in the OECD.[54]

---

50. The total numbers of patent applications and of patent grants around the world in 2007 were around respectively 1.85 million and 0.76 million, in which patent applicants, or patent owners, from Japan, the US and the EU accounted for around 70%. Although there has been an increase in patent filings in developing countries, non-resident applicants accounted for the largest share of total filings in these countries. Moreover, of around USD 110 billion of 2004 receipts for international licensing, the receipts going to the US, the EU, and Japan accounted for more than 90%. See Nguyen, fn. 44 *supra*, 706.
51. B.N. Roin, 'Unpatentable Drugs and the Standards of Patentability'. 87 *Tex L Rev* 503, 547–548 (2009).
52. WHO, 'Public Health, Innovation and Intellectual Property: Global Strategy and Plan of Action – Proposed Time Frames and Estimated Funding Needs', (21 Jan. 2009) <http://apps.who.int/gb/ebwha/pdf_files/EB124/B124_16Add2-en.pdf>, para. 8.
53. WHO, *Public Health, Innovation and Intellectual Property Rights* (Geneva: WHO. 2006), 15.
54. F. Abbott & J.H Reichman, 'Access to Essential Medicines: Lessons Learned since the Doha Declaration on the TRIPS Agreement and Public Health, and Policy Options for the European Union' June 2007 PE 381.392 (prepared for the International Trade Committee, European Parliament). <www.law.duke.edu/news/pdf/a-r_directorate_report.pdf>.

Reducing innovators' monopoly rents in a developing country market, normally a small one, by using competition law, may not result in such a large cost in lost innovation as it might if the rents reduced are those derived from a developed country market. Further, the information disclosure function of an intellectual property system is not very important to developing countries. Domestic firms in developing countries, in theory, may obtain information from the patent disclosures protected in developed countries. It is true from an economic perspective that restricting the market power of patent protection by competition law in small technology-importing economies will 'lead to static gain, while the dynamic loss from discouraging innovation or less information disclosure would be close to zero'.[55] It is reasonable to conclude that poor countries 'should not have to pay as large a share of those research costs' as wealthy countries.[56]

US and EU case law generally shows that domestic competition law can intervene where consumer harms are derived from the anti-competitive exercise of IPRs and exceed economic incentives for innovation.[57] It is, therefore, unreasonable and unfair from either a legal or an economic perspective if developing countries cannot appropriately apply their competition law to remedy anti-competitive practices relating to refusal to license, or derivative conduct, while developed countries are permitted to do so. However, due to the complexity and difficulty of proving abuses of a dominant position by right holders, competition law should not be used by itself where other flexibilities, permitted under the TRIPS Agreement and national intellectual property law, such as public health and public security grounds, may be applicable. The intervention of competition law in IPR abuses in developing countries, as in developed countries, should be as a last resort to protect consumer welfare, competition, and innovation. For all that, in these cases, the adverse effects on foreign investment (if any) appear to be minimal.

Second, it should not be presumed that IPRs create (or even increase) market power when determining the dominant position of a right holder.[58] The way in which the complainants and SACC defined the relevant product market and dominant position in those two cases may pose a threat to right holders. The argument there seemed to be that the mere holding of a patent conferred monopoly on the patent holder in the patented product market. However, this concern is unsubstantiated. SACC and/or the complainants in those two cases followed the view that each active ingredient/substance formed a relevant market

55. R. Aoki & J. Small, 'Compulsory Licensing of Technology and the Essential Facilities Doctrine', 16 *Info Econ & Pol'y* 13, 13–19 (2004).
56. J.H. Barton, 'Statement to the United States Senate Finance Committee Hearing on International Enforcement of Intellectual Property Rights and American Competitiveness', (15 Jul. 2008) <http://finance.senate.gov/imo/media/doc/071508jbtest.pdf>.
57. Nguyen, fn. 25 *supra*, 59–160.
58. In the US, see *Illinois Tool Works Inc. v. Independent Ink, Inc.*, 547 U.S. 28, 45–46 (2006). In the EU, see Case 24/67, *Parke, Davis and Co. v. Probel*, [1968] ECR 55, 72; Case 40/70, *Sirena v. Eda*, [1971] ECR 69, paras 5 and 16; Case 78/70, *Deutsche Grammophon v. Metro*, [1971] ECR 487, paras 11–16.

on its own on the basis of the lack of substitutability (or interchangeability) between the various ARV active ingredients, rather than on the fact that each active ingredient was the subject of a patent. Similarly, the US Supreme Court in *Eastman Kodak*[59] held that a single brand of a product or service can be a relevant product market on the basis of lack of substitutability available to consumers. It is correct for competition authorities to use the principle that a product market definition depends on the substitutability of the relevant product with other comparable products from the consumer perspective regardless of whether or not a product is subject to intellectual property protection. This may be well illustrated by the *AstraZeneca* judgment in the EU.[60]

Third, regarding IPR-related excessive pricing and refusal to license under competition law, many developing countries, and even the EU,[61] stipulate in their domestic competition law and/or intellectual property law that excessive pricing of IPR-embodied products is detrimental to consumers and constitutes an IPR abuse. Excessive pricing, from the perspective of consumers, may be regarded as one of the most noticeable and blatant abuses of a dominant position. It transfers wealth from consumers to dominant firms and causes inefficiency. However, there are three reasons which may prevent competition authorities from intervening in excessive pricing. First, the prohibition of excessive pricing may reduce incentive to innovate and invest in R&D *ex ante*. Second, it is difficult to implement in practice, because it is difficult to identify 'excessiveness'. It is not easy for antitrust courts to directly police price issues; and there is no safe harbour for the pricing practices of firms.[62] Third, this prohibition may be redundant because excessive pricing is self-correcting by the invisible hand of the market.

For the first reason, as presented above, the application of competition law in the majority of developing countries to the field of technology transfer hardly affects innovation in general. Relating to the third reason, the trend of maximizing profit in the market makes the self-correcting nature infeasible.[63] Regarding the second reason, there is still controversy over the criteria required to establish such anti-competitive conduct as excessive pricing. However, if competent authorities can prove that the price of a technology or technology-embodied product in the domestic market of a developing country is unreasonably higher (after deducting reasonable costs, i.e. distribution costs) than the price of a like technology or product in a developed country or in another developing country that has economic

59. *Eastman Kodak Co. v. Image Technical Services, Inc.*, 504 U.S. 451, 481–482 (1992).
60. *AstraZeneca*, fn. 10 *supra*, paras 147–221 (relevant market) and 239–294 (dominant position). It should be noted that substitutability/interchangeability is not 'assessed solely in relation to the objective characteristics of the products and services at issue, but the competitive conditions and the structure of supply and demand on the market must also be taken into consideration'. Case T-427/08, *CEAHR v. Commission*, Judgment of 15 Dec. 2010, not yet reported, para. 67.
61. Nguyen, *TRIPS Agreement*, fn. 25 *supra*, 78–82.
62. *Pacific Bell Telephone Co. v. Linkline Communications, Inc.*, 129 S. Ct. 1109, 1121 (2009); *Verizon Communications Inc. v. Law Offices of Curtis V. Trinko*, 540 U.S. 398, 408 (2004).
63. A. Ezrachi & D. Gilo, 'Are Excessive Prices Really Self-Correcting?', 5 *J Comp L & Econ* 249 (2009).

conditions similar to the former developing country, it appears obvious that the right holder is engaging in excessive pricing in the first developing country market. In such case, in addition to permitting parallel imports (an issue to be decided under domestic legislation),[64] competition authorities can (in combination with intellectual property and/or regulatory authorities) require the right holder (if it has dominance) to reduce the price or grant compulsory licenses. If the unreasonable difference in prices between these two countries is not established, competition authorities could combine both a cost-based and a demand-side approach in order to determine the economic value, and then compare it with the market price. In any event, the costs of innovation should be fully taken into account.

It is worth noting that the excessive pricing in question is an exploitative practice that constitutes an abuse of a dominant position. It thus violates competition law.[65] In contrast, mere high pricing violates neither competition law nor intellectual property law. IPRs granted under domestic intellectual property law gives right holders an opportunity to 'tax the economy' so as to maximize their profits; but profit maximizing by the right holders should not go as far as 'private avarice' if they really hold a dominant position in the relevant market.[66] The exercise of IPRs on the one hand and the enforcement of competition law on the other hand should be reasonably related to the goals of public policy in the context of balancing social costs and innovations.

Briefly, the IPR-related competition cases in South Africa show that a developing country can properly use domestic competition law to promote access to technology, particularly pharmaceutical technology.

3.2 Abbott's Withdrawal of Medicine Registration Applications in Thailand

Thailand recently took advantage of the public health-related flexibilities of Article 31 of the TRIPS Agreement to authorize compulsory licenses for pharmaceutical patents to increase access to medicines. In January 2007, Thailand granted compulsory licensing on the ground of public use for a patented ARV, which is a combination of Lopinavir and Ritonavir, owned and sold by Abbott Laboratories under the trade name Kaletra. In response to the compulsory licensing, Abbott, in March 2007, withdrew registration of seven different medicines, including Aluvia which is a new heat-stabilized version of Kaletra, with the Food and Drug Administration of Thailand. In April 2007, HIV/AIDS treatment activists made a complaint to the Competition Commission of Thailand, alleging that Abbott's withdrawal of

64. Article 6 of the TRIPS Agreement states that 'nothing in this Agreement shall be used to address the issue of the exhaustion' of IPRs.
65. European Commission's Guidance on the Commission's enforcement priorities in applying Art. 82 of the EC Treaty to abusive exclusionary conduct by dominant undertakings, OJ 2009 C45/7, para. 7.
66. P. Carstensen, 'Post-Sale Restraints via Patent Licensing: A "Seedcentric" Perspective', 16 *Fordham Intel Prop, Media & Ent L J* 1079 (2006).

registration applications for medicines, especially Aluvia, in Thailand was a violation of the Competition Act, in particular Sections 25(3) and 28. However, in December 2007, the Trade Competition Commission concluded that the withdrawal did not violate the Competition Act.[67]

The Trade Competition Commission, on the basis of its notification on dominance, found that Abbott's turnover in the Thai market was below the threshold to hold a dominant position. Moreover, Abbott had not yet obtained the required certificate of product registration which is what actually made products unavailable in the market. Therefore, Abbott's conduct was concluded not to constitute a refusal to supply. In addition, the Commission found that the Abbott's withdrawal was not intended to cause harm to Thai consumers. It argued that medicine registration was subject to the decision of the Food and Drug Administration. Moreover, due to the need for medical prescriptions, there would never be direct orders from Thai consumers to the Abbott's head office in the US. As a consequence, the conduct in question was regarded as not violating Section 28.[68]

Although the Abbott case was not successful from the competition law perspective, it may reflect the relationship between the application of intellectual property law and competition law on access to technology. The fact that access to the patent held by Abbott had been achieved via compulsory licensing under Thai intellectual property law should be taken into account. If the patent holder was also attacked under competition law, this could cause concern among the international community as well as to the patent holder. The TRIPS Agreement recognizes the built-in exceptions to, and limitations of, IPRs. A developing country, through its domestic intellectual property law, may use those exceptions and limitations. If that country is not able to apply those exceptions and limitations, it may find another solution through its domestic competition law, since the enforcement of competition law requires a variety of legal and economic analyses to avoid defects in enforcement. This confirms the observation that it is better for a developing country to use domestic competition law as a last resort.

3.3 Philips Package Licensing in Taiwan

3.3.1 Background

Philips Electronics, a Dutch company, and two Japanese companies, Sony and Taiyo-Yuden, together owned around 100 patents relating to the technology for manufacturing recordable compact discs (CD-Rs) under technical standards jointly created by Philips and Sony (the so-called Orange Book). These three companies organized a patent pool, whereby the two Japanese companies licensed their patent rights to Philips, with Philips, acting as an exclusive licensor, bundling all the other

67. Nguyen, fn. 44 *supra*, 710–712; Flynn, fn. 44 *supra*, 466.
68. Decision of 27 Dec. 2007 of the Thai Trade Competition Commission on the complaint against Abbott (on file with the author).

patent rights with its patent rights, for licensing to other CD-R producers. The patent pool arrangement stipulated the royalties to be paid as 3% of the net selling price with a minimum of 10 JPY per licensed product. Philips, Sony, and Taiyo-Yuden would take respectively 60%, 25%, and 15% of the royalties received from the package licensing.[69] At the time of signing the license agreements between Philips and Taiwanese CD-R producers, CD-R market prices were high; but later the prices fell substantially, and the 10 JPY royalty per licensed product accounted for at least 20%–30% of CD-R selling prices. The Taiwanese CD-R producers, after the failing to reach agreement with Philips for an adjustment of the amount of royalties, filed a competition complaint with the Fair Trade Commission of Taiwan (TFTC).

After its investigations, TFTC in 2001 and 2002 concluded that the conduct of Philips, Sony, and Taiyo-Yuden, including joint and package licensing, methods of setting royalties, and refusal to disclose important trading information relating to the license agreements, violated the Fair Trade Act of Taiwan. First, by joining the patent pool and package licensing arrangement, these companies had engaged in a horizontal anti-competitive concerted action preventing CD-R producers from freely choosing licensors and from negotiating royalties with licensors in breach of Article 14 of the Act. Second, due to the patent pool arrangement, these three companies were regarded as a monopolistic firm in the CD-R patent licensing market.[70] Consequently, the fact of not adjusting the method of calculating royalties, irrespective of substantial market changes, constituted conduct carried out by a monopolistic firm that improperly set and maintain prices, violated Article 10(2) of the Act. Finally, in negotiations to reach license agreements with Taiwanese licensees, Philips refused to supply key trading information, including detailed information regarding the content of the licensed patents, their scope, and their duration; the license agreements also contained a non-challenge clause. These collective actions of Philips, Sony and Taiyo-Yuden violated Article 10(4) of the Fair Trade Act, which prohibits other abusive conduct by a monopolistic firm.

This TFTC decision was appealed. In August 2005, the Taipei High Administrative Court reversed the decision, finding that the patent pool was not a concerted action because the members of the pool did not compete with each other. TFTC then appealed to the Supreme Administrative Court of Taiwan. In April 2007, the Supreme Administrative Court affirmed the Taipei High Administrative Court, ruling that the patent pool and package licensing arrangements established by Philips, Sony and Taiyo-Yuden were complementary rather than substitutable. They could not result in a concerted action. Further, package licensing may

69. TFTC Decision of 11 Jan. 2001 (Disposition (90) Kung Ch'u Tzu No. 021), 'A complaint alleged that the CD product patent licensing practices in Chinese Taipei by Koninklijke Philips Electronics, N.V. (the Netherlands), Sony Corporation (Japan), and Taiyo-Yuden Co., Ltd (Japan) were in violation of the Fair Trade Law'. <www.apeccp.org.tw/doc/Taipei/Case/D0575201.htm>.

70. According to Art. 5-1(3) of the Fair Trade Act of Taiwan, if the combined market share of three firms in the relevant market totals 75% or more, these firms are regarded as a monopolistic firm.

give rise to pro-competitive benefits such as one-stop technology shopping and transaction cost reductions.[71]

3.3.2 Other Cases Involved

Interestingly, the Philips package-licensing case first brought under the competition law of Taiwan gave rise to a compulsory licensing case under patent law in Taiwan, an IPR-related antitrust case in the US, and an IPR case in the EU under WTO law.[72]

Due to the substantial fall in CD-R prices leading to very high royalties as noted, some producers in Taiwan, including Gigastorage, stopped paying royalties, in addition to filing the competition complaint against Philips with TFTC. Philips responded by terminating its license agreement with Gigastorage in April 2001. However, Gigastorage continued to produce CD-Rs; and in June 2001, it applied to the Intellectual Property Office of Taiwan (TIPO) for a compulsory license of the Philips-owned patents. According to Article 76 of the Patent Act of Taiwan, one of the grounds under which TIPO can grant compulsory licensing is 'an applicant's failure to reach a licensing agreement with the patentee concerned under reasonable commercial terms and conditions within a considerable period of time'. On this premise, TIPO, in July 2004, decided to grant a compulsory license to Gigastorage for CD-R-related technology patents owned by Philips.[73] While appealing this decision before the Taipei High Administrative Court, Philips also filed two complaints respectively in the US and the EU. In the US, Philips brought suit against Gigastorage with the US International Trade Commission (ITC), alleging that Gigastorage was violating Section 337(a)(1)(B) of the Tariff Act of 1930,[74] by importing into the US certain CD-Rs (and rewritable compact discs: CD-RWs) which infringed Philips's patents. In the EU, Philips submitted a complaint to the European Commission, alleging that Article 76 of the Patent Act of Taiwan infringed Article 28 of the TRIPS Agreement, by permitting compulsory licensing which was not justified by, and even violated, Article 31 of the Agreement.[75]

71. See Nguyen, *TRIPS Agreement*, fn 25 *supra*, 204–210; OECD, 'Competition Law and Policy in Chinese Taipei' (April 2006). <www.oecd.org/dataoecd/18/22/36875564.pdf>.
72. See *US Philips Corp. v. ITC*, 424 F.3d 1179, 1193 (Fed. Cir. 2005), *cert. denied*, 547 U.S. 1207 (2006); European Commission, 'Report to the Trade Barriers Regulation Committee: Examination Procedure Concerning an Obstacle to Trade, within the Meaning of Council Regulation (EC) No. 3286/94, Consisting of Measures Adopted by the Separate Customs Territory Of Taiwan, Penghu, Kinmen and Matsu Affecting Patent Protection in Respect of Recordable Compact Discs' (30 Jan. 2008). <http://trade.ec.europa.eu/doclib/docs/2008/january/tradoc_137633.pdf>.
73. The compulsory licensing decision was upheld by the Committee of Appeal of the Ministry of Economic Affairs of Taiwan on 16 Jun. 2006.
74. 19 U.S.C. § 1337(a) (1)(B).
75. Philips complained under Council Regulation (EC) No. 3286/94, OJ 1994 L 349/71, as last amended by Council Regulation (EC) No. 356/95, OJ 1995 L 41/3, laying down Community procedures in the field of the common commercial policy in order to ensure the exercise of the

Regarding Philips's complaint in the US, it was indisputable that Gigastorage infringed Philips's patents. However, Gigastorage counterclaimed that Philips's conduct of package licensing constituted patent misuse and an illegal anti-competitive practice. The ITC observed that Philips's package licensing contained both licenses to patents essential to CD-R manufacturing and licenses to patents not essential to that activity. Therefore, the ITC contended that the patent package licensing in question was per se patent misuse, because it 'did not give prospective licensees the option of licensing individual patents (presumably for a lower fee) rather than licensing one or more of the patent packages as a whole'. Further, the ITC argued that the package licensing also constituted patent misuse under the rule of reason. The ITC reasoned that including such non-essential patents in the licensing packages could foreclose alternative technologies and injure competitors, who seek to license such alternative technologies to parties in need of the licenses of Philips's essential patents under which, 'the anticompetitive effects of including non-essential patents in the packages of so-called essential patents outweighed the pro-competitive effects of that practice'.[76] However, package licensing provides pro-competitive benefits, particularly the reduction of licensing transaction costs; so the Federal Circuit applied the rule of reason when reviewing this case. This Court held that the ITC's rule of reason analysis was flawed, because there was no evidence on foreclosure of competition, and the benefits of package licensing, regarding the reduction of transaction costs had not been addressed. The Court remanded the ITC's decision.[77] The ITC, after re-examining the case in accordance with the Federal Circuit's ruling and the effect of the Supreme Court's judgment in *Illinois Tool Works* in 2006,[78] rejected Gigastorage's patent misuse defence and entered an exclusion order prohibiting the unlicensed importation of infringing CD-Rs in February 2007.[79] Philips and Gigastorage subsequently reached a settlement of the US litigation.[80]

It is worth noting that on the other side of the Atlantic, the European Commission launched a competition investigation on a similar CD-R package licensing administered by Philips in the EU in 2003. Faced with this investigation, Philips agreed to discontinue the package licensing arrangement with Sony and Taiyo-Yuden. It also agreed to offer individual licenses, the 'Philips Only License Agreement', on revised terms. Philips further agreed to reduce the royalty rates. As a result, the Commission closed the investigation in February 2006.[81]

Community's rights under international trade rules, in particular those established under the auspices of the WTO.

76. *US Philips Corp. v. ITC*, fn. 72 *supra*, 1183–1184. However, the ITC did not address issues relating to the reasonableness of a patent pool, royalty fixing according to the patent pool, or the royalty structure as TFTC did.
77. *Ibid.*, 1188–1189.
78. *Illinois Tool Works Inc. v. Independent Ink, Inc.*, 547 U.S. 28 (2006).
79. US ITC General Exclusion Order and Order to Cease and Desist of 5 Feb. 2007, Investigation No. 337-TA-474.
80. European Commission, 'Report to the Trade Barriers Regulation Committee', fn. 72 *supra*, para. 24.
81. See European Commission press release IP/06/139, (9 Feb. 2006).

Regarding Philips's complaint in the EU, the European Commission, after an investigation, concluded that Article 76 of the Patent Act of Taiwan, which permits the grant of compulsory licensing in circumstances where there has been no more than a failure to reach an agreement on reasonable commercial terms and conditions within a reasonable period of time, was inconsistent with Article 28 of the TRIPS Agreement.[82] According to the Commission, a mere refusal to license by a patent holder cannot lead to compulsory licensing. By analysing the right to exclude of patent holders under EU law as well as other jurisdictions in the context of the WTO, the Commission observed:

> [T]he requirement of attempted negotiations was seen as a procedural requirement [rather than] a substantial condition permitting the grant of a compulsory license ... At the time of conclusion of the [TRIPS Agreement] most (if not all) jurisdictions had a doctrine that a refusal to deal by a patent holder is a perfectly permissible activity, that might only be considered illegal where the patent owner was in a position of market power and the refusal was linked to [an] abuse of that market power.[83]

Further, the Commission also contended that the interpretation of various procedural requirements relating to the decision of the authorities of Taiwan granting compulsory licensing in the case in question were inconsistent with Article 31 of the TRIPS Agreement.[84] As a result, the Commission threatened to bring the case to the WTO dispute settlement system.

In March 2008, the Taipei High Administrative Court reversed the TIPO decision. This Court held that the interpretation and determination of 'reasonable commercial terms and conditions' should take into account not only the amount of royalties but also the profit margin of parties, shared risks, the scope and period of the license and market competition. However, when granting the compulsory license to Gigastorage in question, TIPO based its decision on the calculation of royalties alone.[85] TIPO contacted the European Commission seeking to settle the EU's concerns and decided not to appeal the court's ruling.[86] At the end of 2008, TIPO proposed amending the compulsory licensing-related provisions of the Patent Act of Taiwan. According to this proposal, compulsory licensing would only be granted on three grounds, namely: (i) national emergencies, (ii) use of a patent for enhancement of public welfare (and not for profit) or a dependent patent,

---

82. European Commission, 'Report to the Trade Barriers Regulation Committee', fn. 72 *supra*, para. 107.
83. *Ibid.*, paras 102–103.
84. *Ibid.*, paras 145, 154 & 202.
85. C. Lin, 'Filling in Gaps of the TRIPS Agreement: Reflections on the Taiwan-Philips CD-R Compulsory Licence Case', 3 *Asian Journal of WTO and International Health Law and Policy* 577 (2008).
86. European Commission, 'General Overview of Active WTO Dispute Settlement Cases Involving the EU as Complainant or Defendant and of Active Cases under the Trade Barriers Regulation (9 Nov. 2010). <http://trade.ec.europa.eu/doclib/docs/2007/may/tradoc_134652.pdf>.

and (iii) as a remedy against a competition infringement by a patent holder.[87] As of January 2011, this amendment however has not yet been adopted.

3.3.3 Remarks

The facts of the Philips package licensing case, as well as the compulsory licensing case in Taiwan, involve licensees having to pay high royalties, despite the appreciable reduction of selling prices, and the licensors resulting in their receiving greatly increased total licensing revenues due to growth in demand in the CD-R market. The main concern of TFTC appears to be the licensors' refusal to renegotiate lower royalty rates, although the focus of the case in question was the licensors' abuse of market power through the patent pool rather than the very high royalties as such. However, it appears that the licensees had an unreasonable business strategy regarding royalty calculation when they signed license agreements with Philips. They did not anticipate the substantial reduction in CD-R selling prices. As a result, their royalties which were reasonable and affordable at the time of signing the license agreements, only later became too costly and unaffordable. The licensees wanted to escape the consequences of their contractual obligations relating to royalties by use of competition law. However, the final result demonstrates that this is a very difficult strategy. One may argue that due to the strong (even dominant) position of Philips, the licensees were faced with a take-all-or-nothing licensing offer. They were forced to accept all terms and conditions offered by Philips, including the royalty calculation. But this, at least from the perspective of profit predictability, merely reflects a risky decision made by the licensees. Competition law does not exist to help firms correct mistakes in their business decision-making. The mere fact that royalties to be paid to a patent holder have not been reduced in proportion to the decrease of market prices for patent-embodied products does not constitute an abuse of a dominant position in the relevant market.

Regarding the interpretation of competition law, as the Supreme Administrative Court of Taiwan ruled, TFTC, when evaluating the patent pool and package licensing established by Philips and the two Japanese companies, did not take into account other factors affecting competition in the CD-R technology market. TFTC did not balance anti-competitive effects, on which it mainly relied, with the pro-competitive benefits, which the Court highlighted. Meanwhile, in its Guidelines on technology licensing arrangements, TFTC states that it follows the simple rule of reason by considering many factors in addition to the reasonableness of the provisions of a licensing arrangement.[88] Should TFTC not respect the rule of reason when scrutinizing anti-competitive restrictions on technology transfer, it is inevitable that its decision will be vacated and remanded on appeal.

87. TIPO, 'TIPO Holds Public Hearing for Compulsory Licensing Draft Amendment on 18 August 2008'. <www.tipo.gov.tw/en/News_NewsContent.aspx?NewsID=3018>.
88. TFTC Guidelines on Technology Licensing Arrangement (s. 4). <www.apeccp.org.tw/doc/Taipei/Decision/ctdec001.htm>.

The Philips package licensing case and the TIPO decision to grant compulsory licensing indicate that there was no cooperation between the authorities responsible for competition law and IPRs in Taiwan. One of the grounds for granting compulsory licenses under Article 76 of the Patent Act of Taiwan is to remedy the anti-competitive conduct of a patent holder. At the time TIPO considered the application submitted by Gigastorage for compulsory licensing, TFTC had already concluded that Philips, together with the two Japanese companies, violated competition law. However, TIPO did not invoke this decision, regardless of the fact that the decision was under appeal. If TIPO had combined the competition ground with the ground relating to the applicant's failure to reach a license agreement under reasonable commercial terms and conditions, TIPO's decision to grant compulsory licensing of Philips's patents to Gigastorage would have been more supportable and it would have been more difficult for the EU to protest the decision under the TRIPS Agreement and WTO dispute settlement mechanism. Moreover, it is worth noting that the license agreements between Philips and the licensees in Taiwan had been registered with TIPO.[89] If TFTC had had the cooperation of TIPO, TFTC (and TIPO) might have noticed the debatable nature of these agreements under competition law and could have begun investigations earlier, which might then have focused on the abuse of IPRs through package licensing rather than through high royalties.

The final outcome of the Philips package licensing case and the grant of compulsory licensing in question in Taiwan shows the influence of US and EU views on Taiwanese courts' rulings. It seems that the Supreme Administrative Court of Taiwan followed the holding of the US Federal Circuit in *US Philips v. ITC*, as well as the final decision of the ITC, with respect to the application of the rule of reason to package licensing in the context of the Fair Trade Act of Taiwan and the explanation of the term 'concerted action' in package licensing. Regarding the compulsory licensing issue, perhaps under pressure from the EU to bring the dispute to the WTO, the Taipei High Administrative Court concluded that the concept 'reasonable commercial terms and conditions' comprised many aspects as royalty rates alone are not effective to determine this concept. It appears that the decision of the Taipei High Administrative Court and the proposal to amend the provisions for granting compulsory licenses in the Patent Act of Taiwan are consistent with both competition law and the TRIPS Agreement.

In any event, the decisions and rulings relating to the Philips package licensing case elaborate the application of competition law to technology licensing in Taiwan, with respect to both vertical agreements between Philips and Taiwanese licensees and horizontal agreements between Philips and the two Japanese companies. This helps to pave the way for TFTC to have investigated Philips's other anti-competitive practices relating to the package licensing. Additionally, this case and the related cases reflect the international dimension of technology transfer-related competition issues and the need for international cooperation on IPRs and competition law.[90]

89. European Commission, 'Report to the Trade Barriers Regulation Committee', fn. 72 *supra*, paras 139–140.
90. Nguyen, fn. 25 *supra*, 210–212.

4 OPTIONS FOR DEVELOPING COUNTRIES

The competition flexibilities in the TRIPS Agreement pave the way for developing countries to promote access to technology. The international community has increasingly supported their applying such flexibilities to reach this objective. The above analysis of some typical cases in South Africa, Thailand and Taiwan shows that promoting access to technology by competition law is feasible although requiring a well-established competition law infrastructure.

It depends on developing countries to coordinate domestic competition law and intellectual property protection in the light of the TRIPS Agreement and international supports to 'maximize expected social welfare'.[91] Competition law is still a domestic issue while intellectual property protection is 'globalized'. Therefore, vigorous but responsible domestic competition law enforcement will constitute an important component of the creation of a more pro-competitive environment for access to technology in developing countries. However, this principle is neither easy to apply in practice nor a panacea addressing all of the major issues. It must be remembered that competition law is antitrust. It is neither anti-IPR nor anti-trade. The enforcement of domestic competition law to promote access to technology in general should take into account the legitimate rights of IPR holders regardless of their nationality. It should not hinder international trade liberalization and adversely affect the minimum standards of intellectual property protection under the TRIPS Agreement.

In brief, the role of competition law in the field of technology transfer should be neither over-looked nor over-estimated. Developing countries need to 'exercise their sovereignty over their domestic markets, and elaborate their own "tailor-made" national competition laws and enforce them effectively'.[92] These countries can, and deserve, to utilize domestic competition law in combination with domestic intellectual property law, all of which should be interpreted in the light of human right issues and the Millennium Development Goals of the United Nations,[93] in order to promote access to technology. In such competition law enforcement, close cooperation between competition, intellectual property and regulatory authorities, as well as that between national competent authorities and international forums, should be established. Consequently, competition law should be regarded as one of the most important legal tools to be taken into account in constructing development policies in general and promoting access to technology in particular at domestic and international levels.

---

91. M.J. Meurer, 'Patent Litigation, Licensing, Nonobviousness, and Antitrust', in *Antitrust Law and Economics*, ed. K.L. Hylton (Cheltenham: Edward Elgar, 2010), 270.
92. Supachai, fn. 29 *supra*.
93. See United Nations, *The Millennium Development Goals Report 2010* (New York: United Nations, 2010).

Chapter 5

Intellectual Property Rights, Technology Transfer and Development: The Case of Compulsory Licensing

Keith E. Maskus

The period since the mid-1990s has seen the most extensive reforms in intellectual property rights (IPR) in history. The primary reason has been the requirement, through the Agreement on Trade Related Aspects of Intellectual Property Rights (TRIPS) of the World Trade Organization (WTO), that emerging and developing economies adopt more stringent minimum standards of protection. These mandated reforms involve patents, copyrights, trademarks, plant variety rights, geographical indications, and even elements of trade secrets (or unfair competition law). Additional impetus for change has come from bilateral trade agreements negotiated by the United States and the European Union and various initiatives at the World Intellectual Property Organization (WIPO). A number of rapidly developing economies also have engaged in reforms because of rising domestic economic interests in doing so.[1]

1. An extensive discussion is provided in Keith E. Maskus, *Private Rights and Public Problems: The Global Economics of Intellectual Property in the 21st Century* (Washington: Peterson Institute for International Economics, 2011), manuscript in preparation. See also Jayashree Watal, *Intellectual Property Rights in the WTO and Developing Countries* (The Hague: Kluwer Law International 2001), 48 et seq., among many other references.

Hans Henrik Lidgard, Jeffery Atik & Tu Thanh Nguyen, *Sustainable Technology Transfer*, pp. 89–120.

Figure 5.1 Changes in the GP Index by Income Quartiles

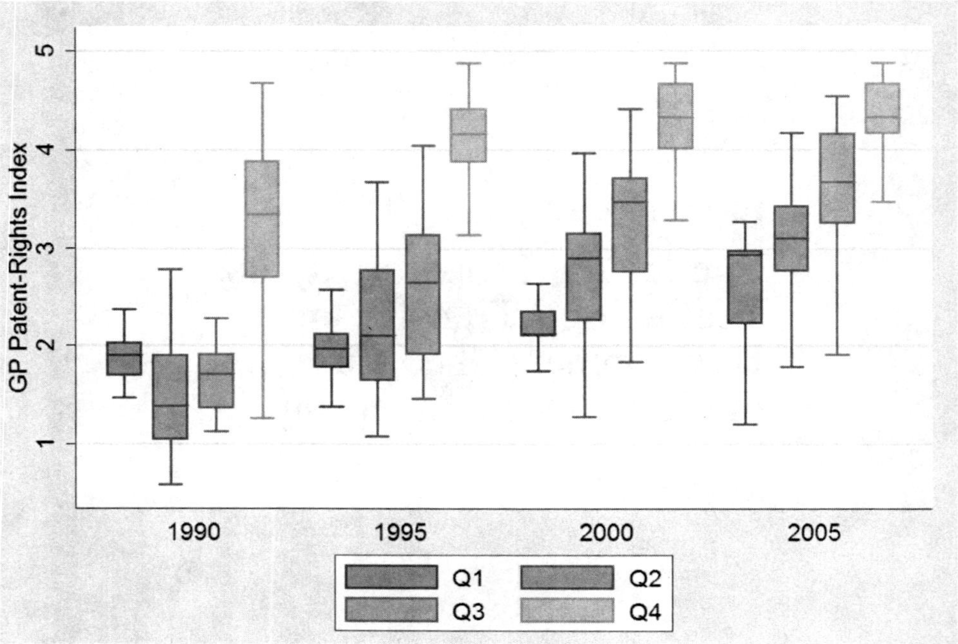

That there have been significant reforms is demonstrated in Figure 5.1, which depicts increases since 1990 in the famous Ginarte-Park (GP) index of legal patent rights.[2] This index varies from zero to five, with a higher number indicating the presence of more legal protection (e.g., longer duration or greater scope) for patents. In this diagram the dimensions of each box reflect the gap between the 25th and 75th percentile of each country group and the horizontal line is the mean value. Country groups are listed by income quartiles, based on real (PPP-adjusted) GDP per capita in US dollars, expressed in year 2000 prices.[3] As may be seen, in the earliest year there was a 'U shape' in patent protection, as discussed in the literature.[4] Countries in the second and third quartiles had protection indexes far below those of the rich economies in the fourth quartile. By 2005, however, there was a clearly rising relationship between patent rights and average incomes. The primary point is that all of the three lower quartiles significantly reformed their legal regimes. Those in the second and third quartiles – which are the economies with the most rapid industrialization

2. For a description of this index see Walter G. Park, 'International Patent Protection: 1960–2005', 37 *Res Pol'y* 761 (2008).
3. Maskus, fn. 1 *supra*, offers details.
4. Keith E. Maskus & Mohan Penubarti, 'How Trade-Related Are Intellectual Property Rights?', 39 *Journal of International Economics* 227 (1995), and Yongmin Chen & Thitima Puttitanun, 'Intellectual Property Rights and Innovation in Developing Countries', 78 *J Deve Econ* 474 (2005).

underway and represent the most significant imitation threats – more than doubled their average indexes. For its part, China's GP index rose from 1.33 to 4.08, while Brazil's rose from 1.28 to 3.59. Moreover, supplemental measures of the *effective* protection of IPRs (involving perceptions of enforcement activity in addition to legal reforms) indicate that global firms see a much stronger international system now than 20 years ago.[5] In brief, the concerted efforts of the United States, the EU and other developed economies to enhance global IPRs have paid off handsomely, especially among the larger emerging economies.

One would expect that major policy transformations of this kind would have important and detectable effects on economic activity. The primary potential benefit for developing economies, as noted in the TRIPS Agreement itself, is for enhanced inward flows of technology transfer (TT), both in terms of volume and products and technologies aimed specifically at their needs. I review the available evidence on this point later in this paper. This evidence is not definitive but it does suggest that stronger IPRs are significantly boosting TT into the larger, middle-income countries. There is little evidence of any impact, positive or negative, in the poorer developing countries. Put differently, IPR reforms are likely to enhance industrial transformation and economic growth within those economies where enterprises are capable of effectively deploying and improving new technologies. This structural transformation will not be without cost, however, as greater competitive pressure is faced by less efficient domestic enterprises. Further, the processes of technology transfer, adaptation, and innovation are highly complex. Simply putting in stronger patent and copyright rules in developing economies is not by itself likely to generate technological change and growth. Sustainable access to global technologies, and the ability to shift an economy toward more innovation, requires an overall economic environment conducive to such processes.

The complex interrelationships among IPR, technology transfer, and growth raise a number of other concerns as well. In developing economies where local competition is weak or may be stifled by entry regulations, stronger patents raise the likelihood of monopolistic practices, restrictive licensing arrangements and diminished access to new goods. In welfare terms this problem is most acute in areas where access is important for treating public needs, such as delivery of essential medicines and mitigation of environmental damages. If such needs go unmet governments may be tempted to resolve the market limitations via public intervention.

One particular intervention that is widely available in the law is to issue a compulsory license, also called a non-voluntary license. Stated narrowly, a compulsory license (CL) is an order by public authorities to permit the use by government or third parties of the subject matter of a patent without the authorization of the patent holder. The concept is broader, however. For example, through the

5. This statement is based on the subjective indexes published in World Economic Forum, *Global Competitiveness Report* (Geneva: World Economic Forum, various years).

fair-use doctrine or similar regulations, copyrighted material may be the subject of implicit compulsory licenses, which do not require a formal public order. Many countries permit unauthorized research and experimental use of technologies protected by patents and other forms of industrial property, though this use does not extend to production or marketing.[6]

This action is taken when the perceived public interest in wider use outweighs the private interests of the exclusive rights owner. In the case of a patent, the users of a CL typically offer some form of market-based royalty to compensate for the order. The CL is also generally non-exclusive and permits the original IPR holder to continue to sell. As described later, CLs may be issued for a variety of public-policy purposes, including providing for government use, dealing with medical emergencies, counteracting market power, and offsetting misuse of intellectual property.

To my knowledge there are no formal statistical analyses in the literature of the effectiveness of CLs in increasing TT in developing countries. This is not surprising, because the ability of CLs actually to implement domestic production capacity is heavily dependent on other market and policy circumstances. Moreover, the most important role of CLs may not be in acquiring technology for domestic firms as much as presenting IPR owners with a competitive threat that may induce them to increase access, license voluntarily, or reduce prices and fees. On the other hand, the CL may push firms to exit markets altogether.

In this chapter, I consider the basic economics of compulsory licensing as a means of inducing and sustaining technology transfer. In the next section I briefly overview recent literature on the role of IPR reforms in attracting technology transfer. I turn in section 2 to a description of the global rules governing the issuance of compulsory licenses and the substantial policy flexibility available to governments wishing to use them. In section 3, I discuss how compulsory licenses have been used in the past in key nations. In section 4 I briefly review available statistical evidence. In section 5, I analyse fundamental economic issues involved in compulsory licensing, including the conditions under which the policy can make sense and the complications facing its use. Against the potential benefits are a number of administrative and international costs to be factored into those decisions. I draw some policy lessons in the concluding section.

1 IPR AND TECHNOLOGY TRANSFER

Available literature suggests that developing countries rarely experienced a notable increase in domestic innovation efforts, as measured by patent applications and

6. Even more expansive definitions exist. For instance, as noted later in the chapter the US Supreme Court, in *eBay Inc. v. MercExchange*, 547 US 388 (2006) recently denied the routine availability of preliminary injunctions in cases of alleged patent infringement, converting them into de facto CLs.

R&D expenditures, after undertaking significant patent reforms.[7] This is not surprising, however, since pressure to reform tends to come from international firms long before domestic economic interests are sufficiently innovative to prefer stronger IPR. The global innovators have an interest in stronger protection in those markets where they wish to operate, whether through exports, foreign domestic investment (FDI), or licensing and joint ventures.

Each of these modes of operation involves transactions in existing technologies (perhaps patented in the developed economies) rather than investments in local R&D. In that regard, economists generally find that any short-term stimulus of IPR on economic activity is through safeguarding the returns from contracting in technology, especially across borders.[8]

Indeed, there is persuasive evidence that IPR reforms attract inward TT through a variety of channels. Rather than review what is now a lengthy literature I will just mention a few recent and important studies. First, regarding international trade, one careful statistical analysis considered similar developing countries that adopted relatively stronger or weaker patent reforms in the pre- and post-TRIPS era.[9] The author finds that a country that enacted more rigorous patent laws, compared to an otherwise highly similar country that did not, attracted significantly more imports of high-technology products from developed economies. Applying this finding to changes in the TRIPS Agreement, she suggests that patent reforms added as much as 8.6% in volume terms to developing-country imports of such goods. As has been noted many times, such imports embody a considerable amount of new technical information that spills over into higher domestic productivity.[10]

Next, licensing contracts also seem sensitive to patent reforms. In a recent study of firm-level data the authors show that variations across countries and over time in the index of patent rights have a significantly positive impact on licensing receipts of headquarters firms, and that this was due to a higher volume of

7. The classic reference is Josh Lerner, *Patent Protection and Innovation over 150 Years* (Cambridge, MA: National Bureau of Economic Research Working Paper no. 8977, 2002), 12 et seq. See also M. Sakakibara & Lee Branstetter, 'Do Stronger Patents Induce More Innovation? Evidence from the 1988 Japanese Patent Reforms', 32 *RAND J Econ* 77 (2001); Patricia H. Schneider, 'International Trade, Economic Growth and Intellectual Property Rights: A Panel Data Study of Developed and Developing Countries', 78 *J Devel Econ* 529 (2005); and Yi Qian, 'Do National Patent Laws Stimulate Domestic Innovation in a Global Patenting Environment? A Cross-Country Analysis of Pharmaceutical Patent Protection, 1978–2002', 89 *Rev Econ Stat* 4369(2007).
8. See Keith E. Maskus, *Intellectual Property Rights in the Global Economy* (Washington, DC: Institute for International Economics, 2000), 136–141; Keith E. Maskus, *Encouraging International Technology Transfer* (Geneva: International Centre for Trade and Sustainable Development, Project on IPRs and Sustainable Development, Issue Paper 7, 2004), 24–25; and Walter G. Park, 'Intellectual Property Rights and International Innovation', in *Intellectual Property, Growth and Trade*, ed. Keith E. Maskus (Amsterdam: Elsevier, 2008) 289.
9. Olena Ivus, 'Do Stronger Patent Rights Raise High-Tech Exports to the Developing World?', 81 *Journal of International Economics* 38 (2010).
10. David T. Coe, Elhanan Helpman & Alexander W. Hoffmaister, 'North-South R&D Spillovers', 107 *Econ J* 134 (1997).

licensing.[11] However, the impact was quantitatively much larger if licensees were located in developed countries than in developing countries. A more detailed study of data from the overseas operations of US-based MNEs finds that royalty payments, affiliate R&D and foreign patent applications rose significantly in 16 large emerging economies after patent laws were strengthened between 1982 and 1999.[12] These effects were concentrated in affiliates of US companies that were highly patent-intensive in their domestic operations.

Finally, using the same data set and IPR reform episodes, these authors discover that FDI-related activities of US MNEs rose significantly in countries after policy changes.[13] There was a significant expansion of domestic industrial production by affiliates and of intra-firm exports to reforming nations. All of that is consistent with an enhanced flow of TT through licensing and FDI.

These findings need to be qualified in a number of dimensions. First, none of this evidence pertains to activities in the poorest countries and LDCs. If anything, IPR reforms in those countries have no effect on market-mediated TT, which depends on many other factors.

Second, the fact that market-oriented channels of TT seem to expand in the wake of IPR reforms does not imply that overall diffusion is enhanced if more informal means of learning are reduced. There is evidence that, at least within the Organisation for Economic Co-operation and Development (OECD) countries, reading patent applications is a significant form of gaining access to information about frontier technologies and seems to have a significant effect on productivity growth.[14] However, this form of diffusion is not of great relevance in countries where domestic enterprises and research institutions do not have sufficient capabilities to deploy information from these published sources into their own programs. This problem is heightened when patent applications do not require inventors to reveal enough information as to enable domestic learning and imitation. In any case, the statistical validity of this patent-mediated TT channel has yet to be demonstrated even in emerging and technically capable countries, much less in developing and least developed countries.

Third, the primary form of learning spillovers is imitation and adaptation by domestic firms. There are several channels through which this can happen, including reverse engineering, accessing and improving slightly out-dated and unprotected technologies, the mobility of technical personnel among firms, and other factors. These processes were surely a significant route to technological

11. Walter G. Park & Douglas Lippoldt, 'International Licensing and the Strengthening of Intellectual Property Rights in Developing Countries during the 1990s', 40 *OECD Econ Stud* 7 (2005).
12. Lee G. Branstetter, Raymond Fisman & C. Fritz Foley, 'Do Stronger Intellectual Property Rights Increase Technology Transfer? Empirical Evidence from US Firm-Level Panel Data', 121 *Q Jl of Econ* 321 (2006).
13. Lee G. Branstetter et al., *Intellectual Property Rights, Imitation and Foreign Direct Investment: Theory and Evidence* (Cambridge, MA: National Bureau of Economic Research Working Paper no. 13033, 2007): 15 et seq.
14. Jonathan Eaton & Samuel J. Kortum, 'Trade in Ideas: Patenting and Productivity in the OECD', 40 *Journal of International Economics* 251 (1996).

transformation in such countries as the United States, Japan, Korea, and Taiwan. They underpin Chinese industrial growth today. In each of these countries, except arguably China, significant IPR reforms came after these mechanisms helped build innovative domestic enterprises. For its part, China has adopted a legal framework that strikes a reasonable balance between the exclusivity needs of major domestic and international innovative enterprises and the diffusion needs of imitative and follow-on enterprises.[15] Nonetheless, some features of its newest patent law clearly reserve policy space for limiting patent scope (see below) while the country's enforcement efforts are sufficiently lax that the effective policy favours local technology spillovers.

Perhaps the most important question is whether the strengthened global IPR regime actually limits such learning and spillovers through raising the costs of imitation. This could happen if patent laws establish low thresholds for patentability and broad scope, including unreasonable restrictions against compulsory licensing, and if copyright laws rigorously restrain unauthorized use of educational and scientific materials and reverse engineering of software. Unfortunately, there are virtually no micro-econometric empirical studies to cite on the question, nor of the role of CLs in this process.[16]

2 HISTORY OF GLOBAL REGULATION AND TRIPS

2.1 General History and Industrial Property

Non-voluntary licensing orders have existed in the IPR laws of the major developed economies for many decades. Indeed, the right to issue such licenses was recognized in the Paris Convention (1883) in order to prevent the abuses that might be taken in terms of exercising a patent. The primary such abuse was failure to work, or to insufficiently work, a patent. The CL could be issued after four years from the date of filing a patent application, or three years after the date of its grant, and had to be non-exclusive and non-transferable by the licensee. It could not be issued if the patentee had legitimate reasons for his actions.[17] The meaning of the

15. See Keith E. Maskus, 'Intellectual Property Rights in the WTO Accession Package: Assessing China's Reforms', in *China and the WTO: Accession, Policy Reform, and Poverty Reduction Strategies*, ed. Deepak Bhattasali, Shantong Li & William J. Martin (Oxford: Oxford University Press, 2004), 49.
16. This issue is discussed in depth in Keith E. Maskus, *Private Rights and Public Problems: The Global Economics of Intellectual Property in the 21st Century* (Washington, DC: Peterson Institute for International Economics, 2011 in press). No simple answer can be given, since the prospects for increasing formal ITT while sustaining imitation processes depend heavily on each country's socio-economic characteristics. The essential point is that nations need to implement sound development and innovation policies to complement the effects of intellectual property rights.
17. Paris Convention for the Protection of Industrial Property of 20 Mar. 1883, as revised, 828 U.N.T.S. 305, Art. 5, as set out in revisions from the Conference of The Hague in 1925.

phrase 'legitimate reasons' was not specified and left much to the discretion of national authorities.

The Paris Convention was silent as to the meaning of the working requirement and most countries limited it to domestic production, either by the patentee or its licensee.[18] The implication is that refusal to supply or license within a reasonable time period was an abuse of exclusive rights that needed to be remedied. In that regard, compulsory licensing was seen as less punitive to rights holders than the prior approach, in which patents could be forfeited in the event of non-working. With a period of priority, non-exclusive licenses, and the ability of patent holders to offer reasons for non-working, CLs seemed to strike a better balance and were adopted in numerous countries, even in nations with no prior law in the matter.

This insistence on local working was widely seen as an appropriate means to encourage the development of national industry.[19] Moreover, countries soon saw a wide scope for issuing CLs in the public interest, particularly in medicines and foodstuffs. Indeed, by 1958 at least 11 developed countries (Belgium, Canada, Denmark, Finland, France, Germany, Israel, Japan, Norway, Netherlands and Sweden) and several developing countries reserved the right in matters of public interest to issue CLs at any time without waiting for the periods listed in the Paris Convention to elapse.[20] In brief, much of the developed world saw the need to use CLs for reasons that went well beyond the abuse of monopoly rights. The revisions of the Paris Convention taken at the Lisbon Conference in 1958 clarified that CLs could be issued without delay in almost any case involving the public interest; the prescribed periods of delay in Article 5 would apply only to cases of non-working.[21] It should be noted that the public interest was interpreted quite broadly in this context and CLs could be issued for selling patented goods at unreasonably high prices or importing patented goods at such low prices that domestic firms could not compete.

The provisions of the Lisbon Conference represented the most permissive approach to CLs in global rules-setting and it is instructive that they existed within the developed world. All of those countries, including the United States, were anxious to promote domestic industrialization. Many saw CLs as a means of promoting or requiring technology transfer in cases where they could not otherwise induce foreign enterprises to produce locally. By the 1960s most of these nations also had liberal rules governing public-interest use. For example, nearly all

18. See Jerome H. Reichman & Catherine Hasenzahl, *Non-voluntary Licensing of Patented Inventions: Historical Perspective, Legal Framework under TRIPS, and an Overview of the Practice in Canada and the USA* (Geneva: International Centre for Trade and Sustainable Development, Project on IPRs and Sustainable Development, Issue Paper no. 5, 2003), 10–13. In principle, domestic needs could be satisfied by sufficient import quantities but few nations recognized this channel as working. Language in the TRIPS Agreement suggests that imports now satisfy any legal working requirement but this issue remains controversial.
19. Stephen P. Ladas, *Patents, Trademarks and Related Rights: National and International Protection* (Cambridge, MA: Harvard University Press, 1975), 516.
20. *Ibid.*
21. Reichman & Hasenzahl, fn. 18 *supra*, 12.

member states of the EU still retain laws permitting the issuance of CLs in cases of public interest, as do other developed economies in Europe.[22] By consensus this provision covers the needs of national defence, environmental problems, energy supplies, worker safety, and access to new medicines.

A further attempt was made to expand the global use of CLs in the Conference to revise the Paris Convention, which lasted from 1979 to 1985. A key component of what was termed the 'New International Economic Order', many developing countries pushed for a general reduction in the minimum standards regulating patents and related industrial property.[23] However, industrial interests in the major developed economies, especially the United States, were determined to elevate those standards. This tension caused the negotiations to collapse and set the stage for moving the global regulation of IPRs into the WTO.

Nonetheless, by the early 1990s over 100 countries had provisions for CLs in their domestic patent laws. Among the reasons listed for issuing CLs were non-working, refusal to deal or license, abusive or anti-competitive practices, the public interest, government use, and access to pharmaceuticals and food. Some also listed CLs as a means of effectuating 'dependent patents', or improvements on technology that required use of the prior patent.[24] Such provisions existed in the legal regimes of both developed and developing economies.

2.2 The TRIPS Agreement and CLs

With this background a major aim of US, EU and Japanese negotiators at the negotiations leading to the TRIPS Agreement at the Uruguay Round was to place significant limitations on the permissible use of CLs. The relevant TRIPS provisions and the question of remaining policy space in this context have been widely discussed in the literature and only require a brief treatment here.[25]

An initial observation is that TRIPS itself is built on the objective of promoting economic development, including through technology transfer. Article 7 states that 'The protection and enforcement of intellectual property rights should contribute to the promotion of technological innovation and to the transfer and dissemination of technology, to the mutual advantage of producers and users of technological knowledge and in a manner conducive to social and economic welfare, and to a

22. *Ibid.*
23. A comprehensive history and analysis is offered in Susan K. Sell, *Power and Ideas: North-South Politics of Intellectual Property and Antitrust* (Albany: State University of New York Press, 1998), 107 et seq.
24. Reichman & Hasenzahl, fn. 18 *supra*, 12; and Carlos M. Correa, *Intellectual Property Rights and the Use of Compulsory Licenses: Options for Developing Countries* (Geneva: South Centre T.R.A.D.E working paper no. 5, 1999), 4.
25. See Correa, fn. 24: 15 et seq., Maskus, *Intellectual Property Rights*, fn. 8 *supra*, 171–224, Watal, *supra* fn. 1: 291 et seq., Reichman & Hasenzahl, fn. 13 *supra*, 13–18, and Carolyn Deere, *The Implementation Game: The TRIPS Agreement and the Global Politics of Intellectual Property Reform in Developing Countries* (Oxford: Oxford University Press, 2009), 74–98.

balance of rights and obligations'.[26] This rather convoluted statement can be interpreted in a number of ways but in the current context, it suggests that developing countries may deploy limitations and exceptions on patents and other IPR in order to promote welfare-enhancing TT.

In this regard, Article 8 lists general principles under which countries may issue exceptions from the overall IPR obligations. Specifically, 'Members may . . . adopt measures necessary to protect public health and nutrition and to promote the public interest in sectors of vital importance to their socio-economic and technological development . . . '.[27] Further, appropriate measures may be taken to prevent the abuse of IPR or the use of practices by rights holders that '. . . unreasonably restrain trade or adversely affect the international transfer of technology'.[28] While these principles do not refer specifically to CLs, they suggest that the two broad areas in which such licenses may be deployed involve matters of vital public interest and the abuse of IPR, especially patents. It is important to note that these principles are subject to a requirement that they be consistent with the provisions of TRIPS.

The specific language regarding CLs arises in Articles 27.1, 30 and 31, which pertain solely to patents. The relevant portion of Article 27.1 states that ' . . . patents shall be available and patent rights enjoyable without discrimination as to the place of invention, the field of technology and whether products are imported or locally produced.'[29] On its surface this provision seems to eliminate the use of CLs in the case of non-working, where working is defined as local production or licensing rather than just imports. However, a number of countries continue to define the absence of local production as an abuse of patent rights, which is a different legal question. In this context, the United States withdrew its WTO complaint against Brazil's assertion of working requirements in pharmaceuticals in 2001, suggesting they may be sustained at least in some circumstances.[30] Whether this definition of abuse can stand in light of Article 27.1 remains controversial.

Article 30 is a general statement that governments may provide 'limited exceptions' to exclusive patent rights so long as they do so in a way that meets a three-pronged test. Specifically, such exceptions may not 'unreasonably prejudice' the legitimate interests of rights holders, may not 'unreasonably conflict' with normal exploitation of the patent and must also take account of the interests of third parties.[31] This seems to be a narrow provision as written since exceptions must be limited and meet the other criteria set out.[32] However,

26. Agreement on Trade-Related Aspects of Intellectual Property Rights, (Annex 1C to Agreement Establishing the World Trade Organization, 33 I.L.M. 1197, (1994), Art. 7 (TRIPS Agreement).
27. TRIPS Agreement, Art. 8.1.
28. *Ibid.*, Art. 8.2.
29. *Ibid.*, Art. 27.1.
30. See 'Brazil-Measures Affecting Patent Protection, Notification of Mutually Agreed Solution', WTO document WT/DS199/4 (2001).
31. TRIPS Agreement, Art. 30.
32. Christopher A. Cotropia, 'Compulsory Licensing under TRIPS and the Supreme Court of the United States' Decision in *eBay v. MercExchange*', in *Patent Law and Theory: A Handbook of*

Article 30 offers no specific limitation on the use of CLs per se. Interestingly, attempts to limit the scope of this exception during negotiations ran afoul of legal provisions in the major developed economies, including the United States, where the courts continue to use CLs as components of remedies in antitrust cases.[33]

Next, Article 31 lists requirements that must be met to issue a compulsory license. It is important to note that Article 31 refers to CLs issued for purposes other than in Article 30 and the provisions it sets out therefore do not apply to the general-exceptions CLs under that article. The uses mentioned are those by the government and those authorized on behalf of third parties (i.e. compulsory licenses) but not limited to these. In such cases, a number of conditions are imposed by TRIPS.[34] Use must be authorized on a case-by-case basis, limited in scope and duration, non-exclusive and non-assignable (by the licensee), and it must pay adequate compensation to the rights holder based on the patent's economic value. Furthermore, such uses are permitted only after prior efforts of potential licensees to gain access to the technology on commercial terms have failed (except in cases of national emergency or public non-commercial use), are subject to judicial review, and may be terminated after the conditions giving to its rise have ceased to exist. CLs may be issued in the case of dependent patents subject to certain limitations. Finally, under Article 31.f authorized use must be 'predominantly for the supply of the domestic market', in essence preventing CLs for export. This last limitation was important to multinational firms that did not want to see CLs used as industrial policy that could support global export competition for their patented goods.

Article 31 was negotiated presumably without concern for its potential implications on access to medicines in poor countries. At the outset of TRIPS, the United States took the view that it precluded the possibility of CLs in pharmaceuticals. However, South Africa and Brazil in their legal reforms asserted the rights to issue non-voluntary licenses in cases of national emergency. Further, it soon became evident that the limitations in Article 31.f requiring that a CL be limited to predominantly supply the domestic market would make it effectively impossible for poor countries with limited production capacity to issue an authorization that could meet their needs. WTO member states recognized in the 2001 Doha Ministerial Declaration on the TRIPS Agreement and Public Health that CLs were permissible and that countries had the '. . . freedom to determine the grounds on which such licenses are granted.'[35] It also extended the required implementation of patent rules covering medicines until 2016 for the least developed countries. Finally, states recognized that relief from Article 31.f should be provided in essential patented medicines, resulting in the 2005 Amendment to the TRIPS Agreement.[36] This

---

Contemporary Research, ed. Toshiko Takenaka & Rainer Moufang (London: Edward Elgar Publishing Ltd, 2009), 557.

33. Reichman & Hasenzahl, fn. 13 *supra*, 14.
34. TRIPS Agreement, Art. 31.
35. Doha Declaration on TRIPS and Public Health, WTO document WT/MIN(01)/DEC/2 (2001).
36. Amendment of the TRIPS Agreement, WTO document WT/L/641 (2005).

amendment inserted Article 31.*bis*, under which countries with export capacity may issue a CL to supply the market of a least developed country or other developing country wishing to use the system. Thus, for example, generic producers in China, India, Canada and other countries may petition their governments for a CL to produce a medicine specifically for distribution in an eligible WTO member state.

To summarize this discussion, a close reading of TRIPS supports the view that developing countries are left with considerable legal room to manoeuvre in the issuance of compulsory licenses. They may be authorized in cases of national emergencies and government non-commercial use without first negotiating with patent holders. They are similarly available to discipline anti-competitive abuses of IPR, including attempts to restrain technology transfer. The use of CLs is generally bound by the requirements of Article 31 but many legal experts believe that these rules may be used flexibly within the limits of each nation's legal regime.[37] For example, authorities may take actions to ensure that licensees under a CL order may continue to produce or be compensated in the event of a cessation of the order. This is important to avoid cases where such licensees would not undertake investments for fear of being forced out of the market after some period of time. For another, it is evident that a CL may be issued in cases where a patented drug or other product of importance for health or nutrition is priced at levels higher than authorities see as consistent with their needs.[38] Finally, least developed countries may take advantage of generic supplies of medicines produced in other countries if those governments agree to issue a CL for exports.

With this flexibility it may be surprising that relatively few cases of CLs have actually been issued in the developing world. As discussed below, both Thailand and Brazil have used them to acquire production rights in patented anti-retroviral drugs, but other cases are scarce. Only Rwanda has taken advantage of Article 31.*bis* under the 2005 Amendment, when in 2008 it imported a combination AIDS drug from Apotex, a Canadian generics producer. Otherwise, there appear to be few cases of formal CL uses in the developing world for any reason, including industrial technology transfer. This puzzle suggests that there must be other constraints on CL processes, arising from economic and political factors. Those factors are discussed in section 5 below.

2.3 Copyrights

Before departing this section, however, it is important to note that compulsory licenses may effectively apply to other forms of IPR beyond patents and industrial property. Most prominent is the doctrine of 'fair use' (a US construct) or similar provisions under which copyrighted material may be used without authorization.

37. Reichman & Hasenzahl, fn. 13 *supra*, 14–15.
38. Note also that TRIPS places no explicit restrictions on drug price controls or other elements of public health policy.

This provision is not formally equivalent to CLs in the sense discussed above, for no licenses need to be issued under fair use. Indeed, countries generally do not have a regime for CLs in their copyright laws and rely instead on such doctrinal limitations and exceptions. Fair-use rules govern general circumstances under which copyrighted material may be taken for personal use without formal authorization. If a copyright holder believes her rights to have been infringed and the copier asserts fair use it is up to the courts to decide the validity of the two claims.

The TRIPS Agreement is largely silent on the issue of fair use except to replicate the language of the Berne Convention.[39] Specifically, Article 13 states that 'Members shall confine limitations or exceptions to exclusive rights to certain special cases which do not conflict with a normal exploitation of the work and do not unreasonably prejudice the legitimate interests of the right holder'.[40] Implicit in this statement is a three-step test (special cases, normal exploitation, and unreasonable prejudice) for determining the allowable scope of limitations and exceptions.

In traditional copyright doctrine (the Berne Convention) such limitations may be either uncompensated or compensated.[41] Quotations in book reviews, criticisms, other commentaries, and scholarly works are examples of the former. So are limited extracts and illustrations for teaching. Countries may also permit reproduction by the press and broadcasters of current periodicals and related materials if the copyright owner does not reserve the right to do so herself. Compensated limitations exist in certain cases where copyright owners in broadcasts and sound recordings may not wish to transmit or offer these services to specific locations. Second users may then distribute the materials without authorization (if permitted by law) but must pay some reasonable compensation at rates determined by competent officials. This provision underlies practices in the US, Canada and other countries to permit the transmission of broadcast materials by local providers to rural areas without requisite cable or wireless systems or other means of receiving them from the copyright owners.[42]

While these are important public-purpose limitations to copyright, they generally refer to allowable use once a legal copy of an item has been procured. Because citizens in developing countries may have more fundamental access problems – legally procuring a textbook, film, or broadcast and having those materials available in local languages – they may be of limited practical relevance in development terms. Rather, for purposes of science and education developing

---

39. Berne Convention for the Protection of Literary and Artistic Works September 9, 1886, as last revised at Paris on 24 Jul. 1971, 1161 U.N.T.S. 30.
40. TRIPS Agreement, Art. 13.
41. For an extensive discussion, see Ruth L. Okediji, *The International Copyright System: Limitations, Exceptions and Public Interest Considerations for Developing Countries* (Geneva: International Centre for Trade and Sustainable Development, Project on IPRs and Sustainable Development, Issue Paper no. 15, 2006): 5 et seq.
42. Copyright systems also permit broadcasters, bars and restaurants to play music or other materials under collective licenses purchased from collection agencies in order to overcome the transactions costs of individual licensing. This is not a compulsory license, however.

countries may wish to extend fair use to copying and translating materials, presumably with compensation.

However, the Berne Convention and TRIPS clarify that the creators have the right to prevent reproduction and to authorize translations. This problem was explicitly recognized in the 1971 Appendix to the Berne Convention,[43] which set up a complex compulsory licensing system for bulk access to, and translations of, undersupplied or overpriced items. The procedures involve significant waiting periods for reproduction and translation licenses and use is broadly limited to educational, teaching and research purposes.[44] Potential translators must show evidence of having attempted to achieve a license. Authors are given a grace period beyond these terms and, if they choose to withdraw a protected work from circulation in a country it cannot be the subject of a CL. These residual access problems are sufficiently difficult that very few countries have acceded to the Berne Appendix. Nonetheless, it was incorporated into TRIPS and remains the only mechanism for bulk access available in developing nations.

Some scholars argue that the general provision in TRIPS, Article 40 that permits taking action on competition grounds against abuses of IPRs could be used to issue compulsory licenses.[45] However, this interpretation rarely has entered the antitrust practices of the developed countries. Further, a claim that limited supplies and high prices of copyrighted material support issuance of a CL might be problematic to sustain if it became a WTO dispute. One reason is the relatively narrow scope of copyright protection (copyright pertains to expressions rather than ideas), which has historically caused it to escape competition purview. It is difficult to identify licensing abuses in the copyright context, for example.[46] Further, given the general absence of resources and expertise for competition enforcement in the poorer developing economies it is not likely to be a sustainable means of ensuring access to published content. Finally, it is not clear that a compulsory license would be a legitimate remedy given the requirement for consistency with the rest of TRIPS. In sum, despite the legal possibility that TRIPS opens an avenue for compulsory licensing in copyrights, countries are better advised to deploy fair-use limitations and exceptions. As noted above, this interpretation is consistent with prevailing national approaches to copyright laws.

This analysis suggests that current provisions under the Berne Convention and TRIPS probably do not support mechanisms for sustainable access at low prices to copyrighted content via the use of non-voluntary licenses. Rather, there may be scope for negotiating a supplemental treaty that sets out international conditions for permissible limitations and exceptions. Indeed, there has been limited progress at WIPO toward a global agreement regarding fair-use exceptions of materials to

43. Appendix: Special Provisions Regarding Developing Countries (1971).
44. Okediji, fn. 36 *supra*.
45. See Ruth L. Okediji, 'Sustainable Access to Copyrighted Digital Information Works in Developing Countries', in *International Public Goods & Transfer of Technology Under a Globalized Intellectual Property Regime*, ed. Keith E. Maskus & Jerome H. Reichman (Cambridge: Cambridge University Press, 2005).
46. Okediji, fn. 41 *supra*, 20.

increase access for the visually impaired.[47] However, as important as that instrument may be it does not address general access issues in poor nations. I return to this question in the final section.

A final important issue in the copyright area refers to the treatment of digital content. The TRIPS Agreement did not come to grips with the complex issues of digital distribution and the internet, including file sharing and technological means of defeating electronic locks, watermarks, and other forms of digital rights management (DRM). These problems were taken up initially in the 1996 WIPO Copyright Treaty (WCT) and the WIPO Performance and Phonograms Treaty (WPPT).[48] The treaties stipulate that authors, composers, artists and performers have exclusive rights to authorize the communication of their works to the public via wired or wireless means. Countries that accede to them must also provide legal protection and remedies against the circumvention of DRM systems.[49]

These obligations are stated broadly and leave precise solutions to contracting parties. In fact, countries have taken different approaches, particularly as the abilities to copy and download and share files on the internet have expanded. For example, under the Digital Millennium Copyright Act the United States essentially forbids any attempt to circumvent DRM. For their part, many developing countries have not signed or implemented the WCT and WPPT and are, therefore, not bound by them. Some countries that have implemented the agreements have taken advantage of the limitations and exceptions they permit. For example, in July 2010 Brazil put forward a copyright reform proposal that greatly differs from the US approach.[50] Article 107 of the proposal permits circumvention in cases of fair use and for content that is otherwise in the public domain. Put differently, Brazil does not recognize DRM approaches that extend implicit protection beyond the bounds of its copyright law. The proposal even establishes penalties for actions that hinder or prevent users from accessing their fair-use rights. Legal opinion seems split on whether this approach is fully compliant with the WIPO provisions.

To summarize, the WCT and WPPT treaties leave discretion to national authorities to define acceptable limits on copyright under the fair-use doctrine of the Berne Convention. Further, both treaties also state that countries may select their own regimes as regards exhaustion of copyrights in software, databases, recordings and other digital products, so that markets may stay open to parallel imports.

---

47. See 'No Decision on WIPO Treaty for Blind Persons Misses ''Golden Opportunity''', *Intellectual Property Watch*, 26 Jun. 2010.
48. WIPO Copyright Treaty, 20 Dec., 36 I.L.M. 65 (1997), and WIPO Performances and Phonograms Treaty 20 Dec. 1996, 36 I.L.M. 76 (1997).
49. As of March 2010 the WCT had 88 contracting parties, with 79 having put it into legal effect, while the WPPT 86 contracting parties and its provisions were in force in 75.
50. See 'Brazil's Approach on Anti-Circumvention: Penalties for Hindering Fair Dealing', <www.michaelgeist.ca/content/view/5180/125/>.

3 HISTORICAL USE OF COMPULSORY LICENSING

A brief review of how compulsory licenses have been used in major developed countries is instructive about their potential for encouraging TT or at least technology diffusion. Again, all of the developed economies have scope within their IPR laws and competition regulations to issue CLs for a variety of public-interest reasons.

3.1 The United States

The United States has essentially two doctrines under which it deploys CLs: a public, non-commercial use exception for patented technologies and orders to license as remedies for anti-competitive behaviour. Both have existed for a long time and remain in use. There is statutory foundation for the former approach, under which public-use licenses have frequently been issued to ensure that critical defence technologies may be deployed if necessary, subject to full compensation.[51] Less frequently, the government has taken patent rights in order to support development and infrastructure goals, such as dams and electricity projects.

Much less frequently, US authorities have issued, or threatened to issue, a CL in order to expand supplies of critical medicines onto the market. The most recent, and highly controversial, example came in 2001 when Bayer was threatened with losing its exclusive rights if it did not expand its supplies of Cipro to offset the anthrax scare. In the end, Bayer did so and retained its patent.[52]

Further, under terms of the Bayh-Dole Act of 1980, the federal government has 'march-in rights' to grant a CL to a third party for technologies that were developed with the assistance of public funds and are patented, and where the patent owner fails to make the product 'available to the public on reasonable terms'.[53] This reservation is aimed primarily at basic defence technologies but may also be invoked if patented medicines resulting from publicly funded research must be licensed to meet unsatisfied health or safety needs. While this provision exists in the law it has rarely if ever been used.[54]

This brief summary suggests that there are certainly provisions within US law for public non-commercial use CLs and these provisions have been undertaken with some regularity except in the case of Bayh-Dole. However, the United States

51. Reichman & Hasenzahl, fn. 18 *supra*, 21–22. The statute in question is Title 28, s. 1498 (latest revision 1 Feb. 2010), which relates to manufacture for US government use.
52. This pressure was raised under the provisions of s. 1498, meaning the United States intended to permit manufacture for its own government distribution. Note that Canada also acted and actually suspended Bayer's patent for a few days. See David B. Resnik & Kenneth A. De Ville, 'Bioterrorism and Patent Rights: Compulsory Licensure and the Case of Cipro', 2 *Am J Bioethics* 29 (2002).
53. Bayh-Dole Act, Pub. L. No. 96-517 (1980) codified at 35 USC. §§ 200–212, ss 200–203.
54. See David Halperin, *The Bayh-Dole Act and March-In Rights* (Washington DC: paper prepared for Consumer Project on Technology, 2001), 5.

generally has made much less use of public-interest CLs in order to gain rights on behalf of domestic firms. The principle appears to be that public-use exceptions are vital but using CLs to promote industrial policy is an unacceptable relaxation of patent rights.

Beyond public non-commercial use the United States has no statutory basis for issuing CLs on behalf of third parties to remedy case of insufficiently worked patents, dependent patents, excessive prices, or other reasons. Rather, US policy is generally hostile to CLs, viewing them as unwarranted limitations on monopoly rights needed to induce investments in innovation.

However, compulsory licenses have been issued frequently by the US courts as components of antitrust findings to discipline misuse (the US term) of patents in ways that amount to violations of antitrust laws.[55] American courts became sceptical about the market power implicit in patent grants in the 1930s and continued in that vein through the 1970s. For example, in 1944 the Supreme Court ruled in *Mercoid Corporation v. Mid-Continent Investment Company* that the plaintiff had inappropriately extended the patent on a machinery switch to complementary components in a tied arrangement, an unwarranted attempt to broaden the patent beyond its grant. In 1965 the court held in *Walker Process v. Food Machinery Corp.* that a firm that fraudulently obtains or asserts its patent is in violation of antitrust laws and subject to treble damages. In the 1970s less than half the litigated patents were held to be valid in US courts, mostly on antitrust grounds.

During the 1980s and the Reagan Administration, however, US policy shifted sharply in favour of patent rights and against the principle that IPR could be readily misused, even on antitrust grounds. The Court of Appeals for the Federal Circuit, founded in 1982 to hear appellate IPR cases, frequently overturned findings of patent invalidity and limited the application of competition principles to such cases.[56] The legal presumption in the United States became essentially that patents are granted as a statutory monopoly right and that rights owners may exercise them, or refuse to do so, at their volition. Recent Supreme Court cases, such as *eBay v. MercExchange*, have moderated this presumption to a considerable degree but have not reinstated a tendency toward vacating patents.

In the cases listed above the judicial remedy essentially was to invalidate a patent that was asserted in anti-competitive or fraudulent ways. This is not the same as issuing a CL, which courts rarely did. However, in their antitrust actions the Federal Trade Commission and Antitrust Division of the Justice Department have ordered CLs as components of consent decrees in a variety of instances to promote competition. There are many such cases but three will make the point. In 1999 the FTC charged Intel with abusing its dominance in the microprocessors industry by withholding key technical information from customers who used or made competing products or demanding unreasonable licensing terms.[57]

55. Reichman & Hasenzahl, fn. 18 *supra*, 21.
56. Keith E. Maskus, *Reforming US Patents: Getting the Incentives Right* (Washington, DC: Council on Foreign Relations, Special Reports series, Report no. 19, 2006), 8.
57. Details are at <www.ftc.gov/opa/1999/03/intelcom.shtm>.

In essence, Intel was refusing to deal in an attempt to coerce three major customers into licensing their patented innovations back to the company. Intel essentially justified this strategy as an assertion of its intellectual property rights. The FTC order prohibited the company from withholding such information and refusing to sell to customers on the basis of an IPR dispute. In essence, this was a CL to disclose sufficient information to customers that the use of microprocessor technology would diffuse through the sector.

A second case came in 2001 when the US Federal Trade Commission (FTC) ordered Dow Chemical, in the process of a merger with Union Carbide, to license patents and other IPR related to polyethylenes to BP Amoco and Exxon.[58] This order was designed to prevent dominant market power in the polyethylenes market as a result of the merger. In a third case, the Antitrust Division sued 3D Systems to block it from acquiring DTM and gaining a dominant position in certain prototyping technologies.[59] In the settlement, the Justice Department ordered 3D Systems to license 178 US, Canadian and Mexican patents to a third party, which would compete in the American market.

To summarize this discussion, US policy on CLs has followed three tracks. First, the government has an active public-use exception that it asserts in the cases of national defence and public infrastructure projects. Second, at least prior to the 1980s the courts were relatively hostile to the market power in patents and often vacated them on the basis of fraudulent behaviour or misuse. Third, in their antitrust actions public authorities frequently order sharing of technology or CLs of patents to promote competition through information diffusion. It should be noted that in order to minimize uncertainty facing business the FTC has issued guidelines on IPR licensing practices that are in a 'safe harbour' and effectively immune from antitrust enforcement.[60]

As noted above, various limitations and exceptions to copyrights amount to compulsory licenses. In the United States, uncompensated CLs exist under the fair-use doctrine, which sets out legal unauthorized uses of legally procured copies. Thus, attributed quotations, limited copying for education and research purposes, criticism, parody, and certain forms of news reporting all come under fair use, though any such use may be subject to litigation by the copyright holder. Uses beyond these levels require acquisition of licenses.

On occasion US authorities also issue a CL to supply remote areas with broadcasts and cable or wireless service. These actions require royalties paid at rates set by negotiation or public authorities. Regarding digital file downloading, file sharing and circumvention, these activities are illegal without authorization of the copyright owners.

---

58. Details are at <www.ftc.gov/opa/2001/02/dowunion.shtm>.
59. Details are at <www.justice.gov/atr/public/speeches/200285.htm>.
60. See 'Antitrust Guidelines for the Licensing of Intellectual Property'. <www.justice.gov/atr/public/guidelines/0558.htm.>. Guidelines have also been issued by the European Commission, the Korean Fair Trade Commission, Japan Fair Trade Commission, and other jurisdictions.

3.2 Canada

While the US approach is aimed at competition and public use, the traditional Canadian policy on CLs had two objectives – industrial development and consumer access in medicines and food.[61] Thus, until the 1930s Canada's patent law required local production or licensing on reasonable terms within two years of a patent grant. Failure to achieve that would result in revocation of the patent. This rigorous working requirement was changed in the Patent Law of 1935 to treat failure to work as an abuse of exclusive rights, the remedy to which was a CL at a reasonable royalty. This policy essentially remained in place until Canada implemented NAFTA in the early 1990s. That agreement permits importation to satisfy working requirements and Canada withdrew its local production rule from its definition of IPR abuse.

Observers point out that, despite this liberal approach to CLs, it did not often result in CLs being issued because of the high cost and lengthy periods of proceedings, the costs of litigation on appeal, and the practice of many patentees to correct the abuse when made aware of the challenge.[62] Of the 96 applications made for a CL between 1935 and 1989, 17 obtained a non-exclusive license, 15 were denied, and at least 57 were withdrawn or abandoned.[63] These figures point to the difficulty of demonstrating abuse and sustaining costly proceedings, even in a developed country with a traditional antipathy toward patents.

In contrast, Canada was much more active in using CLs to support its generic foods and medicines industries and promote consumer access. Between 1969 and 1992, over 1,000 applications for a CL to produce or import were filed in medicines, with 613 applications granted.[64] By 1993, however, this policy had become intensely controversial. A prominent commission noted that the CLs policy had saved consumers some USD 211 million in 1983 but that the mandated 4% royalty paid to patent owners was insufficient compensation for the R&D efforts. The commission also advocated increasing the royalty rate on CLs to 14% and offering even higher royalties to firms that relocated some of their R&D activity to Canada. This interesting history points to the possibility of shifting the conception of CLs away from access toward industrial policy. Whether this policy was effective in encouraging local R&D was the subject of much debate. One prominent study found that neither CLs nor the R&D bonus helped develop a research-oriented pharmaceutical industry in Canada, though the former did support the

61. Reichman & Hasenzahl, fn. 18 *supra*, 20.
62. For details consult David Vaver and Beverley McLachlin, *Intellectual Property Law: Copyrights, Patents and Trademarks (Essentials of Canadian Law)* (Toronto: Irwin Law, 1997), 170 et seq.; and Donald G. McFetridge, 'Intellectual Property, Technology Diffusion, and Growth in the Canadian Economy', in *Competition Policy and Intellectual Property Rights in the Knowledge-Based Economy*, ed. Robert D. Anderson & Nancy T. Gallini (Calgary: University of Calgary Press, 1998), 65.
63. Vaver & McLachlin, fn. 57 *supra*.
64. McFetridge, fn. 57 *supra*, 82.

generics industry.[65] In any event, Canada abandoned CLs with its 1992 patent reforms to bring its policy into conformity with the United States and other developed economies.[66]

Canada generally takes a more access-oriented approach to limitations in copyright than the United States. It pursues the 'fair dealing' doctrine that exists in a number of Commonwealth countries such as the UK and Australia.[67] In this context Canadian copyright law sets out six criteria for evaluating whether a use lies within the confines of fair dealing. Among these are the intended use (research, study, and the like), the number of copies made and how they were distributed, whether alternative non-copyrighted sources could be used, whether or not the original work was published, and the likely impact on the market for the original work. Here, developing countries might note two particular features of the Canadian law. First, the 'market impact' factor is just one of several to be considered, whereas it is paramount in the United States. Second, there is a wide definition of what constitutes fair research use, including within libraries and universities.

3.3 Other Countries

European Union. While there are significant differences in their approaches to IPR and CLs, other major countries fit into these general parameters. Thus, the EU takes an aggressive stance toward ordering licensing in the context of competition orders.[68] The European Commission is generally rigorous in its treatment of abuses of dominant positions, particularly in the area of IPR. In particular, the EC Treaty (Article 102) states that a firm's conduct can be abusive without being anti-competitive. This is quite different from recent practice in the United States, where exercise of market power is encouraged so long as it is not fraudulent or an abuse of IPR. One clear distinction is that the European Court of Justice (ECJ) frequently has restrained the pricing strategies of firms operating under a patent, including their ability to price-discriminate among customers or locations.[69]

The European authorities also at times compel licensing of a product or critical technology in cases of refusal to supply.[70] For example, in 2004 the ECJ found that the databases of IMS Health, a dominant provider of medical prices and practices, had become an industry standard and should be licensed even though it was protected by copyright and database protection.[71] In a familiar case, the European

65. *Ibid.*
66. *Ibid.*
67. See Giuseppina D'Agostino, *Healing Fair Dealing? A Comparative Copyright Analysis of Canadian Fair Dealing to UK Fair Dealing to US Fair Use* (Toronto: York University, CLPE Research Paper 28), 2007, 3.
68. See Mattias Ganslandt, 'Intellectual Property Rights and Competition Policy', in *Intellectual Property, Growth and Trade*, ed. Keith E. Maskus (Amsterdam: Elsevier, 2008), 233.
69. *Ibid.*
70. *Ibid.*
71. Case C-418/01, *IMS Health GmbH & Co. OHG v. NDC Health GmbH & Co. KG*, [2004] ECR I-5039.

Commission ruled in 2004 that the patents held by Microsoft did not justify its refusal to provide interface information to rival firms, which would allow competing programs to interact with the company's operating system.[72] The remedy involved a significant fine until the information was provided.

Copyright fair use is governed largely by the EU Copyright Directive of 2001, which implemented the WCT and WPPT Treaties at WIPO.[73] Under this harmonization directive, member states must protect copyright owners from circumvention of both DRM and technological measures, where those actions are taken knowingly. In some ways it is more protective than the US DMCA.[74] This is consistent with a generally protective approach to copyrights and databases within the EU. For example, recent legislative reforms in France provide stiff penalties for unauthorized file downloading. As for fair-use exceptions, personal copying for private use is permissible unless it is for 'direct or indirect commercial advantage', which was not defined. Otherwise the provisions are similar to those in the United States.

Japan. Japan's patent law recognizes three general justifications for a CL.[75] One is a general public-interest exception, which may be invoked in cases of national emergency, medical crises, and the like. A second is to effectuate the exploitation of an improvement invention that requires access to an existing patented technology (dependent patents). The third is to remedy non-working of an issued patent through an unreasonable refusal to produce, import or license. These provisions come largely from the language in the Paris Convention and TRIPS. For example, the CL for non-working requires a minimum prescribed waiting period, prior negotiations that failed to achieve a license, and adequate compensation for the non-exclusive license. It should be noted that despite these provisions, CLs are rarely used in Japan. No CLs to remedy non-working have been issued. Further, under an agreement reached in 1994 with the United States, Japan pledged not to issue CLs to permit exploitation of dependent patents.

Regarding the application of competition enforcement to the abuse of IPR, Japan's Fair Trade Commission takes a case by case, broad rule of reason approach within its guidelines.[76] It is possible for the JFTC to order licensing under the country's Anti-Monopoly Act, in the event of an abuse such as refusal to deal, in order to restore competition to the market.

---

72. European Commission Decision of 24/03/2004 relating to a proceeding under Art. 82 of the EC Treaty (Case COMP/C-3/37.792 *Microsoft*), OJ 2007 L32/23.
73. Directive 2001/29/EC of the European Parliament and of the Council of 22 May 2001 on the harmonization of certain aspects of copyright and related rights in the information society, OJ 2001 L 167/10.
74. See Aashit Shah, 'UK's Implementation of the Anti-Circumvention Provisions of the EU Copyright Directive: An Analysis'. *Duke Law and Technology Review* 3 (2004), <www.law.duke.edu/journals/dltr/articles/2004dltr0003.html>.
75. See Tadashi Ishii, *Compulsory License System: History and Present Problems, Osaka Institute of Technology* <www.oit.ac.jp/ip/property_lab/ishii/data/pdf/slide_e_03.pdf>.
76. Organisation for Economic Co-operation and Development, 'Intellectual Property and Competition Policy in the Biotechnology Industry', *Policy Brief*, June 2005, 7.

China. China's patent law offers three avenues to a CL: the public interest under a national emergency or other extraordinary problem, exploitation of an improvement patent, and a refusal to license on reasonable terms.[77] The conditions for issuing CLs are consistent with TRIPS provisions, including permission of judicial review. To date it appears that no CL has been issued under patent legislation passed since China joined the WTO in 2001.

Mention should be made of China's highly controversial 'indigenous innovation policy', first announced in 2006 and still under formulation. In essence, this policy provides substantial preferences in government procurement to firms that develop new products in China in six high-technology areas and that own IPR in China. Firms that develop their IPR outside China, which are most multinational enterprises, may be excluded from the country's massive procurement market. One consequence might be a decision by these enterprises to shift R&D facilities there and license their newest technologies to Chinese firms. While this approach is not a compulsory license it bears potential for inducing quicker technology transfer than international firms might prefer. The procurement policy itself is arguably permissible under the WTO, since China is not yet a signatory to the government procurement agreement. Its implicit effect on IPR may run afoul of non-discrimination rules in TRIPS, though this remains to be seen. In any event, the policy is too new to be assessed in terms of its outcomes for TT.

India. India waited until 2005, the latest dare available under TRIPS, to implement a new patent law. A CL may be issued in national emergencies and to prevent abuse of a monopoly position.[78] India also will grant CLs in order to permit production for exports of medicines to a qualifying country under TRIPS 31bis. The law also requires working 'in the territory of India', suggesting that importation does not satisfy working. A CL may be issued in the case of non-working after three years of patent grant. Non-voluntary licenses may also be authorized where the patented invention is not available to the public at a reasonably affordable price.

3.4 Summary: The Scope for Compulsory Licenses

It is useful to summarize the analysis to this point by stating succinctly the conditions under which CLs may be justified, whether by standard practice or within the TRIPS format.

Patents are the primary focus of compulsory licensing, provisions for which exist in the legislation of all major developed countries. Copyrights are also subject to at least implicit CLs under the fair-use or fair-dealing doctrines. Compulsory licenses exist to resolve the following market failures and public needs.

---

77. Jon Wood & Raj S. Dave, *Compulsory Licensing of Patents in the US, China, Japan, Germany, and India* <www.ipo.org/AM/CM/ContentDisplay.cfm?ContentFileID=6521>.
78. Express Pharma, *Compulsory Licensing under the Indian Patent Law* (2006), <www.expresspharmaonline.com/20060915/management06.shtml>.

Public-interest goals, including national emergencies, medical crises, and domestic shortages of critical public goods, including medicines, medical machinery and diagnostic kits, and foods. In principle, there is no legislative restraint on such uses, including for environmental goals and agricultural needs.

Needs for government use, including by designated firms operating on behalf of the government. Prominent examples include national defence and public infrastructure (a form of 'eminent domain' in intellectual property).

Exploitation of patents on improved technologies, where the exercise of the secondary patents requires access to a prior patent, may require the use of CLs. This case may be of particular importance in developing countries, where the bulk of innovation is adaptive and incremental, with the results aimed at meeting local agricultural or industrial needs.

The need to offset abusive practices in the private exercise of issued patents. Commonly recognized elements of abuse include refusals to supply or license (a form of 'non-working'); restrictive conditions in licensing contracts, such as required technology grant-backs from licensees; and tied sales or licensing provisions that attempt to extend a patent grant in one area to other markets.

It is possible to go beyond these elements in order to promote competition and technology diffusion where widespread use of key technologies is important. Here, competition authorities pay attention to the possible need to break up restrictive patent pools and to require licensing of IPR-protected critical product or technical standards.

In copyrights a limited scope of exceptions for private use and for educational and research purposes may be advanced as fair use.

Those elements are generally seen as acceptable uses of CLs, though there are substantive differences across countries in their legal status and frequency of use. One way to characterize them is that they are necessary limitations on patent monopolies in order to achieve particularly important public goods, including competition. In that regard, CLs aimed at achieving *social goals* or addressing basic *market failures and externalities* are uncontroversial, in principle.

The following justifications have also been made for permissible compulsory licenses (or at least implicit CLs), but are not uniformly accepted around the globe and may be the subject of disputes.

The need to discipline a firm that is selling a patented good on a market at unreasonably high prices, particularly in medicines and agricultural inputs.

Using CLs to force unauthorized technology transfer where the patentee makes reasonable efforts to supply the good.

The public monitoring of technology transfer contracts within FDI, joint venture and licensing transactions, with the intent of favouring particular forms of TT, enforcing performance standards, or requiring that technology be surrendered to domestic firms within a particular time period.

Public procurement policies that have the effect of forcing TT to participate in government contracts.

Permissive policies regarding free downloading and file sharing of copyrighted content.

The reason these policies are less uniformly accepted is that they are aimed at addressing problems that are not necessarily abuses of patent rights or the result of market failures. Rather, they are aimed at supporting strategic industrial policy and maintaining low-cost consumer access, or simply reflect a lack of enforcement resources. As an economic matter, using CLs to pursue such goals is inefficient in comparison with more direct policies, such as price controls and subsidies to R&D in local innovation and adaptation.[79]

To put this difference in perspective, CLs enjoy widespread global support as an available policy tool where they address fundamental social goals and overcome abuses of IPR. They are more controversial where they are aimed at industrial policy goals or counteract activity that is not clearly an abuse of exclusive rights. While this is an easy generalization to make, it is unfortunately vague as a guide to policy because there is significant room for disagreement between the two extremes. Countries may opt for a broad definition of fair use, for example, so long as their actions meet the Berne Convention (and TRIPS) three-step test. The TRIPS Agreement is silent on the question of granting CLs to bring down prices of drugs and agricultural inputs, so long as the action meets the terms of Article 30 and/or 31. Aggressive use of CLs to support a domestic generic drugs industry may be perceived either as an attempt to manage health-care costs or a strategic industrial policy. Perhaps most vexing, what constitutes abuse of IPR is a highly technical question that has widely divergent answers among the major developed economies, and those differences would likely be even larger between developed and developing countries. There is substantial flexibility in this area, subject to the concerns expressed about CLs as a practical matter in the following section.

Before leaving this policy overview a final question should be addressed. Can other forms of industrial property beyond patents be the subject of CLs? In principle, any element that offers exclusive production and distribution rights for a period of time may be the subject of a CL, including utility models, design patents, plant variety rights, and semiconductor designs. However, the scope of these rights in generating market power is far more limited than is the case with 20-year invention patents. As a result they are typically not the subject of CLs and, indeed, many countries exclude some or all of them from legal review.[80] China, for example, does not permit CLs on design patents. Trademarks and related devices, such as collective marks and geographical indications, also rarely support market power and are not necessarily exclusive. These elements are not generally subject

79. See the discussion of local innovation support in Keith E. Maskus & Ruth L. Okediji, *Intellectual Property Rights and Technology Transfer to Address Climate Change: Risks, Opportunities and Policy Options* (Geneva: International Centre for Trade and Sustainable Development, Project on IPRs and Sustainable Development, Issue Paper no. 32, 2011), 25–30.

80. In the case of semiconductor technologies the possibility of CL is mentioned in TRIPS, Art. 31.c, which states that authorized use for such technologies can only be for public, non-commercial use or in cases of anti-competitive practices.

to CLs and TRIPS Article 21 states that '... the compulsory licensing of trademarks shall not be permitted'.[81] However, trademarks may be subject to revocation in the event of non-use for a period of at least three years. Finally, the EU has a sui generis form of patent-like protection for databases. In principle these exclusive rights may be the source of concentrated market power and could be subject to licensing orders. The European Commission and ECJ treat this as an issue for case-by-case competition regulation and no such orders have been issued to date.

As for trade secrets, these are not formal exclusive intellectual property rights. Rather, they are confidential business information that is protected by national law against disclosure via unfair or fraudulent means. Accordingly, they are not ordinarily the subject of CLs. However, in cases where a refusal to make available trade secrets in a way that permits abuse of market dominance, competition authorities may order that the information be made available to competitors. This was a central issue in the case brought by the European Commission against Microsoft, for example.[82] Microsoft argued that its interoperability protocols, which permit other computer programs to work with its operating systems, were trade secrets protected nationally and not subject to compulsory licensing. In 2007, the General Court ruled essentially that it did not matter whether the protocols were trade secrets or formal IPR since the company was using them to support its market dominance by refusing to license them to competitors. The Court ordered that the information be made available within 120 days on reasonable and non-discriminatory terms and also fined Microsoft. This remains a controversial ruling.

4 STATISTICAL EVIDENCE

Despite their potential importance for development purposes, there are very few systematic econometric studies that provide evidence on the effectiveness of CLs. In part, this is because there are no systematic international databases available on cases in which CLs have been granted. It also reflects the simple fact that CLs may not need to be issued to accomplish their objective. Merely threatening to authorize a CL may be sufficient to induce the patent owner to license voluntarily, produce locally, or reduce prices. This is an issue that will be addressed further below.

There are just two studies of which I am aware that closely examine the impact of CLs. The first analysed US firm-level R&D expenditures data in 1975.[83] The author found that 42 companies subjected to CL decrees did not reduce their R&D relative to sales in comparison to other firms of similar size and product mix. However, there was some tendency for the affected firms to patent fewer of their inventions in the future, resorting more frequently to trade secrecy.

81. TRIPS Agreement, Art. 21.
82. See Case T-201/04, *Microsoft Corp. v. Commission*, [2007] ECR II-1115. I am grateful to Robert Schwartz for bringing this case to my attention.
83. F.M. Scherer, *The Economic Effects of Compulsory Licensing*. (New York: New York University, Stern School of Business, 1977), 5 et seq.

The author's view is that CL orders judiciously applied in cases of clear abuse or extension of patent rights beyond granted levels have little or no impact on the rate of technological progress in the United States.

A second paper focused on a historical episode. The one major exception to the US policy of not issuing public-interest CLs came during and after World War I, as studied in a recent analysis.[84] Under terms of the *Trading with the Enemy Act* of 1917 (TWEA) US firms that had contributed to the war effort were permitted to violate patents owned by enterprises in enemy countries. By February 1919 all German patents could be licensed to interested American companies. The authors studied sub-classes of inventions in the chemical industry that were particularly susceptible to this licensing to see if the US firms became more innovative after TWEA. They found a substantial increase in domestic patenting by firms that received these compulsory licenses in comparison to those that did not, with the difference being as much as 20%. This episode is often thought to have been crucial in the twentieth century development of the American chemicals industry. This is a case where what amounted to CLs turned out to be an effective means of acquiring and improving foreign technology, though tariffs and the absence of German competition for a period of time also promoted the domestic industry. It is also worth noting that the German chemicals industry recovered from this policy, suggesting that foreign CL orders in even large markets may not markedly diminish incentives for investments in the originator nations. Of course, such stark episodes, in which the patents of an entire nation are vacated, are not likely to characterize modern opportunities for CLs in developing countries.

It is worth mentioning a third analysis, which is not literally about CLs but has some indirect relevance.[85] The authors considered how the Japanese patent system (JPS) affected Japanese technical progress after World War II, as measured by increases in total factor productivity (TFP). The JPS in place over the estimation period 1960–1993 was designed to encourage incremental and adaptive innovation and diffusion of knowledge into the economy. Mechanisms for doing so included early disclosure of, and opposition proceedings to, patent applications, an extensive system of utility models, and narrow claim requirements in patent applications. There was also an active system of monitoring inward FDI and technology transfer contracts in the early part of the period. That policy discouraged FDI in favour of supporting licensing from abroad. While this was not a formal CL system it bears similarity to the role that non-voluntary licensing might play in encouraging domestic firms to acquire international technology. Specifically, the single-claim requirement, opposition, and licensing monitoring induced international firms to license their patents to domestic suppliers and rival firms in order to enter the Japanese market.

---

84. Petra Moser & Alessandra Voena, 'Compulsory Licensing: Evidence from the *Trading with the Enemy Act*' (Stanford CA: Stanford University working paper, 2008), 3 et seq.
85. Keith E. Maskus & Christine McDaniel. 'Impacts of the Japanese Patent System on Productivity Growth', 11 *Japan and the World Economy* 557 (1999).

The authors found that this system promoted the development of large numbers of utility model applications for incremental inventions, which were based in part on prior applications for invention patents and in part on licensing. In turn, utility models had a strongly positive impact on real TFP growth over the period, while patent applications had a weaker but still positive effect. The conclusion they reached is that utility models were an important source of technical change and information diffusion in Japan, while patent applications had both a direct and an indirect impact on raising productivity. It is interesting to note that as Japan has become a global leader in technology creation, its patent system has shifted dramatically in recent years away from emphasizing learning and diffusion and more toward protecting underlying inventions.

It should be noted that both South Korea and China have pursued technology and innovation policies with similar characteristics. While there are no systematic studies, anecdotal evidence in Korea suggests that its encouragement to domestic firms to license or acquire available international technologies was central in its industrial transformation.[86]

What lessons about CLs and technology transfer might be drawn from this history? First, CLs are likely to be more effective where the domestic market is large and growing rapidly. Second, each of these countries had significant domestic capacity to use and improve the technologies made available. In this context, a sound domestic base of manufacturing, process technologies, and at least a moderate amount of R&D in adaptive innovation seems to be a key in learning from TT. Third, there may be policy alternatives to CLs that can encourage technology transfer, including limitations on patent scope, use of utility models and similar devices to encourage domestic adaptive innovation, and programs to promote technology licensing. However, to the extent that the last approach involves intrusive monitoring, mandated licensing, and performance requirements they may be challenged by foreign trade partners. It is not clear that they are the most sensible policy approach in any event.

5 ECONOMIC TRADEOFFS

Perhaps the most remarkable thing about compulsory licenses, particularly in developing countries, is that they are so little used. It is, in fact, difficult to locate cases where any emerging or developing country granted a CL for reasons other than ensuring low-cost access to patented medicines on their markets, as noted below, and even those situations are scarce. Only Rwanda has used the terms of TRIPS Article 31.f.*bis* to procure medicines from another country by asking the exporting government to grant a CL. That CLs are almost never used is even more

86. Linsu Kim, *Technology Transfer and Intellectual Property rights: Lessons from Korea's Experience* (Geneva: International Centre for Trade and Sustainable Development, Project on IPRs and Sustainable Development, Issue Paper no. 2, 2003): 3 et seq.

remarkable in light of the substantial remaining flexibility that countries may deploy under terms of TRIPS, as discussed above.

The economic logic underlying non-exclusive CLs as a means of disciplining abuses of market power associated with IPR is well established. They can also address market externalities arising from an unwillingness of patent owners to license a technology needed for a critical standard. And CLs are one channel through which technology may be acquired, whether for such social needs as adequate medical supplies or to support development of domestic industry. It is also possible that CLs may be deployed so effectively that patent owners simply choose to change their behaviour, expanding supply or reducing costs, in order to avoid losing their exclusivity. Why, then, are they granted so infrequently?

There are two essential answers to this question, which are particularly relevant in the poorest countries. The first is that CLs may be unnecessary or may not do much good. They may not be necessary for the simple fact that very few global technologies are actually patented in poor countries with small markets.[87] In those cases there is nothing to issue a CL for and local markets are left to buying unpatented older technologies or reverse engineering or other forms of imitation to acquire knowledge. This prospect is of little comfort overall, however. If a useful product or process is not patented in a particular location it effectively signals disinterest in supplying that market because of its small size or weak business climate. In turn, it is unlikely that local firms would be willing to undertake the significant costs of reverse engineering for the domestic market. This situation underscores the reason that CLs may not do much good: there may be no domestic private or even public (State-owned) enterprises that are capable of producing a product under such an order.

The alternative is to purchase the product from a larger market, where a variety of substitute technologies may be available. However, these are exactly the countries where global firms wish to protect their intellectual property, meaning that key technologies may be patented and therefore unavailable except at high cost. The point here is that CLs may not be an effective solution to the most fundamental problems limiting technology transfer: limited income and market size and weak investment climates.

The second basic answer is that in comparison with the potential benefits of CLs there may be even greater economic and reputational costs, even in countries with larger markets and better domestic production capacity. These costs must be reckoned with in any calculation of how to establish a CL regime and whether actually to grant non-voluntary licenses in any particular case. Following is a listing of those potential costs.

87. For example, patent scoping exercises find that few climate-change mitigation or adaptation technologies are patented in poor developing countries, as reviewed in Keith E. Maskus & Ruth L. Okediji, *Intellectual Property Rights and Technology Transfer to Address Climate Change: Risks, Opportunities and Policy Options* (Geneva: International Centre for Trade and Sustainable Development, Project on IPRs and Sustainable Development, Issue Paper no. 32, 2010), 8–11.

First, establishing an effective CL system is technically demanding and requires significant legal and administrative resources, even given the flexibility inherent in TRIPS. Dealing with anti-competitive abuses, the primary justification for CLs, requires establishing sophisticated competition laws and regulations. Many developing nations have implemented or reformed competition laws in recent years but few have established norms and guidelines against which to recognize and discipline abusive practices in the use of IPR. Beyond the legal complexities in this question lie certain practical questions. What are reasonable royalty rates in different sectors and technologies and on what market should they be based? Which administrative or judicial regime should mediate in setting these royalties? What is the appropriate duration for any particular CL? Under what set of market circumstances should the CL be vacated? Should the venue for judicial review be different from those engaged in setting the terms of CLs? What mechanisms need to be in place to ensure that compulsory licenses meet the terms of TRIPS in cases where those provisions are possibly binding? This issue may explain the near absence of recourse to CLs for medicines under the 2005 TRIPS Amendment, which sets out a series of technical hurdles to its use.

All of these questions would tax the technical and legal expertise of the major developing countries. They are likely to require more administrative resources than the poor countries, faced with more pressing development problems, are willing to invest. This situation would be compounded where particular CL regimes were seen by foreign authorities to exceed TRIPS bounds, which could result in costly litigation at the WTO or elsewhere.

Second, CLs may not be the panacea for technology transfer and development that they are sometimes depicted to be. As noted above, while it is possible to grant a license to patent rights it is generally not possible to order the surrender of confidential business secrets, which may be essential to the domestic grantee in terms of actually learning the technology.[88] Further, the CL must be non-exclusive, which means a potential license recipient may have relatively little incentive to serve the market and improve the technology. And, unless the courts or competition authorities choose to prevent it on legitimate grounds, the patent owner may simply acquire the grantee and reduce the number of suppliers on the market.[89] There are other means by which the patent owner might frustrate the TT intent of the license, including interfering with the grantee's access to key imported inputs.

A further difficulty is that the domestic producer operating under the CL may itself become monopolistic, particularly in small markets with little competition.[90] A variant of this problem is that the laws in numerous developing countries support monopoly distributors of goods, which may make it difficult for the grantee to

88. Maskus, *Encouraging International Technology Transfer*, fn. 8 *supra*, 10.
89. Reichman and Hasenzahl, fn. 16 *supra*, 24.
90. *Ibid.*

penetrate the market.[91] More generally, to the extent that business services are monopolized the effectiveness of CLs in encouraging acquisition of technologies and sustainable domestic production is liable to be stunted.

Finally, frequent resort to CLs where they may not be warranted under typical practice is likely to make a country seem less hospitable to FDI and market-based licensing. This will have a negative, albeit unknown in terms of volume, impact on general flows of inward technology transfer, even as it may expand TT directly. More broadly, global corporations and the governments of developed economies (and, increasingly, such countries as South Korea and China that are exporting technologies) may react vocally and depict countries that issue CLs as worsening their investment climates. In a global environment in which all countries seek greater access to global technologies this reputational impact may be the largest concern of all.

Despite these problems there are cases where granting CLs makes sense from a development or public-health standpoint. For example, licenses were issued by Thailand in 2006 and 2007 on two anti-retroviral drugs (Efavirenz and the combination Lopinavir-Ritonavir) and one heart medication (Clopidogrel) and by Brazil in 2007 on Efavirenz. The patent owners, prominently Abbott Laboratories, reacted by threatening to leave the markets and painting these countries as unreliable business locations. However, supplies seem not to have diminished in the wake of these actions. Whether smaller countries would have the clout and expertise to replicate these episodes remains an open question.

6 POLICY LESSONS

From this long overview and analysis it is possible to draw the following conclusions for policymakers. First, under certain circumstances compulsory licenses are an important tool in generating technology transfer. They can be especially useful in cases where IPR are frustrating governments in their needs to increase supplies of critical public goods. However, they come with a number of technical complications and costs that cannot be ignored. Accordingly, the granting of CLs should be done on a clearly justifiable, case-by-case basis. In short, compulsory licensing should be a component of each country's IPR and development policy, but used judiciously.

In this context, developing countries have a number of legislative and regulatory choices to make. First, how extensive should their government-use and public-interest exemptions be? There are no real legal restraints imposed by the global system and nations have a wide range of historical experience and legislative models to choose from, including India's recent patent law. Second, what should determine the contours of their treatment of fair use in copyrights? Again, this area is largely unconstrained, except by the three-step test. A wide

91. See Keith E. Maskus & Mohamed Lahouel, 'Competition Policy and Intellectual Property Rights in Developing Countries', 23 *The World Economy* 595 (2000).

exception for technical literature and scientific data in order to raise sustainable access to materials for education and research seems particularly appropriate here.[92] Due consideration should be given to the permissible scope of digital copying for libraries and educational purposes. Further, there is no global requirement that countries adopt database protection other than by copyrights, which may be held subject to fair-use limitations. Following the EU model in this regard could materially retard access to scientific and educational materials.

More demanding technically is the question of how to integrate CLs with principles determining actionable abuse of patents. Developing countries need to establish or refine competition laws and policies in ways that are appropriate for their economic environments but pay attention to the needs of patent owners. An excessively wide definition of abuse will encounter significant opposition from global firms and foreign governments, not to mention domestic enterprises that may be caught by it. For this purpose, countries can scrutinize the policies set out in the developed world, where competition law and judicial decisions have established important principles and means of enforcement. They can also benefit from consulting competition legislation adopted recently in India, Brazil, China and elsewhere, which tend to be more development-oriented. One basic approach is to publish a set of guidelines that offer a 'safe harbour' to particular forms of licensing behaviour.

Ultimately, however, competition guidelines in detecting patent abuse are unlikely to be effective in attracting technology transfer unless they are accompanied by a transparent and enforceable IPR system. The standards selected can be appropriate to individual country circumstances so long as they meet TRIPS minimum requirements. Thus, they can be more focused on development needs than those found in the rich nations. Examples in the patent domain include exemptions of certain technologies from eligibility, use of utility models and design patents, effective opposition systems, and a broad research exemption.[93]

In fact, developing countries in principle could move beyond the confines of traditional IPR and entertain more access-regarding innovation models. There is, for example, considerable scope for supporting incremental innovation and design within industries where it is important to share knowledge by means of liability rules.[94] Such systems, which essentially involve open use by those in a knowledge pool in return for compensation at mutually agreed rates, could complement or

92. See Jerome H. Reichman & Paul F. Uhlir, 'A Contractually Reconstructed Research Commons for Scientific Data in a Highly Protectionist Intellectual Property Environment', Duke University Law School, 66 *L & Contemp Probs* 315 (2003). The Max Planck Institute recently proposed such an exemption, under which research materials could be used for a wide array of scientific purposes, including, with attribution, the use of data in published and on-line articles for experimentation by researchers. See Jerome H. Reichman, 'Intellectual Property in the Twenty-First Century: Will the Developing Countries Lead or Follow?', 4 *Hous L Rev* 115 (2009).
93. Reichman, fn. 92 *supra*, 1115.
94. See Jerome H. Reichman & Tracy Lewis, 'Using Liability Rules to Stimulate Local Innovation in Developing Countries: Application to Traditional Knowledge', in *International Public*

even replace mechanisms based on exclusive rights, such as utility models. While this approach is not a compulsory licensing regime it could prove more attractive to international and local firms developing smaller innovations for the domestic economy.

A final observation is implicit in much of the prior analysis. Any legislative approach to compulsory licensing, and IPR and competition law more generally, is not likely to induce much technology transfer or local innovation in the absence of reasonable domestic capacities to use and adapt technologies. For example, evidence suggests that patents are more effective at attracting TT where domestic enterprises already have at least modest R&D programs underway.[95] In that regard, the policy challenge to achieve sustainable technology transfer is really three-fold: use compulsory licenses as necessary, surround them in an appropriate IPR system, and improve the national innovation profile.

Goods and Transfer of Technology under a Globalized Intellectual Property Regime, ed. Keith E. Maskus & Jerome H. Reichman (Cambridge: Cambridge University Press, 2005), 337.

95. Maskus, *Encouraging International Technology Transfer*, fn. 8 *supra*, 23.

Chapter 6
ACTA and the Destabilization of TRIPS

Jeffery Atik

1 FROM TRIPS TO TRIPS PLUS

1.1 THE UNRAVELLING OF THE GRAND BARGAIN

A significant divide has developed between the large portion of the WTO membership which sees little or no benefit from the TRIPS[1] undertakings and the small number of advanced economy members who seek ever-greater intellectual property (IP) standards. This chasm – between those who now judge TRIPS to be overreaching and those who find TRIPS to be inadequate – raises doubt as to whether a TRIPS consensus continues to exist. The movement of position described here is both relative and absolute: relative, in that IP debates are fiercer now than at the time of the foundation of the WTO; and absolute, in that both sides are taking increasingly hardened positions with respect to the original terms of TRIPS. Our story is one of two regrets: that TRIPS did not go far enough, that TRIPS went too far. In particular, the call for more IP protection, known as 'TRIPS Plus', has led to even greater fragmentation of the global intellectual property system. To ask for

1. Agreement on Trade-Related Aspects of Intellectual Property Rights (TRIPS). The TRIPS Agreement is Annex 1C to the Marrakesh Agreement Establishing the World Trade Organization, done at Marrakesh 15 Apr. 1994, 33 I.L.M. 1125, 869 U.N.T.S. 299.

Hans Henrik Lidgard, Jeffery Atik & Tu Thanh Nguyen, *Sustainable Technology Transfer*, pp. 121–147.

more, when the other side wishes to offer less, is a path to a potential breakdown. By pursuing TRIPS Plus, through new free trade agreements, and in particular through the negotiation and implementation of the 2010 Anti-Counterfeiting Trade Agreement (ACTA) is to risk losing the moral force behind TRIPS. ACTA may prove strangely counterproductive. Instead of achieving higher IP standards, it may lead to an increasingly resented, and hence less effective, observation of TRIPS.

One of the most unexpected features of the World Trade Organization (WTO) system was its integration of new intellectual property mandates. Most of the WTO infrastructure is derived from the prior General Agreement on Tariffs and Trade (GATT), and its substance is largely built on foundations set by GATT and by the related GATT-era side agreements. By and large, these commitments were negative in character: they sought to limit national freedom-of-action in order to dismantle trade barriers. During the Uruguay Round that led to the establishment of the WTO, GATT-like discipline was extended into theretofore-excluded fields of trade – such as agriculture and textiles – and for the first time to services and some investment controls. IP was not an obvious field for inclusion in the WTO, as it involved imposing on the WTO membership positive requirements.

The Grand Bargain refers to the central set of commitments made by the founding WTO membership that led to the incorporation of TRIPS into the international trade system. Had TRIPS been presented as a stand-alone treaty, it would have been unlikely that any but a small group of countries would have undertaken its obligations. The globalization of intellectual property standards greatly favours the small number of countries which host ownership of the greater part of valuable intellectual property. Most countries have few owners of internationally valued IP; for these countries (both poor and mid-level countries) intellectual property recognition creates obligations to make significant outbound wealth transfers. Only a few counties are net recipients of royalties. Unlike trade in goods which in theory benefits all participating parties, the recognition of IP rights benefits a discrete few. As a matter of economic self-interest, most countries would prefer a weak, if non-existent national intellectual property regime.

The promoters of TRIPS (chiefly the United States and Europe) were able to persuade the greater part of the WTO community to accept minimum standards of intellectual property protection by offering market access in other sectors, notably agriculture and textiles. This linkage, between recognition of IP rights and access to markets for developing world products, a peculiar result of the Uruguay Round, constitutes the Grand Bargain.

At a decade and a half of distance from the foundation of the WTO, many WTO members have come to regret the making of the Grand Bargain. On the one hand, a strident block of anti-IP countries, led by Brazil and India, oppose any expansion of IP obligations at the global level and consistently urge minimizing interpretations of TRIPS commitments within WTO dispute settlement. Further, effective market access in the newly covered sectors (agriculture and textiles) have disappointed the developing block. With little realized benefits, the mandates of IP recognition (and the costs and headaches associated with IP enforcement) are presently seen as an even greater burden on the larger WTO membership.

On the other hand, the knowledge economy country block, led by the United States, Europe and Japan, are pushing for greater IP protection than is required by TRIPS, with an emphasis on IP enforcement, in various international settings. The dissatisfaction felt by advanced countries is that TRIPS did not go far enough. Between these two species of regret, it is clear there would be no Grand Bargain reached today; that is, there would be no global consensus reached to erect TRIPS.

While TRIPS extended basic IP protection throughout the WTO area, it has – in the view of the United States, Europe and Japan – largely failed to bring about meaningful protection for rights holders from the scourge of counterfeiters and other unauthorized users.[2] TRIPS was certainly a step in the right direction, the advanced economies believe, as all WTO members were required to provide basic IP rights. Post-TRIPS experience has shown, however, a large enforcement gap. The increase in international trade facilitated by the construction and the expansion of the WTO regime (including the important accession of China to the WTO) has been accompanied by growth in IP piracy. In the absence of meaningful enforcement of the rights established by TRIPS, the TRIPS achievements seem illusory.

The United States – prodded by its powerful IP rights holder lobbies (principally Hollywood and the pharmaceutical industry) – has urged a meaningful extension of TRIPS, with an emphasis on effective enforcement.[3] For reasons to be discussed here, further enhancement to TRIPS has been effectively blocked (at least for the time being) within the WTO structure. Notwithstanding this hostile environment, the United States promulgated a set of reforms to TRIPS it would like to see adopted. And it has cobbled together an alliance of pro-IP States supportive of certain enhancements to the corpus of mandatory IP protections. This agenda (or wish list) has become known as 'TRIPS Plus'.[4]

---

2. Alan M. Anderson & Bobak Razavi, 'The Globalization of Intellectual Property Rights: TRIPS, BITS, and the Search for Uniform Protection', 38 *Ga J Int'l & Comp L* 265, 273–274 (Winter 2010).
3. In 2002 this view was incorporated into 'Special 301' annual review of the US Trade Representative (USTR), examining in detail the adequacy and effectiveness of intellectual property protection in approximately 72 countries. Office of the United States Trade Representative, '2002 Special 301 Report' (2002) (USTR Special 301 Report), 1–2. ('[T]he consensus view of the world community [is] that the vital framework of protection under existing treaties, including the TRIPS Agreement, should be supplemented to eliminate any remaining gaps in copyright protection on the Internet that could impede the development of electronic commerce.') See also 2007 Annual Report of the President of the United States on the Trade Agreements Program, 58. ('At the February 2007 meeting of the TRIPS Council, the United States made a presentation on its experience with border enforcement measures. A number of Members have resisted a substantive discussion of enforcement in the TRIPS Council.') USTR Reports may be found at the USTR website <http://ustraderep.gov/Document_Library/Reports_Publications/Section_Index.html>.
4. Peter K. Yu, 'TRIPS and Its Discontents', 10 *Marq Int'l Prop L Rev* 370, 383 (2006). See also Margot Kaminski, 'The Origins and Impact of the Anti-Counterfeiting Trade Agreement (ACTA)', 34 *Yale J Int'l L* 247, 249 (2009). ('Maximalist countries have been pushing for a TRIPS-plus regime, adding new requirements to the TRIPS standards through bilateral and regional trade agreements.')

1.2 Regime Shift Redux

The WTO structure, which provided for the triumphant extension of basic IP coverage throughout most of the globe, has become an increasingly unreceptive site for any expansionist IP agenda.[5] This is partly due to disappointment perceived by the larger part of the WTO membership (that is, the WTO's less developed members) with the implementation of the Grand Bargain.[6] The unhappiness due to the non-achievement of sought-for market access is a political fact that has increased the general hostility felt by much of the world to the recognition of IP rights.

Further, experience under TRIPS has brought forward the under-addressed resources question: from the time of the TRIPS negotiation, it had never been clear how an effective level of IP enforcement was to be funded in many poorer countries. TRIPS constitutes an 'unfunded mandate' for IP enforcement. While TRIPS calls for the provision of technical assistance (in Article 67[7]), it leaves funding to each member's own resources. Here too is a potential collision: any call for greater IP enforcement inevitably further strains the administrative resources of most countries with little or no compensating capture of value, as most of this gain passes to foreign interests. As Chow has noted, counterfeiters can be powerfully connected – and are inevitably local.[8] Their economic gains largely remain within national confines.

The United States made a strategic decision to largely abandon the WTO as the chief site for effecting its expansive IP ambitions – engaging in what Laurence Helfer might recognize as a further 'regime shift'.[9] Rather, the United States has sought to incrementally bind various countries to IP commitments that exceed the TRIPS minimum standards (that is, TRIPS Plus) in various bilateral trade agreements, beginning with the United States – Australia free trade agreement.[10]

---

5. See generally Paul Drahos, 'Trade-Offs and Trade Linkages: TRIPs in a Negotiating Context'. Notes from a talk given at Quaker House Geneva (12 Sep. 2000), <www.quno.org/economic-issues/intellectual-property/intellectualLinks.htm#OCCASIONAL>.
6. David Vivas-Eugui, 'Regional and Bilateral Agreements and a TRIPS-Plus World: The Free Trade Area of the Americas (FTAA)', TRIPS Issue Papers no. 1, Quaker United Nations Office (QUNO) (August 2003), 14. <www.quno.org/economicissues/intellectual-property/intellectualLinks.htm#ISSUES>.
7. TRIPS Art. 67 provides:

   In order to facilitate the implementation of this Agreement, developed country Members shall provide, on request and on mutually agreed terms and conditions, technical and financial cooperation in favour of developing and least developed country Members. Such cooperation shall include assistance in the preparation of laws and regulations on the protection and enforcement of intellectual property rights as well as on the prevention of their abuse, and shall include support regarding the establishment or reinforcement of domestic offices and agencies relevant to these matters, including the training of personnel.
8. Daniel Chow, 'Anti-Counterfeiting Strategies of Multi-National Companies in China: How a Flawed Approach Is Making Counterfeiting Worse', 41 *Geo J Int'l L* 749, 750–751 (2010).
9. See generally, Laurence R. Helfer, 'Regime Shifting: The TRIPS Agreement and the New Dynamics of International Intellectual Property Lawmaking', 29 *Yale J Int'l L* 1 (Winter 2004).
10. United States-Australia Free Trade Agreement, Signed 3 Aug. 2004, entered into force 1 Jan. 2005. P.L. 108-206.

The trajectory of regime shifts is worth noting. Helfer describes the shift from the UN-based WIPO institution, where developing world concerns about IP are given weight, to the newly minted WTO, where IP burdens were to be compensated by valued concessions in other areas. After the two-sided disappointments with TRIPS (failure to realize anticipated market access gains by the developing world and the enforcement gap perceived by the advanced economies), the WTO/TRIPS field is no longer promising. For the IP expansionists the new institutional horizon for pursuing TRIPS Plus includes bilateral agreements as well as regional and plurilateral structures.

The United States-Australia Free Trade Agreement is generally considered to be the first exercise in the US pursuit of TRIPS Plus. The presence of the extensive IP provisions found in the United States-Australia FTA likely does not signal deep US concerns about the parlous state of IP protection in Australia. The US was not solely functioning as a traditional *demandeur* of Australia in these negotiations (or vice versa). Rather, the United States-Australia FTA might be better seen as a laboratory project (or exemplar), setting out and exercising post-TRIPS state-of-the-art enforcement obligations in an important bilateral setting.[11] Australia, under this telling, was merely the convenient counterparty for its expression.

The US expanded its communication of desired IP enforcement program in subsequent free trade agreements entered with Korea, Morocco and Singapore. While the terms of the IP provisions of these agreements vary, each represents an accorded articulation of US goals. And the Morocco agreement supports the US position that TRIPS Plus obligations may be subscribed to by a developing country.

1.3 The Road to ACTA

Having achieved some success in persuading various trading partners to accede to enhanced IP enforcement, the US, together with Japan, launched the

11. See Ruth L. Okediji, 'Back to Bilateralism? Pendulum Swings in International Intellectual Property Protection', 1 *U Ottawa L & Tech J* 125, 128–129 (2003–2004):

> Despite the hope that the TRIPS Agreement would diminish the use of bilateralism to secure international protection for intellectual property, post-TRIPS bilateralism remains the dominant policy of the United States. Intellectual property bilateralism is evident in trade negotiations such as the recent FTAA negotiations, in the continued use of Section 301 of the Trade Act of 1974 ... to secure extra-TRIPS or 'TRIPS-plus' commitments from other countries, as well as in pure bilateral intellectual property agreements.
>
> ***
>
> The new bilateralism is clearly a tool to effectuate the benefits of forum shifting ... to overcome limitations imposed by the TRIPS Agreement, ... and to sustain the expansion of intellectual property rights at the expense of the public interest both in developed and developing countries.

Ibid., 140–141. (Footnotes omitted.)

Anti-Counterfeiting Trade Agreement (ACTA).[12] The initial impetus for ACTA came with a 2005 anti-counterfeiting proposal by Japan closely followed by a similar proposal from the US, to which the European Union joined, in October 2007.[13] ACTA was the first plurilateral[14] (as opposed to bilateral) effort to expand the range of TRIPS Plus coverage.[15] The choice of the initial state parties (that is, the founding ACTA negotiating parties) was critical: these were by-and-large 'like-minded' States; States which hosted significant stocks of IP. Japan was the original partner in the ACTA initiative; thereafter the European Union, Japan, Canada, Switzerland, and Korea joined the negotiations.

The ACTA negotiations concluded in December 2010 and it is, as of this writing, open for signature. ACTA represents a deliberate 'regime shift' away from the WTO structure (which houses TRIPS).[16] Clearly, expanding intellectual property protections is not in the pro-development spirit of the current Doha Round of WTO talks. It is not – and likely could not become – part of the agenda to be considered by the WTO membership within the Doha Round. Any negotiations within the WTO would necessarily involve substantial representation from the newly industrialized, developing and less-developed blocks, and hence significant pushback against expansion of intellectual property rights. The like-mindedness that can be found in a smaller, club-like setting may favour the formation of substantively more attractive (and ambitious) IP norms than might result in a more cosmopolitan setting.

---

12. Anti-Counterfeiting Trade Agreement dated 15 Nov. 2010 among Australia, Austria, Belgium, Bulgaria, Canada, Cyprus, the Czech Republic, Denmark, Estonia, Finland, France, Germany, Greece, Hungary, Ireland, Italy, Japan, the Republic of Korea, Latvia, Lithuania, Luxembourg, Malta, Morocco, the Netherlands, New Zealand, Poland, Portugal, Romania, Singapore, Slovakia, Slovenia, Spain, Sweden, Switzerland, the United Kingdom, the United Mexican States, the United States, and the European Union (ACTA). Final version, Subject to Legal Review. <www.ustr.gov/webfm_send/2379>.
13. Margot Kaminski, 'The Origins and Potential Impact of the Anti-Counterfeiting Trade Agreement (ACTA)', 34 *Yale L J Int'l L* 247, 250–251 (Winter 2009) (Kaminski). See also Michael Blakeney & Louise Blakeney, 'Stealth Legislation? Negotiating the Anti-Counterfeiting Trade Agreement (ACTA)', 16(4), *Int'l T L R*, 85, 90–91, 201 (2010). Blakeney and Blakeney connect these negotiations with the Agreement on Measures to Discourage the Importation of Counterfeit Goods, GATT Doc. No. L/5382 (18 Oct. 1982), a US sponsored informal plurilateral anti-counterfeiting agreement. *Ibid.*, 87. Accord Emily Ayoob, Note, 'Recent Development: The Anti-Counterfeiting Trade Agreement', 28 *Cardozo Arts & Ent L J* 175, 178–179 (2010).
14. In WTO parlance, a plurilateral agreement is a multilateral agreement that binds some but not all WTO members. The term 'multilateral' is reserved for those undertakings that bind the complete WTO membership.
15. See Peter K. Yu, 'A Tale of Development Agendas', 35 *Ohio N U L Rev* 465, 555 (2009) (Tale of Development Agendas). (Developed countries' push for establishing ACTA is a forum shifting mechanism to achieve 'the maximalists' abiding goal of ratcheting up IP protection and enforcement worldwide.' quoting Susan K. Sell, 'The Global IP Upward Ratchet, Anti-Counterfeiting and Piracy Enforcement Efforts: The State of Play 4' (IQsensato, Occasional Papers No. 1, 2008).).
16. Charles R. McManis, 'The Proposed Anti-Counterfeiting Trade Agreement (ACTA): Two Tales of a Treaty', 4 *Hous L Rev* 1237 (2009).

The removal of the latest phase of global intellectual property law-making from the WTO results in part from a determination to provide for faster tracking than a WTO process could provide. It also involves (with some risk) a de-linking of intellectual property concerns from other issues within the WTO remit. And it provides splendid isolation from the inconvenient presence of the many WTO members which have been consistently critical of TRIPS, including one economic superpower (China) where enforcement of intellectual property protection is generally perceived to be deficient.

Perhaps more telling than the list of those 'like-minded' states participating in ACTA negotiations are the identities of the powerful states (within the WTO) that were excluded from the ACTA process: Brazil, India, and China. There are, of course, significant concerns expressed about counterfeiting and piracy in all these States.[17] That apart, the consistent hostility displayed by India and Brazil within the WTO regime to the expansion of IP rights may be a more plausible explanation for their exclusion from the ACTA negotiating process.[18]

The US makes the rather puzzling assertion that counterfeiting and piracy (that is, non-recognition and/or non-enforcement of IP rights) 'undermines legitimate trade and the sustainable development of the world economy'.[19] One wonders if a

17. See e.g., Lawrence A. Kogan, 'Brazil's IP Opportunism Threatens U.S. Private Property Rights', 38 *U Miami Inter-Am L Rev* 1, 7–8 (Fall 2006), European Commission's annual report on EU Customs Enforcement of Intellectual Property Rights, 'Results at the Border 2009', (22 Jul. 2010) (EU 2009–2010 Customs Action Plan). ('Overall, China continued to be the main source country from where goods suspected of infringing an IPR were sent to the EU (64% of the total amount of articles).') Executive Summary, 2. <http://ec.europa.eu/taxation_customs/customs/customs_controls/counterfeit_piracy/statistics/index_en.htm>, USTR 'Special Section 301 Report' (30 Apr. 2010), 9–12 Executive Summary, Section I Trends in Counterfeiting and Piracy, Internet and Digital Piracy (Counterfeit goods and internet piracy originating in BRIC countries).
18. See e.g., World Trade Organization (WTO) Council for Trade-Related Aspects of Intellectual Property Rights (TRIPS Council), Minutes of Meeting from 18 to 22 Jun. 2001, Special Discussion on Intellectual Property and Access to Medicines, IP/C/M/31 (10 Jul. 2001) (Statements of Brazil and India on sponsorship of TRIPS and Public Health IP/C/296), Peter K. Yu, 'Six Secret (And Now Open) Fears of ACTA', (describing Argentina's, Brazil's, China's and India's, strong opposition to an EU initiative to push for further discussions of IPR enforcement issues, in the June 2006 TRIPS Council Meeting), (forthcoming 64 *SMU L Rev* 2011) (20 Oct. 2010), 10–11 (Six Open Fears). <http://papers.ssrn.com/sol3/papers.cfm?abstract_id=1624813>, Peter K. Yu, 'Access to Medicines, BRICS Alliances, and Collective Action', 34 *Am J L & Med* 345, 349–352 (2008) (Detailing some of Brazil's and India's 'pushback to developed countries' IPR initiatives at the WTO); Shashank P. Kumar, 'Border Enforcement of Intellectual Property Rights against In-Transit Generic Pharmaceuticals: An Analysis of Characters and Consistency, 32 *E.I.P.R.* 506 (2010). (Analysis of India's and Brazil's requests at the WTO for consultations regarding seizures of in-transit drugs – WT/DS408/19 May 2010 – against background of India's and Brazil's assertions before the WTO General Council 3–4 Feb. 2009 that Council Regulation EC 1383/2003 Concerning customs action against goods suspected of infringing certain intellectual property rights and the measures to be taken against goods found to have infringed such rights, OJ 2009 L 196/7-14, violates GATT Art. V and TRIPS.)
19. USTR Press Release, 'Statement by the Anti-Counterfeiting Trade Agreement (ACTA) negotiating partners regarding the recent round of ACTA negotiations in Lucerne, Switzerland' (5 Jul. 2010). <www.ustr.gov/about-us/press-office/press-releases/2010/june/office-us-trade-representative-releases-statement-act>.

single official of the developing world would agree with the notion that fighting IP counterfeiting is the key to achieving 'sustainable development'. There seems to be little here beyond vulgar sloganeering: the connection between the control of counterfeiting and piracy and economic and social development has yet to be persuasively demonstrated. Indeed the only true believers in the substantiality of this linkage seem to be IP rights holders, for which this credence is most convenient.

The objectives of the US in the ACTA negotiations were fairly well-defined. ACTA, in the view of the US, was not to be a forum for the enlargement of additional substantive intellectual property rights intended to become subject to international mandates. ACTA's sole focus should have been the establishment of effective enforcement standards for the set of rights mandated by TRIPS and other international instruments.

According to the Office of the United States Trade Representative (USTR), those interested in 'understanding the US approach' to ACTA should view the IP-specific provisions of the recent free trade agreements negotiated by the US with Australia, Korea, Morocco and Singapore.[20] While this assertion hardly limited the range of negotiating outcomes the US might have sought within the ACTA process, it does give content to US ambitions.

If one uses the provisions of the various FTAs cited by USTR as guides to the intentions of the US in the ACTA negotiations, one discovers that ACTA largely confines itself to the enforcement of IP rights. Enhanced IP enforcement is only part of the TRIPS Plus agenda of the US.[21] Each of the four exemplary FTAs contains additional IP provisions (that is, commitments binding on their respective signatories) that go beyond the minimum standards imposed by TRIPS. It seems to have been a political calculation to confine ACTA to enforcement issues only. This may reflect a sober realization of the maximum progress feasible at the plurilateral level at the time; expansion of substantive rights might have engendered more fierce opposition. Or it may reflect a consensus within the US IP right holder community that in fact the most serious challenges to their interests arise in the area of enforcement. The final ACTA text reflects both achievements and setbacks with respect to the announced ambitions of the US.

ACTA is hardly the endpoint of the TRIPS Plus 'regime shift' outlined here. By moving from TRIPS, and then from bilateral agreements, on to ACTA, more general trade issues have been decoupled from IP and IP enforcement concerns. ACTA is a far more specialized agreement than any bilateral trade agreement, or even more so than TRIPS, given that TRIPS is part and parcel of the WTO 'single undertaking'. Having completed ACTA, and having in the process established new 'best practices' for IP enforcement standards, further shifting may return these newly minted norms to mixed subject agreements. In later sections of this chapter,

---

20. See Letter from Ron Kirk, United States Trade Representative, to Ron Wyden, U.S. Senator (28 Jan. 2010), 2. <www.ustr.gov/webfm_send/1700>.
21. See IPEC 2010 Joint Strategic Plan for Intellectual Property Enforcement, 14–15, 'Promote Enforcement of U.S. Intellectual Property Rights through Trade Policy Tools'.

a prediction will be made that ACTA provisions will be replicated in future trade agreements, including regional agreements, to which the United States will be party. Specifically, ACTA norms are likely to take root in the currently forming Trans-Pacific Partnership agreement and will thereby become binding against countries that were not able to contribute to the formation of those norms due to their exclusion from the ACTA process.

2 ACTA AND HEALTH ANXIETIES

2.1 Revisiting the Essential Medicines Controversy

The process by which ACTA was formed was itself extremely controversial. There had been a general awareness of ACTA during the course of its negotiation, and the identity of the founding states participating in the sessions was known. Beyond the mere fact of ACTA's preparation (involving a general statement of ACTA's intended coverage and various declarations of national target objectives) very little of substance leaked from the earliest sessions. This almost complete lack of transparency, particularly during the earliest rounds, fuelled the imagination of various actors (including WTO members outside the ACTA club and representatives of civil society) opposed to further expansion of IP protections. During ACTA's formation, many writings were published, criticizing possible provisions of ACTA that may or may not have been on the negotiating table.[22]

Specifically, there was considerable anxiety manifested concerning the possibility that resort to the ACTA process would be used in order to reverse the hard-won gains within TRIPS resolving the essential medicines controversy. Shortly after TRIPS took effect, the AIDS pandemic exploded in large parts of the developing world.[23] Critics of TRIPS seized on the unwillingness of the proprietors of various patented pharmaceuticals to make those medicines available on affordable terms. The withholding of essential medicines by patent holder was highlighted in order to reveal TRIPS' adverse effect on the ability of WTO members to achieve appropriate public health responses. Aggressive actions by patent holders within the WTO[24] and

22. See e.g., cases cited in fn. 2 *supra*, T. Jesse Goff, 'Regulation of Digital Copyrights and Trademarks at the U.S. Border: How the Proposed Anti-Counterfeiting Trade Agreement and the Enacted U.S. Pro-IP Act Will Destabilize the Current System', 16 *Sw J Int'l L* 207 (2010), Henning Grosse Ruse-Khan, Thomas Jaeger & Robert Kordic, 'The Role of Atypical Acts in EU External Trade and Intellectual Property Policy', 21 *Eur J Int'l L* 901 (2010), 75 Academics' Letter to President Obama (28 Oct. 2010) (Questioning Constitutionality of ACTA), <www.wcl.american.edu/pijip/go/blog-post/academic-sign-on-letter-to-obama-on-acta>.
23. Hans Henrik Lidgard & Jeffery Atik, 'Facilitating Compulsory Licensing under TRIPS in Response to the AIDS Crisis in Developing Countries', in *Corporate and Employment Perspectives in a Global Business Environment*, ed. Roger Blanpain & Boel Flodgren (The Hague: Kluwer Law International, 2006), 49–64.
24. WTO Brazil-Measures Affecting Patent Protection, Request for Consultations by the United States, dated 30 May 2000, WT/DS199/1 G/L/385 IP/D/23 (8 Jun. 2000).

in national courts[25] to protect their exclusivities incited adverse public reaction, leading to the abandonment of these proceedings. Pharmaceutical industry interests beat a retreat, opening the way for significant clarifications and reform within TRIPS.

Advocates for access to health care for the impoverished communities afflicted by the AIDS pandemic were able to score a series of important rollbacks of the patent mandates contained in TRIPS. The first of these was the Doha Declaration,[26] restating the availability of certain flexibilities within TRIPS and committing the WTO membership to recognition of public health concerns in the administration of national IP systems. Further, additional extensions of the TRIPS transition period for least developed country WTO members were conceded.[27] And TRIPS itself was amended, reducing the effect of Article 31(f) which had categorically banned the use of compulsory licenses for the production of generic pharmaceuticals for export. New TRIPS Article 31*bis* will permanently implement the August 30, 2003 Decision,[28] by which compulsory licenses will be available, notwithstanding Article 31(f), for the production of generic pharmaceuticals for export to countries which lack the manufacturing capacity to produce those medicines locally.[29]

There was rampant fear that ACTA would be a vehicle by which these important concessions could be effectively withdrawn.[30] Much would depend upon the scope of the final ACTA text. Vague locutions, such as 'piracy' or 'counterfeiting', might be extended to cover the manufacture of or commercialization of generic pharmaceuticals. New enforcement mandates within ACTA might create new opportunities for patent holders to delay or harass firms shipping generic medicines.

---

25. Case 4183/9, *Pharmaceutical company lawsuit (forty-two applicants) against the Government of South Africa (ten respondents)* (Constitutional Court 1998), <http://cptech.org/ip/health/sa/pharmasuit.html>.
26. WTO, Ministerial Conference Fourth Session Doha, 9–14 Nov. 2001WT/MIN (01)/DEC/2 41 I.L.M. (2002) (Doha Declaration).
27. WTO Council for Trade-Related Aspects of Intellectual Property Right Minutes of Meetings Held 25–26 and 28 Oct., 29 Nov. and 6 Dec. 2006 IP/C/M/49 (31 Jan. 2006). See also, WTO TRIPS Council, 'Maldives – Extension of the Transition Period under Article 66.1 of the TRIPS Agreement', Document IP/C/35 (June 2005).
28. WTO General Council Decision of 6 Dec. 2005, Amendment of the TRIPS Agreement, Attachment to Protocol Amending the TRIPS Agreement, WT/L641.
29. See Decision the WTO General Council of 6 Dec. 2005 and accompanying protocol, providing for the insertion of new Art. 31*bis* in TRIPS.
30. See e.g., American University's Program on Information Justice and Intellectual Property ACTA Communiqué, 'International Experts Find that Pending Anti-Counterfeiting Trade Agreement Threatens Public Interests' (23 Jun. 2010). <www.wcl.american.edu/pijip/go/acta-communique>. Six Open Fears, 84, Henning Grosse Ruse-Khan, 'A Trade Agreement Creating Barriers to International Trade? ACTA Border Measures and Goods in Transit, Max Planck Institute for Intellectual Property', Competition & Tax Law Research Paper Series No. 10-10 (forthcoming *Am U Int'l L Rev*), <http://papers.ssrn.com/sol3/papers.cfm?abstract_id=1706567&>.

2.2 The EU Transhipment Case

Since 2008, in what has been perceived as a rear-guard attack on the new understandings with respect to the public health flexibilities in TRIPS, Dutch authorities have seized various shipments of generic pharmaceuticals produced in India and Brazil and destined for developing countries. Presumably, the manufacture of these medicines in India or Brazil and their commercialization in their intended markets would not violate any relevant patent rights in those respective territories. Patent rights arise on a national basis and are confined to national territory. But upon entry of these generic medicines into any national territory, such as the Netherlands, where patent rights obtain, those rights become applicable.

The controversial Dutch seizures involved instances of transhipments, where the presence of the offending products within the national territory was temporary and often happenstance. These cases do not involve the making, use or sale of products subject of the Dutch patent within the Netherlands – though admittedly they do involve the import of such products.[31] Given the particular transport links between important generic producing countries (such as India and Brazil) and many developing country markets for those generics, it may be impractical if not impossible to avoid transhipment through ports (such as Dutch international ports and airports) where patent rights arise.

Formally, the Dutch authorities (and the relevant patent owners) appear to be within their rights. Given the territorial nature of the patent system, a Dutch patent owner should be able to take action against infringing goods 'imported' into Dutch national territory. But exercising these rights, in instances of transhipments (where no patent rights are violated in either the country of origin or destination) seems mean spirited, at least in instances where the trade is one sheltered by the new understandings within TRIPS.

On May 11, 2010 India commenced a dispute settlement process in Geneva with respect to the WTO-compatibility of these seizures,[32] followed on May 12, 2010 by a similar complaint by Brazil.[33] In the end, both India and Brazil appear to have abandoned these complaints, though what, if any, commitment they may have extracted from the EU (and the Netherlands) has not been revealed.[34] Thus, the WTO-compatibility of the seizures remains unresolved.

ACTA was thought to be a 'second front' for the contest as to the legitimacy of these seizures. An expansive definition of 'piracy' or 'counterfeit goods' might

31. TRIPS Art. 28 assures patent owners the rights 'to prevent third parties not having the owner's consent from the acts of: making, using, offering for sale, selling, or importing' products that are covered by the patent.
32. European Union and a Member State – Seizure of Generic Drugs in Transit, DS 408.
33. *Ibid.*, DS 409.
34. See, Peter K. Yu, 'TRIPS Enforcement and Developing Countries' (forthcoming *Am U Int'l L Rev*) (2011). <http://papers.ssrn.com/sol3/papers.cfm?abstract_id=1736030&##>, 23. (Citing BBC report of dispute resolution between India and the EU involving amendment of EU regulations to provide that generic pharmaceuticals in transit will only be inspected in cases of counterfeiting.)

conceivably have covered generic medicines temporarily passing through ports where relevant patent rights obtain. Moreover, ACTA might have mandated such seizures upon the initiative of the officials, resolving any doubt as to the legitimacy of the actions of the Dutch authorities.

The first generally leaked version of ACTA seemed to confirm these fears. First, early ACTA drafts covered all categories of intellectual property covered by TRIPS.[35] Thus, early ACTA drafts reached beyond copyright and trademark violations to potentially include patent infringements as well. Second, draft versions of ACTA contained a specific provision regarding transhipments as part of its broader treatment of border measures. Bracketed language indicated that at least some ACTA founders desired IP enforcement action to be mandatory (as opposed to discretionary) with respect to transhipped goods.[36]

In its final version, ACTA appears less threatening. While the final version of ACTA continues to cover the same categories of IP as does TRIPS[37] – and thus includes patents within its scope – the key terms 'piracy' and 'counterfeit' are restricted to copyright and trademark respectively.[38] Further, the transhipment provisions provide that authorities may (and not that they shall) have the authority to enforce national IP.[39] And ACTA's Border Measures section (which includes ACTA Article 16(2)) does not apply to patents and trade secrets.[40] While ACTA does not impede seizures such as those made by Dutch authorities of transhipped generic medicines from India or Brazil, neither does it mandate those seizures. But it hardly declares such seizures to be inconsistent with TRIPS norms, as modified with respect to trade in essential medicines – the position that India and Brazil had sought in their WTO requests for consultations.

---

35. It is now widely understood that broad ACTA coverage was sought by the European Union in order to assure that protection of geographic indications would be enhanced.
36. The leaked version of 1 Jul. 2010 stamped 'U.S. Confidential Modified Handling Authorized* EU Restricted' contained mandatory injunctive provisions in Art. 2.2 with bracketed alternate references to 'may', <www.publicknowledge.org/files/docs/ACTA_consolidatedtext.pdf>.
37. In ACTA Art. 5 (General Definitions), intellectual property is defined as 'all categories of intellectual property that are the subject of Sections 1 through 7 of Part II of the TRIPS Agreement'. This establishes the potential coverage of ACTA, thus including patents, copyrights, trademark, geographical indications and other forms of intellectual property.
38. The relevant ACTA terms are 'counterfeit trademark goods' and 'pirated copyright goods'. ACTA Art. 5. Here the ACTA definitions provide the link between the notion of counterfeiting with trademark violation and the notion of piracy with copyright violation. A good that infringes a patent (but not a trademark or copyright) is neither counterfeit nor pirated.
39. ACTA Art. 16(2) provides that:

    A Party may adopt or maintain procedures with respect to suspect in-transit goods or in other situations where the goods are under customs control under which:

    (a) its customs authorities may act upon their own initiative to suspend the release of, or to detain, suspect goods; and
    (b) where appropriate, a right holder may request its competent authorities to suspend the release of, or to detain, suspect goods.

40. Article 16 appears in s. 3 (Border Measures) of ACTA. A footnote states that '[t]he Parties agree that patents and protection of undisclosed information do not fall within the scope of this Section.'

2.3 ACTA and Health

As noted above, ACTA has had little effect on the delicate balance reached within TRIPS on access to essential medicines. ACTA includes language which specifically refers to these understandings, furthering the interpretation that ACTA shall leave the essential medicines outcomes undisturbed. In the preamble to ACTA, the parties recognize 'the principles set forth in the *Doha Declaration on the TRIPS Agreement and Public Health*, adopted on 14 November 2001, at the Fourth WTO Ministerial Conference'.

With respect to transhipments, ACTA supports the view that seizures of generic medicines that violate national patents are permitted, but does not mandate such seizures. Whether Dutch or other national authorities will undertake future seizures of generic drugs passing through transhipment nodes, either on their own initiative or upon request of patent owners, remains to be seen, but ACTA neither mandates nor forbids this.

On the trademark front, some concern has arisen with respect to the possible exercise of trademarks that closely resemble the scientific names of pharmaceutically active ingredients. Commercialization of generics could be impeded on the theory that the generic name too closely resembles the registered trademark.[41] While there may be grounds for apprehension here, the problem seems to lie more with the prospect of the maladministration of trademark registration than as a creature of ACTA.

3 THE UNEASY COEXISTENCE OF ACTA AND TRIPS

3.1 The Force of ACTA's Norms

Upon ACTA's entry into force,[42] the global intellectual property system will change profoundly. ACTA, by its terms, manifests global ambitions. Although negotiated by a small number of 'like-minded' countries, ACTA is open for eventual adhesion by the entire WTO membership.[43] This distinguishes ACTA from other TRIPS Plus initiatives, which have involved bilateral and regional instruments. It may be that this invitation to the broad WTO membership is cosmetic; it is

41. As a hypothetical example, consider Lipitor, the world's biggest 'blockbuster drug', patented and produced by Pfizer. Lipitor's scientific name is Atorvastatin. Pfizer's patent will expire in June 2011. Imagine if Pfizer had marketed Lipitor under the tradename Atarva. Conceivably, there would be a concern that Pfizer could use its Atarva trademark (established in the public mind during the period of patent validity) to interfere with generic producers' ability to market generic Atorvastatin.
42. ACTA Art. 40 provides that ACTA enters into four 'thirty days after the date of deposit of the sixth instrument of ratification, acceptance or approval' (update status).
43. ACTA Art. 39 provides that ACTA shall remain open for signature 'by any other WTO Members the participants may agree to by consensus, from 31 March 2011 to 31 March 2013'.

difficult to imagine the hard core of anti-IP members within the WTO, such as India or Brazil, volunteering for ACTA adhesion.

Many of the global substantive IP agreements – such as Berne and Paris – formally exist independently of the WTO/TRIPS structure, but are accommodated by TRIPS.[44] More recent substantive agreements, such as the WIPO copyright and performance rights treaties, have been layered on top of the TRIPS obligations with little controversy. ACTA is distinctive in that it focuses on IP enforcement, as opposed to expanding the scope of internationally recognized IP rights. ACTA also provides for a deepening of commitments of public resources and facilities to assist IP right holders in asserting their rights.

ACTA can be seen as a blueprint for eventual TRIPS reform and renovation. It serves as a laboratory for building and testing novel legal formulations and tests. In many areas, multilateral treaties have suffered from obsolescence, as conditions overtake the understandings that were foundational to their original adoption. Amending multilateral treaties through negotiations is fraught with difficulties (as the recent experience with the insertion of Article 31*bis* into TRIPS demonstrates). Parallel construction of external instruments is a common solution to the problem of obsolescence and norm rigidity in treaties. Of course it is hardly a given that TRIPS is – after 15 years of effectiveness – suffering from obsolescence.

ACTA may provide norms in the event of a return to unilateral action. Prior to the adoption of TRIPS, the US notoriously applied unilateral trade sanctions against countries that did not (in the estimation of the US) provide adequate protection for holders of intellectual property rights.[45] The WTO Agreements supposedly ended this practice, by declaring (in Article 23 of the Dispute Settlement Understanding[46]) that the WTO's dispute settlement mechanism was to be the

44. TRIPS Agreement Arts 1(3) and 3.
45. In 1999 the EU, after consultations with the US requested the establishment of a WTO Panel to consider whether the US's practice under its 'Special 301' legislation violated the GATT and TRIPS. (WT/DS152/11).Under US law the USTR is required to take action to enforce trade agreements as well as against other 'unreasonable' acts which are unjustifiable or burden US trade, even if permitted under treaty arrangements including TRIPS. See e.g. 19 U.S.C. § 2411(d)(3) (B) (i) (II).

    On 8 Nov. 1999 the DSU Panel held that, in light of statements made at the time of the Uruguay Round, s. 301 did not violate WTO commitments. See, Report of Panel, United States – Sections 301–310 of the Trade Act of 1974, 8 Nov. 1999, WT/DS152/R.
46. Article 23 of the WTO Dispute Settlement Understanding provides, in part:

    1. When Members seek the redress of a violation of obligations or other nullification or impairment of benefits under the covered agreements or an impediment to the attainment of any objective of the covered agreements, they shall have recourse to, and abide by, the rules and procedures of this Understanding.
    2. In such cases, Members shall:
        (a) not make a determination to the effect that a violation has occurred, that benefits have been nullified or impaired or that the attainment of any objective of the covered agreements has been impeded, except through recourse to dispute settlement in accordance with the rules and procedures of this Understanding, and shall make any such determination consistent with the findings contained in the panel or Appellate Body report adopted by the DSB or an arbitration award rendered under this Understanding.

exclusive forum for trade conflicts and providing no cover for trade sanctions purportedly justified by norms unilaterally asserted by a particular member.

DSU Article 23 calls for a general renunciation of unilateral trade measures in the event of the non-observation of a WTO undertaking. It is not clear whether this WTO prohibition on unilateral responses applies to instances of non-observation of norms established in an instrument that is outside the WTO system – even when the non-observed norm resembles a norm found within a WTO agreement. ACTA is outside the WTO structure, and so violation of ACTA commitments may not lead to WTO dispute settlement. It is less clear whether the WTO's prohibition on unilateral measure applies to ACTA norms. Presumably there would be less objection to the taking of a unilateral measure against a WTO-member ACTA signatory (on the theory that the ACTA norms were lawful undertakings of that WTO member) – although this is formally in doubt. There would likely be fierce objection were an ACTA signatory (like the US) to take a unilateral measure against a WTO-member non-ACTA signatory (like China, India or Brazil) based on non-observation of an ACTA norm that arguably represents an emerging international standard.

ACTA also serves as a template for a slow and deliberate expansion of new IP enforcement norms. First, the membership of ACTA may well increase. Any ACTA member may pressure a non-signatory to join the club, either by formal subscription or by an independent assumption of ACTA norms, as part of an exaction of benefits in the conduct of foreign relations. Smaller and poorer countries might even 'volunteer' to join ACTA in order to demonstrate good faith, openness to foreign investment, and general friendship to the core ACTA membership.

Second, ACTA norms may be inserted in the continuing thicket of bilateral and regional agreements. Major ACTA powers (US, Europe and Japan, for example) may view the ACTA package as a minimum commitment in the field of IP, to be the basis of undertakings in new instruments. The result of this process could be a far greater diffusion – and effect – of ACTA norms than the ACTA signatories directly represent.

ACTA will establish new norms. The ambition of the signatories is that these norms will become recognized as state-of-the-art standards – minimum IP enforcement standards for the 21st Century. In this view, they form an expected updating of the mid-90s IP norms found in TRIPS. They also reflect an evolution of circumstances and learning from experience: TRIPS in retrospect was too sanguine in presuming effectiveness of enforcement. Of course this may or may not come to pass. Recent experience provides examples of deliberate norm-generation that faltered (witness the eventual decline of the New International Economic Order initiative to a historical footnote).

ACTA involves a 'regime shift' – at least provisionally – away from the WTO. If they take root, ACTA's norms might one day be incorporated into a reformed TRIPS. For the near-term, however, ACTA stands on its own feet. It thus constitutes a fragmentation of intellectual property law, with the resulting threats to harmonization and legitimizing *a la cartisme*. International lawyers will differentiate

ACTA and non-ACTA (or non-ACTA complying) states. Remedies will be available in one set of states that likely will be absent elsewhere. Further reform will be complicated, in multiplying the potential fora: WTO/TRIPS, ACTA or someplace new.

ACTA will have effects – both formal and informal. Formally it will bind only those countries signatory. But ACTA presence will create strong pressures to conform national law to its norms. If the ACTA promoters are successful in establishing ACTA as the new minimum standard, even as a talking point, much of the battle will be won.

An important point of contrast between TRIPS and ACTA will be the role of WTO dispute settlement mechanisms. Of course ACTA, falling outside of the WTO, will not be directly justiciable within WTO dispute settlement. But WTO dispute settlement is not a closed universe. WTO panels and the Appellate Body have shown openness to the consideration of international law external to the WTO system. It is not inconceivable that ACTA might be given some effect even within WTO dispute settlement. It may, for example, fill gaps or aid interpretation of WTO norms.

Given the general unavailability of WTO dispute settlement, ACTA norms may or may not be faithfully followed by its various signatories. On the one hand, *pacta sunt servanda* may be adequate here (and may also be adequate within WTO law) to compel substantial compliance, even without formal dispute settlement.

3.2 Conflict of Norms

The entry into effect of ACTA raises some issues of first impression as to the compatibility of ACTA with the WTO system. To a degree, WTO agreements generally and TRIPS specifically preclude the establishment of additional undertakings among WTO members. At least parts of ACTA may or may not be preempted by the terms of TRIPS.

The law concerning conflicts of treaty norms is not terribly well-defined. Certain rules are well recognized. Generally, in cases of conflict, the terms of the later-in-time treaty will prevail. The subsequent in time treaty is viewed as having implicitly amended the conflicting term contained in the prior treaty. Likewise, a term from a treaty that addresses an issue with greater specificity will likely prevail over a treaty of general application. One can see, however, that even these two simple rules could, in application, yield conflicting results.

There is greater doubt when the conflict exists between a multilateral treaty and a bilateral, regional or plurilateral treaty (such as ACTA) involving some but not all the signatories of the multilateral treaty. Of course (absent an amendment mechanism so providing) some, but not all, treaty signatories cannot amend a multilateral treaty – at least not with respect to any treaty signatory that is not party to the bilateral, regional and multilateral agreement. No matter how many WTO-member adherents ACTA attracts, it could not be considered as altering any WTO/TRIPS undertakings so long as any WTO members remain outside.

Indeed, it might be the case that even full subscription by the entire WTO membership to ACTA would fail to effect any alteration of the WTO undertakings. A similar issue arose with respect to the European Union's 'constitutional' treaties. A mere agreement among the constituent European Member States cannot effect an amendment of the treaties, such as to avoid resort to the formal amendment mechanisms found within those treaties.

Note the WTO agreements provide for formal amendment by less-than-unanimous consent – this is the process being pursued with respect to the new TRIPS Article 31*bis*. As such, ACTA cannot be argued to serve as a quasi-legislative instrument to impose new obligations on non-signatory WTO members. The more difficult question is whether ACTA can adjust WTO undertakings, including the TRIPS obligations, in relations between ACTA signatories.

TRIPS Article 1.1, in part, addresses this issue:

> Members may, but shall not be obliged to implement in their law more extensive protection that is required by [TRIPS].

Article 1.1 thus serves both as authorization for and limitation on Member States in their national law-making. It may be presumed that this residual legislative authority may be invoked in a treaty undertaking, but Article 1.1 does not so expressly provide. Contrast this with the enabling provision found in Article XXIV of the GATT, where GATT contracting parties (and now WTO members) are expressly permitted to enter free trade agreements and customs unions.

If Article 1.1 is enabling of later-in-time agreements with respect to the protection of intellectual property, it cannot be argued that TRIPS otherwise categorically precludes the entry of ACTA. That is, to use US constitutional terminology, TRIPS does not provide field pre-emption regarding IP rights: the entry of TRIPS cannot be read to prohibit any subsequent treaty intrusion on its subject matter, including undertakings that do not directly conflict with TRIPS norms.

TRIPS may however pre-empt an ACTA undertaking that falls outside the precise contours of the enablement found in Article 1.1. First, Article 1.1 expressly permits 'more extensive protection'; by negative implication, an ACTA term that provided 'less extensive protection' would be inconsistent with, and pre-empted by TRIPS. This reading is consistent with the understanding that TRIPS is a summary of minimum standards.

As Ruse-Khan and Yu have both suggested, TRIPS may be more than a minimum standards treaty;[47] it may indeed, at least in part, provide for maximum standards. Yu argues that TRIPS reflects a carefully negotiated balance with respect to enforcement of intellectual property protection. If he is right, then a WTO-member enactment – or the terms of treaty among certain WTO members such as ACTA – that provides for excessive enforcement of intellectual property rights might be pre-empted by TRIPS.

---

47. See A Trade Agreement Creating Barriers to International Trade? ACTA Border Measures and Goods in Transit, *supra*; TRIPS Enforcement and Developing Countries, *supra*.

3.3 ACTA Norms in WTO Dispute Settlement

WTO panels and the Appellate Body follow the rules of interpretation found in the Vienna Convention to give effect to the various WTO undertakings, including TRIPS.[48] In his recent paper on the WTO panel report in *China-Measures Affecting the Protection and Enforcement of Intellectual Property Rights*,[49] Peter Yu discusses the possibility that a bilateral agreement might serve as subsequent practice in the sense of Article 31.3 of the Vienna Convention.[50] This opens the possibility that ACTA might be used to give content to an otherwise ambiguous term found in TRIPS.

Article 31.3 of the Vienna Convention[51] provides that 'there shall be taken into account, together with the context . . . any subsequent agreement between the parties regarding the interpretation of the treaty or the application of its provisions.' The WTO panel rejected China's argument on the interpretation of 'commercial scale' that had been based on the subsequent United States-Australia Free Trade Agreement, as China was not party to that agreement.

Article 31.3 may make ACTA available (for purposes of subsequent practice) in a WTO dispute between two WTO-member ACTA signatories. Of course, such a finding formally has no power outside the particular dispute between the two parties; any interpretation of a TRIPS norm so informed by ACTA would not be binding in any subsequent dispute, and would not alter the rights and obligations of any other WTO member (including the body of non-ACTA signatories within the WTO).

It is not certain that a WTO dispute panel would strike down a TRIPS inconsistent measure where that imposition of that measure is supported by a ACTA norm. WTO dispute panels and the Appellate Body have seen fit to recognize the existence of, and give effect to, treaty obligations arising outside the WTO system. There had been doubt at one point whether a dispute panel could reach outside the WTO agreements to uphold an otherwise inconsistent measure. There had been a period where much anxiety was expressed over possible conflicts between WTO norms and the provisions of various multilateral environmental agreements (MEAs). WTO decisional law has assuaged much of this concern – as WTO dispute panels and the Appellate Body have shown a willingness to interpret WTO obligations in a manner that does not unnecessarily threaten the effectiveness of MEAs. In the final proceeding in the Shrimp-Turtle dispute, the

48. See GATT Annex 2, Understanding on Rules and Procedures Governing the Settlement of Disputes, Art. 3(2) providing for application of public international law to the DSU. The Vienna Convention on the Law of Treaties rules on treaty interpretation have attained the status of customary or general international law. Report of the Panel, United States – Section 211 Omnibus Appropriations Act of 1998, 3 Jul. 2001 WT/DS176/R (6 Aug. 2001), Sections 8.14–8.15.
49. See fn. 34 *supra*.
50. *Ibid.*, 17–19.
51. Vienna Convention on the Law of Treaties, Done at Vienna on 23 May 1969. Entered into force, 27 Jan. 1980. U. N. T. S. Vol. 1155, 331.

Appellate Body upheld an arguably otherwise inconsistent US measure, based in part, on the negotiation of an international regime for the protection of sea turtles. While there is no WTO decisional law on point, one would imagine similar subordination of WTO norms in the event of a conflict with a norm contained in a multilateral human rights treaty.

It is somewhat uncertain how the WTO would manage a conflict with a norm fixed in a treaty falling within the same general subject matter as one of the WTO agreements: TRIPS after all does establish certain IP enforcement obligations. It is one thing to recognize the fragmented nature of international law – partitioned among various regimes with specialized institutions and specified subject matter – and mediate conflicts accordingly. It is quite another to defer to an agreement occupying the same policy space.

Of course TRIPS does coexist with other international intellectual property agreements, most importantly the WIPO-centred Paris and Berne Conventions. But these are by-and-large earlier in time than TRIPS, and where they were found to be inadequate (Paris says little about minimum patent standards), TRIPS provides enhanced coverage.

Finally, there is arguably a 'rule of reason' in WTO law. There are several textual bases for it. With respect to the GATT, Article XX (General Exceptions) softens the various non-discrimination and market access provisions of the GATT. TRIPS contains exceptions to the substantive patent, copyright and trademark mandates. There is, however, no general exception provision to TRIPS analogous to GATT Article XX, and as such many of the enforcement mandates appear to be unqualified. This absence should not preclude a dispute panel from finding inspiration in the objectives of TRIPS (which, it is often noted, are found not in the preamble but in a substantive article of TRIPS) to introduce some flexibility. ACTA might serve an ambivalent role here, depending on the particular norms involved and the context of an eventual dispute. ACTA might, on the one hand, suggest scenarios where a 'rule of reason' approach is appropriate. It might also be urged to resist any departure from stern application of TRIPS enforcement obligations.

WTO panels and the Appellate Body have shown great deference to the form of arguments promoted by contesting WTO members. To the extent disputing parties bring forth arguments based on provisions on ACTA, panels and the Appellate Body will not be able to escape expressing judgment on the weight (if any) of ACTA within the realm of WTO dispute settlement. The WTO institutions may in the end decide that ACTA has no force at all, either directly or indirectly. But arguments will inevitably be made.

ACTA is likely to be more influential in the hidden zone of WTO dispute settlement that occurs prior to the establishment of a panel. Most WTO disputes, and most TRIPS disputes, are resolved at this point. As these settlements are diplomatic and not juridic, neat divisions of formal coverage matter less. Behind closed doors, there will be little to inhibit a WTO member from insisting on compliance with an ACTA norm, even with respect to a non-ACTA signatory.

In a 'realpolitik' vision more broadly, ACTA norms will be urged against the entire WTO membership. And certain WTO members will resist. There are very powerful countries – Brazil and India of course, but especially China – that may see no need for mandatory enhancements to IP enforcement. But many other WTO members, including many developing countries, may be persuaded to give effect to ACTA norms even if doing so will draw resources away from more pressing state interests. As Yu and others point out, even within the category of law enforcement, there may be other offences that present more compelling claim to scarce national resources.

4 THE EXTENSION OF TRIPS PLUS NORMS TO DEVELOPING COUNTRIES: THE CASE OF VIETNAM

ACTA may serve as an important bridgehead for the eventual transmission of TRIPS Plus norms to the developing world. Additional developing countries are likely to become signatories to ACTA, and hence directly bound to its terms. These may include 'voluntary' adherents, who will join ACTA in order to signal to important constituencies (donor countries, private investors, the International Monetary Fund (IMF)) their political commitment to the maintenance of open market economies. Others may be compelled to join ACTA due to the suasion of powerful States.

Still other developing countries may be required to commit to ACTA's norms through the entry of bilateral agreements with powerful ACTA promoting States, such as the US. Every treaty partner who accepts ACTA and other TRIPS Plus provisions in a free trade agreement creates more difficulty for a subsequent developing country negotiating an FTA with the US in resisting such provisions in its agreement.

Finally, ACTA norms may find their way into broader regional trade agreements. Accession to these agreements may be attractive, if not irresistible, to developing countries desirous of participating in high-volume regional trade notwithstanding the presence of unappealing TRIPS Plus intellectual property norms. Vietnam may provide a telling case illustrating each of these processes.

Vietnam is not an original WTO member, and so did not participate in the Uruguay Rounds negotiations. As such, it did not contribute to – or rely upon – the so-called Grand Bargain, by which the most highly developed members (the US and EU) extracted commitments for the establishment and enforcement of intellectual property provisions (ultimately enshrined in TRIPS) in exchange for market access commitments in textiles and agricultural products sought by the developing world.

Vietnam's first exposure to international intellectual property standards arose during the negotiation and implementation of the Bilateral Trade Agreement with the United States (BTA).[52] It was widely understood that the BTA would be a 'comprehensive' agreement (similar to NAFTA), with coverage of services, investment and intellectual property, as well as conventional trade-in-goods

52. Agreement between the United States of America and The Socialist Republic of Vietnam signed on 13 Jul. 2000, in effect 10 Dec. 2001.

commitments.[53] Moreover, both Vietnam and the US viewed the BTA as a 'stepping stone' to an eventual accession of Vietnam to the WTO.[54] BTA in fact served as merely the first round of a bilateral dialogue on trade, investment, services and intellectual property issues, as the US exacted further commitments from Vietnam in the bilateral agreement concerning Vietnam's accession to the WTO.

As such, it could hardly have been surprising that the US insisted on Vietnam's early implementation of TRIPS standards in the BTA, even as it was clear that WTO membership would be several years in the future. There is no record of resistance by Vietnam to acceptance of TRIPS standards in the BTA negotiations; rather, other areas seemed to consume more bargaining capital and political attention.

In the course of the implementing its obligations under the BTA, Vietnam enacted a TRIPS-consistent Intellectual Property Law in 2001.[55] These internal reforms placed Vietnam in a position where further IP reform enactment would not be required in order for Vietnam to meet the obligations of TRIPS that became binding upon its joining the WTO in 2007.

There was a (perhaps) minor intellectual property provision in the BTA that could conceivably be described as 'TRIPS Plus': a requirement that Vietnam provide protection with respect to certain satellite broadcasts.[56] Beyond this instance, it appeared that the United States was satisfied by Vietnam's commitment to the package of TRIPS norms. During the BTA negotiating process, the US' attention was focused in other sectors, where meaningful commitments were exacted from Vietnam above and beyond WTO requirements.

Not only did BTA require Vietnam to implement TRIPS 'prematurely' (that is, in anticipation of, rather than as a result of, Vietnam's admission to the WTO), it eliminated Vietnam's access to certain phase-ins of TRIPS effectiveness that had been provided to the founding WTO membership. Because of the obligations found in the BTA, and consistent with its accession commitments, Vietnam entered the WTO in 2007 fully committed to the intellectual property obligations imposed by TRIPS. And while there may have been some shortcomings in Vietnam's discharge of its TRIPS obligations, there is no record of major significant discontent by the US or other WTO members with Vietnam with respect to the formal aspects of its system of intellectual property protection.

---

53. See Quy Binh Nguyen, 'Harmonization of Law for Economic Development in Vietnam and Impacts of the US-Vietnam Bilateral Trade Agreement toward This Process and Future US-VN Trade Relations: Part 2 BTA and WTO Negotiations', University of Louisville Legal Studies Research Paper No. 2010-06.
54. See David Gantz, 'Doi Moi, the VBTA and WTO Accession: The Role of Lawyers in Vietnam's No Longer Cautious Embrace of Globalization', 41(3) *Int'l Lawyer*, 873–890 (Fall 2007).
55. Decree No. 06/2001/ND – CP dated 1 Feb. 2001 of the Government to revise and modify some provisions of Decree No. 63/CP on industrial property rights protection (amended by Law No. 50/2005 adopted by the National Assembly of Viet Nam, Legislature XI, 8th session, dated 29 Nov. 2005, as amended by Law No. 36/2009 (The Law on Amendments to the Law on Intellectual Property) adopted by the National Assembly of Viet Nam, Legislature XII, 5th session, dated 19 Jun. 2009).
56. U.S-Vietnam BTA, Art. I (3) (E).

That said, Vietnam is facing a new round of international intellectual property mandates, occasioned by the negotiation of a treaty to which it is not a founding party: the ACTA. While there may have been no particular desire to exclude Vietnam (qua Vietnam) from the ACTA process, the political fact remains that it has been so excluded, and any input it might have provided to the process of norm formation has been cut off. Vietnam will eventually encounter ACTA as it earlier encountered the TRIPS requirements, as a set of fully-baked intellectual property norms, presented on a take-it-or-leave-it (or perhaps simply take-it) basis. There is much in the text of ACTA that would affect Vietnam were these TRIPS Plus norms effectively applied to it. Of course, Vietnam might choose to joint ACTA (ACTA is open for signature to all WTO members) or it might engage in further national intellectual property 'reform' on its own initiative.

Vietnam did not participate in the Uruguay Round negotiations either; as to it, TRIPS was part and parcel of the WTO *acquis* (although, as noted before, Vietnam enacted TRIPS-compliant intellectual property standards in discharge of its obligations under the United States-Vietnam BTA). Vietnam did achieve meaningful market access that was connected to its adoption of intellectual property; IP was part of the quid pro quo that brought Vietnam access to the US market (in the BTA) and to global markets (in its accession to the WTO). Still, the 'bargains' entered by Vietnam through BTA and WTO accession were different from the Uruguay Round Grand Bargain that led to TRIPS.

It may be that Vietnam has suffered less disappointment and regret with the accorded terms of its trade agreements (including WTO accession) than have Brazil and India with respect to the Grand Bargain. Vietnam might thereby feel less predisposed to resist any TRIPS Plus expansion of IP rights than certain other WTO members. Looking forward, however, it appears that Vietnam did not participate in the formation of IP norms that it will be called upon to enact. Rather, it will be required upon to accept a complete package of obligations as part of a broader trade agreement.

Vietnam is one of the participants in the recently launched Trans-Pacific Partnership (TPP) talks with the US with the object of forming a regional Free Trade Agreement.[57] The TPP originally was a free trade agreement among Brunei, New Zealand and Peru.[58] Current TPP negotiating participants include Australia, Chile, and Singapore. The US has existing deep bilateral trade agreements with four of these countries: Australia FTA, Singapore FTA, Chile FTA, and the Peru TPA. Joining and hence reforming TPP thus offers the US an opportunity to

57. See Two Letters, dated 14 Dec. 2009, from Ambassador Ronald Kirk to the Hon. Nancy Pelosi and the Hon. Robert C. Byrd, announcing President Barack Obama's intention to enter into negotiations of a regional Trans-Pacific Agreement (TPP) *inter alia* with the Socialist Republic of Vietnam. <www.ustr.gov/tpp>. See also Australia's, Peru's and Vietnam's intentions to join TPP negotiations. 74 Fed. Reg. 15, 4480–4482 (26 Jan. 2009).

58. Trans Pacific Strategic Economic Partnership Agreement among Brunei Darussalam, the Republic of Chile, New Zealand and the Republic of Singapore, entered into force May 2006. <www.mfat.govt.nz/downloads/trade-agreement/transpacific/main-agreement.pdf>.

rationalize its various commitments into a single instrument. TPP further presents an opportunity to deepen the commitments of each signatory. It creates network efficiencies by creating a large trading area with the US as its hub. It serves as a structure to which additional Pacific countries may adhere. And finally – much as NAFTA did two decades ago – TPP may serve as a laboratory for introducing novel obligations that will eventually be globalized, either through expansion of the WTO structure or establishment of yet another global institutional edifice.

The TPP process includes an intellectual property negotiating group – and hence IP will form part of its coverage – although it is not yet clear exactly what the negotiating objectives of the US will be in this regard. It is hard to imagine, however, that the US would not insist on the incorporation of ACTA standards as a minimum. Note that each TPP participant is a WTO member and as such TRIPS effectiveness within the TPP zone is already assured. Any non-redundant IP provisions found in TPP will necessarily be TRIPS Plus. By inserting the ACTA norms into TPP and other US trade agreements, the US can gradually expand the coverage of its TRIPS Plus desiderata.

In doing so, the US would take advantage of the ratchet nature of a trade concession. Once a country agrees to accept a particular norm within any particular structure (be it a bilateral trade agreement with a powerful country like the US – or via membership in a regional agreement), it then gains little in the redundant application of that norm by subsequent agreements. ACTA states are unlikely to join bargaining coalitions resisting the inclusion of ACTA-inspired TRIPS Plus provisions in other international fora.

The ACTA package of TRIPS Plus norms will likely be transmitted to Vietnam through the TPP process. Several of the most important TPP participants (Australia, Chile, Peru) are ACTA promoters. ACTA will likely become effective prior to the conclusion of the TPP negotiating process – and so acceptance of ACTA's TRIPS Plus norms will be *fait accompli* among a dominant block of TPP negotiating countries.

It is unlikely that any ACTA party participating in the negotiations of TPP would resist an incorporation of ACTA mandates, given that these countries will already be treaty bound to give effect to those mandates. As a practical matter, the only effect that inserting ACTA's mandates into TPP could have would be to extend these requirements to the only two TPP countries that did not participate in ACTA: Bahrain and Vietnam. Of course, this calculus might shift as additional countries adhere (as expected) to either ACTA or TPP.

There is little that Vietnam can do singly, or in concert with Bahrain, to block the incorporation of ACTA TRIPS Plus provisions into TPP. Vietnam would face the certain desire of the US to expand the territorial coverage of ACTA and the indifference of other TPP countries, given their prior concessions within the ACTA process. If this prediction runs course, in the short term the effective ACTA coverage will include the territories of the original ACTA parties plus Vietnam and Bahrain (via TPP) – and perhaps other countries as well, via eventual bilateral or regional free trade agreements dominated (at least with respect to IP policy) by the US.

Through the renewal and expansion of its bilateral and regional trade agreement (or trade promotion agreement) programs the US can little by little increase the number of countries bound to its TRIPS Plus design. Each country that accepts TRIPS Plus – whether through adhesion to ACTA or by acceding to IP commitments contained in other agreements – represents one fewer potential opponent within the WTO community. In the future, when IP returns to the WTO agenda (that is, to the global agenda), the anti-IP leadership (Brazil, India, China) may have few passionate followers, as many Newly Industrialized Countries, Developing Countries or even Least Developed Countries may have bought in to TRIPS Plus in other settings. These countries will not waste political capital – their WTO 'bargaining chips' – resisting the imposition of norms they already face through other instruments.

The transmission of ACTA's TRIPS Plus norms to Vietnam through the TPP process may be just one of many incremental steps that together will realize the US' vision of a post-TRIPS global IP expansion. It is a strategy of dividing the block of potential IP resisters. In this course, the indirect acceptance by Vietnam of the ACTA basket of obligations is the fall of a domino, creating increasing pressure on other countries to fall in line.

The intellectual property rights holders of the US straightforwardly embrace the extension of the 'high standard' (i.e. TRIPS Plus) intellectual property mandates found in the various existing Free Trade Agreements between the US and certain TPP states (Australia, Chile and Peru) to the entire TPP zone. As such, the US is not advancing a call to increase IP standards beyond the TRIPS Plus provisions found in the existing FTAs in TPP. Rather the focus should be an extending these mandates to the three countries participating in TPP that lack a 'modern' FTA: Brunei, New Zealand and Vietnam. Of the three non-FTA countries, only New Zealand will be bound by ACTA. Given this constellation of treaty relations, TPP represents the most direct route for the extension of TRIPS Plus to Vietnam and Brunei – and between the two, Vietnam possesses the market and the economy that matters. Thus, at least with respect to furthering the US' objectives in IP, TPP is in some sense not a regional or plurilateral effort at all, but rather has as its focus further IP reform in Vietnam.

The US media sector has identified specific shortcomings with respect to Vietnam that will be addressed in the TPP. These included asserted 'market access' limitations that prevent US media enterprises from establishing subsidiaries in Vietnam to produce or distribute 'cultural products'. As such, all in-bound licensing is directed to domestically controlled enterprises in Vietnam which have proven ineffective, in the eyes of US licensors, in preventing piracy. Thus, the TPP, though a plurilateral agreement, can serve as a site for obtaining a treaty-based resolution of a dispute between a particular pair of countries.

5 DESTABILIZING TRIPS

The notion of progressive development is found in many fields of international law. It forms, for example, a central part of the mandate of the United Nations

International Law Commission. Progressive development is part of the process of international law-making.

Progressive development is tied to the technique of codification. Codes contain provisions that are either (1) articulations of current settled principles (restatements) or (2) aspirations designed to pull States towards the adoption of new norms. Treaties may contribute to progressive development in much the same way. Of course treaties do bind their signatories. But they may also effect a normative pull on non-signatories, urging these states to move toward an emerging international standard. In certain cases, progressive treaty provisions can be used to base arguments as to the emerging status of customary international law in an area.

Of course one state's vision of 'progress' might seem like backward movement to another. ACTA represents a claim for progressive development in the construction of the global intellectual property system. It is a roadmap for progress in the eyes of its promoters. But ACTA's opponents are unlikely to see it in this light. For developing countries, ACTA represents a further intrusion on national autonomy in the field of IP, greater adverse terms of trade vis-à-vis the developed world, and increased burdens on State authorities.

ACTA also constitutes a regime shift, at least in the short run. The promoters of ACTA declined to use the WTO space for this exercise in law formation, preferring the comfort of a club of the like-minded. Whether ACTA norms are subsequently introduced to the WTO/TRIPS structure remains to be seen.

A deliberate result of the entry of ACTA is a destabilization of TRIPS. ACTA is a critique of TRIPS – its very core signals a diagnosis that TRIPS inadequately addressed the problem of IP enforcement. A kind reading might admit TRIPS as a good faith response in its day – but that experience and the passage of time have revealed latent weakness that need redress. A colder assessment might concede that TRIPS inadequacies were there to be seen at the time of its creation – but that a better job was unavailable, due to the immensity of the larger WTO negotiating project or due to hardened opposition at the time.

The Grand Bargain has crumbled – and tensions surrounding IP enhancement have increased. This essential quid-pro-quo was not achieved in the view of many WTO members (including the self-appointed leaders of the developing world interests in the field of IP, Brazil and India). The TRIPS element of the Grand Bargain has been given effect, of course. Notwithstanding the dismantlement of GATT-era regimes protecting textile and agricultural markets, little market access has been created for developing world products. As such, there is no small amount of bitterness in many quarters of the WTO membership concerning the outcomes realized from the Grand Bargain.

The US, Japan and the European Union have not been satisfied with TRIPS. While there is a grudging satisfaction with its substantive mandates found in TRIPS with respect to intellectual property protection, there remains a deep unhappiness on the part of the WTO members with the largest stock of IP with the weak national enforcement of IP norms in other WTO members.

The US has largely withdrawn from the WTO space to conduct its more expansive program for enhanced intellectual property protection. Its desiderata is widely known as TRIPS Plus. In its first iteration, TRIPS Plus provisions were inserted into various bilateral treaties – chiefly Bilateral Investment Treaties and Free Trade Agreements. The United States-Australia FTA was the first major treaty to display TRIPS Plus features. United States-Australia was followed by free trade agreements with Korea, Morocco and Singapore – all of which contain extensive TRIPS Plus provisions.

Bilateral free trade agreements have not been the unique setting for imposition of TRIPS Plus norms. The US (and to a lesser extend Europe) took advantage of its powerful negotiating position within the WTO to impose some heightened obligations on post-Uruguay entrants to WTO membership. These new WTO members – notably China (but also Vietnam) – became party to sui generis accession agreements which imposed obligations that were not generally part of the WTO *acquis,* including IP obligations.

More recently, the US has diversified its TRIPS Plus program. While bilateral and regional agreements remain an important part of its strategy, the US, together with Europe, has returned to a multilateral – though not universal – approach: the entry of a stand-alone treaty, the ACTA.

ACTA was negotiated among like-minded States. This is not to suggest that there had not been significant disagreement among the parties throughout the process. The negotiating parties, however, constitute a limited subset of the broader WTO membership. Importantly, China, Brazil and India were excluded. And many other WTO members, including Vietnam, did not participate. There is, as such, a certain 'club-like' quality to the ACTA process. There was certainly a consensus within the ACTA Club that 'counterfeiting' is a serious and continuing problem, notwithstanding the presence of TRIPS (and the potential for enforcement of its requirements on WTO members through WTO dispute settlement processes).

Were ACTA to merely provide rules for the ACTA Club, the substantive terms reached within ACTA would have little import to the broader WTO membership. But this is not likely to be the case. Indeed, this was not the plan, Rather, ACTA will serve as a stepping stone toward the broader distribution of certain TRIPS Plus norms.

To create ACTA at this time – a decade and a half after TRIPS – is more than a technical assessment as to TRIPS' ability to attain its own goals. ACTA also represents a shattering of the consensus (to the extent one ever existed) behind TRIPS. After ACTA, one cannot read TRIPS as expressing the common understanding of the world community (or even the WTO membership) as to the minimum standard of IP protection to be conceded by every state. ACTA formalizes a rift between the developed world and many other countries (it is these non-signatory countries where 'counterfeiting' and 'piracy' are perceived to be most prevalent).

With ACTA, the developed world (led by the US) is playing a high-stakes game. It seeks to destabilize TRIPS in order to induce movement. But in so doing, it also necessarily undercuts the prestige, normative pull, and – perhaps – legitimacy of TRIPS. For if one group of WTO members distance themselves from TRIPS (due to its asserted ineffectiveness or obsolescence), so too might another group.

In progressive development, one can imagine states embracing a common conception of progress at different speeds. There may be distance in execution without normative disagreement. But this does not seem to describe the broader context in which ACTA is sited. The WTO membership may be as divided – or more divided – as to IP now than it was at the time of TRIPS. Polarization may lead to a fissure between the pro-IP ACTA proponents and the newly industrialized country, developing country and less-developed country majority. Ironically, what all these countries might share is a judgment that TRIPS is failing – though of course their respective prescriptions for reform would differ.

Of course, the proponents of ACTA are not calling for a dismantlement of TRIPS. A functional assumption of ACTA is that TRIPS will remain in place. But the ACTA promoters cannot disguise their assessment of TRIPS and their prescription for its inadequacies. This move is inherently destabilizing of TRIPS – and perhaps strips TRIPS of some of its authority, prestige and legitimacy. The resort to ACTA also makes clear the breakdown of the momentous (and surprising) consensus that brought us TRIPS at the moment of the creation of the WTO.

Chapter 7

Breaking Down Barriers to Technology Transfer: Reforming WTO Standard-Setting Rules and Establishing an Advisory Facility in Standard-Setting for Developing and Least Developed Countries

Christopher Gibson

This chapter focuses on a subject that might be considered as somewhat specialized or even 'arcane' in the technology transfer debate, but one which I will argue is really quite central to facilitating technology transfer to developing and least developing countries: my topic is technology standard-setting. I have included two concrete proposals.

Much has been written about intellectual property (IP) and its relevance to technology transfer. In the international context, there has been extended debate about the impact of intellectual property rights on economic development, trade, and technology transfer.[1] However, the standard-setting process is also integrally

1. See e.g., John. H. Barton, 'New Trends in Technology Transfer Implications for National and International Policy', International Centre for Trade and Sustainable Development, Issue Paper No. 18 (2007) ('New Trends in Technology Transfer'), Daniel. J. Gervais (ed.), *Intellectual Property, Trade and Development: Strategies to Optimize Economic Development in a TRIPS Plus Era*

Hans Henrik Lidgard, Jeffery Atik & Tu Thanh Nguyen, *Sustainable Technology Transfer*, pp. 149–175.

related to technology transfer, and this carries important implications for developing and least developed countries. Currently, these countries are largely marginalized or excluded from important standard-setting processes. Without 'access' or the knowledge and understanding to productively participate in these standards processes, developing and least developed countries are caught in a difficult and negative cycle, in which their exclusion breeds lack of innovation and continuing technological backwardness.

In this chapter, I approach this problem on two levels. First, I propose changes within the WTO system, in particular, focusing on the rules associated with the WTO's Agreement on Technical Barriers to Trade (TBT Agreement),[2] to facilitate fairness in the international standard-setting where intellectual property rights are involved, and thereby benefitting developing and least developed countries. Second, I also propose a new mechanism to facilitate increased understanding of and participation in the standard-setting process by developing and least developed countries.

1 TECHNOLOGY STANDARD-SETTING, INTELLECTUAL PROPERTY AND ISSUES FOR DEVELOPING COUNTRIES

1.1 Complexity in Technology Standard-Setting

Technology standards play a critical role in trade, investment and technology transfer. To be sure, standards are an integral part of the largely invisible infrastructure of the modern world that makes things work and, in particular, makes them work well together.[3] Technical, health and safety standards of all kinds touch

---

(New York: Oxford University Press, 2007); Keith. E. Maskus & Jerome. H. Reichman (eds), *International Public Goods and Transfer of Technology under a Globalized Intellectual Property Regime* (Cambridge, U.K.: Cambridge University Press, 2005) ('*International Public Goods and Transfer of Technology*'), Carsten Fink & Keith. E. Maskus (eds), *Intellectual Property and Development: Lessons from Recent Economic Research* (Washington, DC and New York: co-publication of the World Bank and Oxford University Press, 2005), Daniel J. Gervais, 'Intellectual Property, Trade & Development: The State of Play', *74 Fordham L R 505* (2005), Denis Borges Barbosa, Margaret Chon & Andrés Moncayo von Hase, 'Slouching towards Development in International Intellectual Property', 2007 *Mich St L. Rev* 71 (2007); Keith E. Maskus, *Intellectual Property Rights in the Global Economy* (Washington, DC: Institute for International Economics, 2000).

2. Agreement on Technical Barriers to Trade, 15 Apr. 1994, Marrakesh Agreement Establishing the World Trade Organization, Annex 1A, Legal Instruments – Results of the Uruguay Round, 1867 U.N.T.S. 410 (TBT Agreement).
3. See Christopher. Gibson, 'Globalization and the Technology Standards Game: Balancing Concerns of Protectionism and Intellectual Property in International Standards', 22 *Berkeley Tech L. J.* 1407 (2007) ('Globalization and the Technology Standards Game'). A standard can be defined simply as 'any set of technical specifications that either provides or is intended to provide a common design for a product or process'. *Ibid.*, 1410 (quoting Mark A. Lemley, 'Intellectual Property Rights and Standard-Setting Organizations', 90 *Cal L Rev* 1889, 1896 (2002) ('Intellectual Property Rights and Standard-Setting Organizations').

all aspects of our daily lives and are essential to modern commerce. A current example is the set of wireless communication standards essential for my handheld Blackberry device. Standards regulate the technologies that will allow my Blackberry to operate and interoperate, in a compatible manner, with wireless communication networks. There are standards for the communications protocols, and standards for the security protocols.[4] All of these standards work in the background, the largely invisible infrastructure that is needed so that my wireless device will communicate when I travel with it within the United States, or from the United States to Vietnam. As a maker of handheld wireless devices such as the BlackBerry, the company, Research in Motion must be aware of, and have access to, the technology standards that will allow these devices to work in a networked environment.

A 1999 OECD study stated that the impact of standards on trade is so widespread that, on purely economic grounds, up to 80% of trade is affected by standards or associated technical regulations.[5] Moreover, in an age of globalization, there is no denying the advantage of using *harmonized* standards where technology and products of all types must conform to the technical and other standards-related requirements in various national or regional markets. So we can ask two questions. First, who sets the standards? Second, how are standards developed?

Today, we have a proliferation of many different types of standard-setting bodies. Standard-setting can take place through accredited national, regional or international standard-setting organizations (SSOs), unaccredited industry-based groups often referred to as 'consortia', or direct government involvement.[6] Much of the substantive standard-setting activity has historically taken place at the national level, which presents a mismatch to the increasing need for harmonization of global standards. A further complicating factor is the distinction that must be made between institutional and de facto standards. Institutional standards are those defined and adopted by some form of SSOs, whereas de facto standards are often based on proprietary designs that win a leading position in the market.[7] In some of the fastest developing commercial sectors, such as information and communications technologies (ICTs), standards are self-regulating in the sense that technologies may

4. For example, BlackBerry devices that are Wi-Fi enabled support the IEEE 802.11a, IEEE® 802.11b, and IEEE 802.11g wireless networking standards. For security protocols, BlackBerry offers two transport encryption options, Advanced Encryption Standard (AES) and Triple Data Encryption Standard (Triple DES), for all data transmitted between a BlackBerry Enterprise Server and BlackBerry smart phones. See Help Center <www.blackberry.com>.
5. Organisation for Economic Co-operation and Development (OECD), Working Party of the Trade Comm., Regulatory Reform and International Standardization, TD/TC/WP(98)36/FINAL, at para. 3 (28 Jan. 1999).
6. See e.g., International Organization for Standardization (ISO), <www.iso.org>, American National Standards Institute (ANSI). <www.ansi.org>, Institute of Electrical and Electronics Engineers (IEEE), <www.ieee.org>, and The Internet Engineering Task Force (IETF), <www.ietf.org>.
7. See WTO Secretariat, *World Trade Report 2005 – Exploring Links Between Trade, Standards and the WTO* 75 (2005), <www.wto.org/english/res_e/booksp_e/anrep_e/wtr05-0_e.pdf> ('World Trade Report 2005').

become standards either through de facto dominance in the marketplace or voluntarily consensus (often implemented through an industry SSO).[8]

The international system for standards is thus complex, due in significant part to the diversity of approaches for standards followed in different parts of the world. Because of this diversity and the motivations of those involved in the standard-setting process, standards can be used as a means of facilitating trade, investment and technology transfer, or *equally*, as a means of protectionism when employed at the country level, or as a tool used to gain competitive advantage at the expense of those who are excluded at the enterprise level.[9] It has become clear that standards can impact winners and losers in the marketplace while shaping national or regional competitiveness.

The WTO devoted its World Trade Report 2005 to an examination of the links between trade, the WTO's legal framework and standards.[10] The Report explained how the standard-setting dynamic can play out in competition between countries:

> To the extent that promoters of competing standards come from different countries and the winner can claim rents from the adoption of their standard, strategic trade policy considerations come into play. A government can try to tip the balance in favour of its national champion by mandating the use of the firm's standard at home. This would be in the hope that an installed base of users would create a strong enough bandwagon effect to convince foreign suppliers to switch to the national firm's standard in other markets.[11]

Once governments become involved – whether through regulation, preferential treatment to domestic firms, or measures that make commercial operations more difficult for foreign firms – the World Trade Organization (WTO) trade rules can be implicated. As the 2005 Report states, the WTO 'deals with the rules of international trade and inevitably has to deal with the role of standards in international trade'.[12]

The development of standards using diverse approaches and by different types of organizations – whether governmental, quasi-governmental, or private – complicates the situation with respect to appropriate regulation at the international level. As a further complicating feature, SSOs operate in an increasing global context for the technology that is the subject of their standards processes. The question arises then how, in the face of this diversity, can there be any common approach on fundamental issues for standard-setting such as inclusion and transparency, let alone a common policy for intellectual property rights? The OECD study mentioned above proposes that '[t]o deal with [these] issues . . ., the players

8. See Globalization and the Technology Standards Game, fn. 3 *supra*, 1412.
9. 'Standards can be leveraged legitimately to advance the objectives of economic development and national competitiveness. At the same time, however, national standards can be used to protect domestic industries from global competition.' See Globalization and the Technology Standards Game, fn. 3 *supra*, 1403–1404.
10. See fn. 7 *supra*.
11. *Ibid.*, 41.
12. World Trade Report 2005, fn. 7 *supra*, xxv.

will be better equipped if they have accurate, transparent information about what goes on in the standardization field. In other words, transparency is a valid goal in its own right'.[13] All of this cautions that from an international perspective, international regulation or related measures concerning standards should not unduly constrain these diverse systems for standards development. However, at the other extreme, ignoring underlying issues, such as intellectual rights in standard-setting, is not a positive alternative either.

From the development perspective, those who do not understand standard-setting or who cannot participate in standard-setting processes are at a significant disadvantage. The diverse fora where standard-setting takes place are incubators for collaboration and innovation. They are a locus where real competitive stakes are played out. The setting of standards can give an advantage to one company over another or to one country's industries over another. Equally, standards in technology can facilitate technology transfer, or they can impede it. An open standard like the Hypertext Transfer Protocol (HTTP) used for networking and data communication in the Internet context has done wonders to promote technology transfer. On the other hand, standards and technical regulations in a given country – for example in relation to testing and certification requirements for technology – can cause delays and raise costs and thus create significant obstacles for technology transfer and trade to that country.[14] If developing countries do not get involved early in the standard-setting process, they can be excluded from participation in important new technologies, or end up using standards that are essentially dictated to them by others. Moreover, in certain cases, those who would need to comply with, and integrate standards into their goods, services or operations may be forced to pay substantial royalties when such standards are encumbered by privately held intellectual property rights, such as patents.[15] This practice has been well-documented in the case of East Asian manufacturers of DVDs and other optical storage devices and the efforts to coordinate strategic industrial policy responses.[16]

While it is arguable whether China should be viewed as a developing or developed country for purposes of its role and responsibilities under WTO

---

13. OECD, Working Party of the Trade Comm., Regulatory Reform and International Standardization, fn. 5 *supra*, ¶ 82.
14. John S. Wilson, 'Standards, Regulation, and Trade: WTO Rules and Developing Country Concerns', in *Development, Trade and the WTO: A Handbook*, ed. Bernard Hoekman, Aaditya Mattoo & Philip English (World Bank, 2002), 431–432.
15. See Globalization and the Technology Standards Game, fn. 3 *supra*, 1404; see also Barton, 'New Trends in Technology Transfer, fn. 1 *supra*, 29. ('Hence the royalty tax paid by the outsiders amounts to a tax on developing nations firms. This was the case, for example, for East Asian manufacturers of DVD devices, who had to pay what seemed to them to be an exorbitant royalty.')
16. See Greg Linden, 'Optical Storage In China: A Study in Strategic Industrial Policy', *Center for Work, Technology and Society*, University of California, Berkeley (while focusing on China, the author discusses the efforts of other East Asian economies as well in relation to royalty payments that are viewed as burdensome by local producers of high-tech goods); see also generally Milan Brahmbhatt and Albert Hu, 'Ideas and Innovation in East Asia', Policy Research Working Paper No. 4403, World Bank (November 2007).

disciplines, it has undoubtedly moved to recognize the importance of technology standards. In particular, 'China [has] placed the development of standards at the center of its national strategy to emerge as a world leader in science and technology by the year 2020'.[17] To achieve this objective China has adopted a long-term standards strategy with the launch of two standards programs in 2002 – a technical standards development strategy, and the establishment of a national technical standards system – with the overarching goal of raising the level of Chinese technical standards so that they may have a leading international status.[18] This kind of focus on standards development can serve as a model for developing countries seeking to advance in the technology transfer context.

In the next section, I discuss how intellectual property rights are critical to technology standards and can play a key role in the standard-setting process. In the long-run, learning not only about intellectual property rights but also about how to participate in the standard-setting process – a goal that might appear challenging for many developing or least developed countries – can mean the difference between acting as innovators and standard setters, rather than merely as standards adopters.

1.2 The Trouble with Intellectual Property and Standards

Intellectual property, which is frequently an issue in discussions involving technology transfer, is also central to the development of standards. The more technical the content in a particular product, service or technology, the more likely that intellectual property rights will be involved. Thus, the need to deal with intellectual property rights – principally patents – often lies at the centre of standard-setting activities. The problem at the international level, however, as discussed below is that the respective frameworks and structures in place for standards, on the one hand, and intellectual property, on the other hand, are not sufficiently integrated. Put another way (more colloquially), 'they don't talk to each other'.

Intellectual property rights and their proper enforcement are often seen as an essential pre-condition to the dissemination of intellectual assets and technology transfer. Due to their intangible nature, intellectual property rights differ from other property rights in that they are often non-rivalrous in use, and non-excludable by

17. *See* Brian J. Delacey et al., 'Government Intervention in Standardization: The Case of WAPI', 25 (2006). <http://ssrn.com/abstract=930930>.
18. China's objectives are to be accomplished in three phases: (i) form a new voluntary technical standards system and enhance the market adaptability of technical standards by 2010; (ii) complete and perfect the technical standards system and raise the level of Chinese technical standards development by 2020; and (iii) ensure that Chinese technical standards hold a pre-eminent and prominent international status by 2050. See Teresa Cendrowska, 'Enabling US-Chinese Cooperation in Standards and Conformity Assessment', *ASTM Standardization News*,<www.astm.org/cgi-bin/SoftCart.exe/SNEWS/APRIL_2005/cendrowska_apr05.html?E+mystore>. (Referring to comments of Liu Fei, director of operations in the Beijing office of the Consortium for Standards Conformity and Assessment.)

private means alone.[19] Given these characteristics, the State needs to play a pivotal role, first in defining the valid legal scope of intellectual property rights, and then in establishing measures for their protection and enforcement.

In the context of technology transfer, intellectual property rights can provide the owner with the legal protection necessary to support entry into a foreign country's marketplace and to maintain a competitive position there. This is particularly true for firms relying on innovation and producing technology-intensive goods and services.[20] When one thinks of international trade law and the related topics of intellectual property and technology transfer, the natural focus at the international level is often on the WTO's TRIPS Agreement. The TRIPS Agreement codifies Member countries' obligations to establish, protect and enforce IP rights within their territories. In doing so, the TRIPS Agreement 'create[s] the conditions necessary for [technology] transfer to take place'.[21] The TRIPS Agreement, however, does not expressly address standards issues.

Nevertheless, in the standard-setting context, intellectual property rights, and particularly patent rights, pose special challenges to SSOs. Patent claims may be essential to the implementation of a standard, but they can be owned by those who have little or no incentive to make their intellectual property rights available on acceptable terms. Following adoption of a standard, owners of intellectual property rights may be unwilling to grant access under licensing terms conducive to economical and wide adoption of the standard. Or, during the development phase, members of a SSO may secretly seek to ensure that any standard adopted will infringe their undisclosed patent claims, in order later to reap a hefty yield of royalties when those rights are revealed to the standard-adopters after the marketplace has become locked-in.

As a result, SOSs should have policies in place for addressing these IP-related issues designed to minimize such problems, reducing the risk that a finished standard will encounter IP-related obstacles while also facilitating its broader implementation. During the 1990s and even into the first years of the new century,

---

19. Christopher S. Gibson, 'Latent Grounds in Investor-State Arbitration: Do International Investment Agreements Provide New Means to Enforce Intellectual Property Rights?', in *Yearbook on International Investment Law & Policy 2009-2010*, ed. Karl. P. Sauvant (New York: Oxford University Press, 2010), 406–407. Intellectual property rights can give an inventor the incentive to invest in the creation and exploitation of new technology through a legally enforceable and, in the case of patent and copyright, time limited right to exclude others. The broader public purpose behind granting such exclusionary rights is to benefit the public through the increase in knowledge and cultural goods, while in the case of trademarks providing a measure of consumer protection.
20. Christopher S. Gibson, 'A Look at the Compulsory License in Investment Arbitration: The Case of Indirect Expropriation', 25 *Am U Int'l L. Rev* 367 (Winter 2010).
21. Carlos M. Correa, 'Can the TRIPS Agreement Foster Technology Transfer to Developing Countries?', in *International Public Goods and Transfer of Technology*, fn. 1 *supra*, 227, 254, 256. As noted above, the TRIPS Agreement also requires that Members must provide incentives to enterprises and institutions in their territories for the purpose of promoting and encouraging technology transfer to least developed country Members. TRIPS Agreement, Art. 66.2.

many standards organizations did not have proper policies in place. Fortunately, much has changed in the last 10 years. A number of different approaches – some with wide and even international acceptance – have been developed to address the issue of intellectual property rights in standard setting.[22] To consider an all-inclusive approach, for example, Mark Lemley in his article 'Ten Things to Do about Patent Holdup of Standards (And One Not to)', provides a laundry list of potentially helpful measures:

> SSOs could get members to agree in advance of the standard to license patent rights on reasonable and nondiscriminatory ('RAND') terms, bind SSO members to that RAND policy by license agreements, require patentees to specify the content of their RAND licenses ex ante, impose penalty defaults to force disclosure, and/or establish a step-down royalty rate procedure.[23]

The requirement to negotiate licenses on reasonable and non-discriminatory terms embodies the well-established RAND approach for standards licensing, which has emerged as 'the majority rule' among standards bodies that have an express patent policy.[24] Without such a policy, the problems identified above, which may occur at the level of the individual SSO, can also redound to cause ripple effects and raise concerns of fairness at the international level.

1.3 Standard-Setting, Intellectual Property and a Test-Case to Explore Unfair Competition

Technology standard-setting provides a test-case to explore the frontier that exists between the proper scope of intellectual property protection, on the one hand, and practices that might be considered as anti-competitive or amounting to unfair competition and an abusive use of IP rights, on the other. Those who control the standard wield significant economic power, and can work to exclude those who might otherwise wish to participate in a relevant technology market. A simple standard for connectors in a specific technology implementation can act as a 'gatekeeper' for firms that do not have rights to use the particular configuration.

Thus, we can look at special cases where intellectual property and standard-setting are involved, such as who gets to participate (as a member) in SSOs in the first place, and then after a standard is set, potential 'refusals to deal' with third parties. Refusals to deal may be justified where there is an independently developed IP element; however, this becomes more of a concern where the IP was developed in conjunction with standard-setting activity. In this context, refusals to deal might in certain circumstances be considered as an act of unfair trade practice. This could be the case where a company violates an SSO's rules by

22. See generally Intellectual Property Rights and Standard-Setting Organizations, fn. 3 *supra*.
23. Mark A. Lemley, 'Ten Things to Do About Patent Holdup of Standards (And One Not To)', 48 *B C L Rev* 149, 155–156 (2007) ('Ten Things to Do About Patent Holdup of Standards').
24. Intellectual Property Rights and Standard-Setting Organizations, fn. 3 *supra*, 1906.

intentionally failing to disclose a patent bearing on a standard in development by the SSO. In addition, setting the level of any royalty rates to be paid also merits special concern, and good work on this issue has already been done.[25] In particular, as noted above, requiring SSO members to specify their royalty rates 'ex ante', before the standard has been set and it is known whether the selected standard is covered by a particular party's patent, can inform the standard-setting process itself and also work to minimize the potential for exorbitant royalty rates sought by parties who know, only after a standard is selected, whether they might be in a commanding position for subsequent negotiations. Finally, preventing access to certain standards that fall within the realm of an 'essential facility' can be an area of special concern, particularly where a given firm has a dominant position in the marketplace.

Article 8(2) of the TRIPS Agreement provides guidance where use of IP rights might be considered abusive:

> 2. Appropriate measures, provided that they are consistent with the provisions of this Agreement, may be needed to prevent the abuse of intellectual property rights by right holders or the resort to practices which unreasonably restrain trade or adversely affect the international transfer of technology.

Article 8(2) highlights the important interface between intellectual property rights and the regulation of anti-competitive and restrictive business practices. While some commentators have said that this article merely provides a 'policy statement that explains the rationale for measures' taken under other articles of the TRIPS Agreement,[26] Article 8(2) may be of special relevance for cases in which abuse of IP rights in the standard-setting process can be demonstrated. It might be of use, for example, when improper or exclusionary dealings with respect to IP in the standard-setting process bring an impact at the international level. For instance, practices that might be considered anti-competitive in one WTO Member State with respect to a technology standard encumbered by IP rights, might justify 'appropriate measures' in another Member State concerning that same standard.

25. Globalization and the Technology Standards Game, fn. 3 *supra*, 1420; Ten Things to Do About Patent Holdup of Standards, fn. 23 *supra*, 154–155.
26. Daniel Gervais, *The TRIPS Agreement: Drafting History and Analysis*, 2nd edn (London: Sweet & Maxwell, 2003), 121–122. Gervais suggests that Art. 8 is essentially a policy statement that explains the rationale for measures taken under Arts 30, 31 and 40 of the TRIPS Agreement. Others have indicated that, from the developing country perspective, Art. 8(2) 'serves as a conspicuous reminder' of what some less developed countries initially considered as the scope of IP-related matters within the proper mandate of the negotiations leading to establishment of the WTO. For example, India indicated in a meeting of the TRIPS Negotiating Group that it was only the restrictive and anti-competitive practices of intellectual property owners that could be considered to be 'trade-related' because they alone distorted or impeded international trade. See Peter K. Yu, 'The Objectives and Principles of the TRIPS Agreement' (2009), 18–19. <www.peteryu.com/correa.pdf> (chapter is abridged and adapted from Peter K. Yu, 'The Objectives and Principles of the TRIPS Agreement', 46 *Houston L Rev* 797 (2009)) ('The Objectives and Principles of the TRIPS Agreement').

In this sense, Article 8(2) can possibly be used 'as a shield to defend a member state's use of flexibilities that have been built into the TRIPS Agreement.'[27] Put another way, although Article 8(2) uses the word 'may' and is constrained by the 'TRIPS-consistency requirement', in the patent area it can help strengthen the basis for exceptions such as that set forth in Article 31 of the Agreement (compulsory licensing), by reinforcing the notion of a more balanced and flexible interpretation of the TRIPS Agreement.[28]

Article 31 of the TRIPS Agreement permits the government grant of a compulsory license by Member countries over IP (patent) rights, when justified in certain extraordinary cases and where the terms and procedures of Article 31 are followed. While Article 31 leaves open the substantive grounds on which a WTO Member may rely to justify a compulsory license, one area that may be pertinent to standard-setting is the possibility of authorizing a compulsory license to correct anti-competitive behavior so that a particular patent does not 'undesirably constrain competition between firms.'[29]

Precedent exists in the domestic context for granting a compulsory license in response to anti-competitive behaviour relating to standards. A compulsory license was used as a remedy in the United States to address patent rights and related concerns of bad faith conduct in the standard-setting context. In the well-known 1995 case, *In re Dell Computer Corp.*, the Federal Trade Commission (FTC) brought a claim against Dell for engaging in unfair competition by participating in a standard-setting process, yet failing to disclose pertinent patent rights it owned, and seeking to enforce those rights once the applicable industry standard was adopted and had began to take hold.[30] Dell's conduct was found to 'constitute unfair methods of competition in or affecting commerce',[31] and the FTC ruled that Dell violated Section 5 of the Federal Trade Commission Act. The FTC banned Dell from ever enforcing its patent again – in effect, imposing a compulsory license without remuneration as the remedy for Dell's anti-competitive behaviour.[32]

In a second well-known United States standards case, *In re Rambus, Inc.*, the FTC again brought a Section 5 antitrust claim against a private company for its conduct in a SSO, claiming that 'through deceptive acts and practices, Rambus obtained monopoly power over the DRAM market.'[33] In particular, the FTC alleged that Rambus had unlawfully monopolized the market for four computer memory technologies incorporated into industry standards for dynamic random

27. *Ibid.*, 24.
28. *Ibid.*, 27–28.
29. Antony Taubman, 'Rethinking TRIPS: "Adequate Remuneration" for Non-Voluntary Patent Licensing', 11 *J Int'l Econ L* 947 (2008). To date, governmental authorities have granted relatively few high-profile compulsory licenses for reasons other than antitrust-competition law concerns.
30. 121 FTC 616, 1996 FTC LEXIS 291 (1996).
31. *Ibid.*
32. The dissenting opinion in the *Dell* case found that the patent disclosure policy was unclear and might have just 'encouraged' rather than 'required' patent disclosure. *Ibid.*
33. *In re Rambus, Inc.*, Docket No. 9302 (FTC 18 Jun. 2002), Public Record Version, 3.

access memory (DRAM) chips. An Administrative Law Judge first ruled in favour of Rambus, but then the full Commission overruled that decision and issued a remedy setting maximum royalty rates for the technology in question, thereby granting what amounted to another form of compulsory license, this one compensated at a FTC-established rate.[34] This did not end the case, however, because the United States Court of Appeals for the District of Columbia unanimously overruled the FTC's decision, holding that the FTC had failed to prove deceitful conduct and, even if Rambus had done so, the FTC failed to sustain the allegation of monopolization.[35]

These examples illustrate how WTO Member countries may, consistent with WTO disciplines, issue remedies or take other measures to address IP-related unfair competition concerns that might arise in the standard-setting context. The examples also serve to show how the TRIPS Agreement regime for intellectual property cannot be considered in isolation from other national policies, such as competition and development policies, but must be integrated into such policies.[36] While the TRIPS Agreement contains several provisions that can be interpreted to hold relevance for standards-setting in cases of potential abusive use of IP rights, the Agreement nevertheless does not directly or expressly address standards themselves. Instead, the WTO's TBT Agreement has detailed provisions governing standards, and this is where my proposal for reform, outlined below, directs its focus.

Finally, viewing these complex issues from the perspective of developing countries, they need to gain understanding of how to participate in standard-setting processes, just as understanding is needed to effectively harness use of the intellectual property system. Moreover, from the IP perspective, developing countries can consider drawing on the flexibilities that have been built into the TRIPS Agreement, where appropriate, in response to standards that might reflect an abusive use of intellectual property rights. Professor John Barton addressed standards and the issues posed for developing countries in his 2007 paper, *New Trends in Technology Transfer: Implications for National and International Policy.* Considering patents and the potentially high royalties that might be required to access the technology they cover, he suggests that '[o]ne possible response for these firms and nations is to become important enough in the product development process that they can set a standard of their own'.[37] He further emphasizes that

> [t]he best long-term strategy is therefore to encourage firms [in developing countries] to become strong enough that they will hold intellectual property rights on aspects of the newest technologies and have a say in setting the global standard, so that they can become royalty recipients rather than royalty payers.[38]

34. FTC Docket No. 9302 (2 Aug. 2006) (commission opinion).
35. *Rambus, Inc. v. Federal Trade Commission*, No. 07-1086 (D.C. Cir. 2008), at 4–5.
36. The Objectives and Principles of the TRIPS Agreement, fn. 26 *supra*, 33.
37. New Trends in Technology Transfer, fn. 1 *supra*, 29.
38. *Ibid.*

Professor Barton has identified a preferred path for stakeholders in developing and least developed countries as they seek to engage in world markets that are globalized and increasingly technological. In the next section, I propose reform of certain standard-setting rules within the WTO framework that should facilitate all stakeholders having a 'fair say' in setting global standards.

2 TBT AGREEMENT AND PROPOSAL FOR REFORM

The WTO's TBT Agreement recognizes the important contribution that international standards can make to trade, investment and technology transfer, but also cautions that standards may be used to create trade barriers.[39] The TBT Agreement in its Preamble identifies several key objectives, including (i) encouraging international standards and conformity assessment systems due to the important contribution they can make by improving efficiency of production and facilitating international trade, and (ii) ensuring that technical regulations and standards, and procedures for assessing conformity with them, do not create unnecessary obstacles to trade.[40] The Preamble also recognizes 'the contribution which international standardization can make to the transfer of technology from developed to developing countries', while acknowledging that developing countries may encounter difficulties and require assistance in the development and application of technical regulations, standards and procedures for conformity assessment.[41]

The TBT Agreement focuses on standards and technical regulations: thus, only measures that fall within the scope of the definitions of a standard or technical regulation are subject to the Agreement's disciplines.[42] The TBT Agreement defines 'standard' as a '[d]ocument approved by a recognized body, that provides, for common and repeated use, rules, guidelines or characteristics for products or related processes and production methods, *with which compliance is not mandatory*.'[43] When the specification is intended to be voluntary in implementation, the term 'standard' is used, whereas specifications with regulatory force (i.e. where compliance is mandatory) are referred to as 'technical regulations'.[44] 'Most of the principles applied by the TBT Agreement to technical regulations also apply

39. TBT Agreement, fn. 2 *supra*, pmbl., paras 2 & 4. The TBT Agreement, unlike its predecessor the Standards Code, is fully integrated into the WTO system with provisions that bind all WTO members and disputes arising under it governed by WTO dispute settlement rules. See WTO Secretariat, *Guide to the Uruguay Round Agreements*, 71.
40. *Ibid.*, pmbl., paras 1 & 3.
41. *Ibid.*
42. Jan M. McDonald, 'Domestic Regulation, International Standards, and Technical Barriers to Trade', 4:2 World Trade Rev 249, 252 (2005) ('Domestic Regulation, International Standards, and Technical Barriers to Trade').
43. *Ibid.* Annex 1 (Terms and Their Definitions for the Purpose of This Agreement), para. 2 (emphasis added).
44. *Ibid.*, para. 1.

to voluntary standards, which are covered by the Code of Good Practice for the Preparation, Adoption and Application of Standards (Annex 3 of the TBT Agreement).'[45] While the underlying principles for the preparation and adoption of technical regulations and standards are the same, the former are normally issued by a governmental body and require mandatory compliance, while the latter are developed by diverse standard-setting bodies for voluntary adoption.[46]

The TBT Agreement emphasizes the international harmonization of standards as a means to facilitate trade by creating conditions for economies of scale, while avoiding unnecessary obstacles to trade and preventing the use of standards for protectionist motives. The Agreement contains a number of provisions to encourage harmonization of technical regulations, standards and conformity assessment procedures. While WTO members are free to decline to adopt any standard at all in a given instance,[47] when they do so they should use relevant international standards as a basis for their technical regulations and national standards, except when an international standard would be ineffective for achieving the legitimate objectives pursued.[48] The TBT Agreement extends this responsibility to central and local governments and non-governmental bodies, although with different levels of required adherence.[49]

Article 2.4 of the TBT Agreement is central to these harmonizing objectives, providing that:

> Where *technical regulations* are required and relevant international standards exist or their completion is imminent, Members *shall* use them, or the relevant parts of them, *as a basis* for their technical regulations except when such international standards or the relevant parts would be ineffective or inappropriate means for the fulfillment of the legitimate objectives pursued, for instance because of fundamental climatic or geographical factors or fundamental technological problems.[50]

45. World Trade Report 2005, fn. 7 *supra*, 131.
46. See Globalization and the Technology Standards Game, fn. 3 *supra*, 1454.
47. Domestic Regulation, International Standards, and Technical Barriers to Trade, fn. 42 *supra*, 260. Jan McDonald comments that:

> Members that have technical regulations in place may have to modify them to fit international standards, while other Members are free to ignore international measures and to keep domestic standards low or non-existent. Indeed, it appears that a Member that has declined to adopt any standard, even the relevant international standard, is still free to challenge the technical regulation of another Member that is more restrictive than the international standard.

Ibid.

48. TBT Agreement, fn. 2 *supra*, Arts 2.4, 4, & Annex 3, para. F.
49. *Ibid.*, Arts 2.4, 3.1 & 4.1 & Annex 3, para. F. Central governments are required (shall) to use relevant international standards except as provided in the TBT Agreement, whereas for local government and non-governmental bodies within their territories, WTO members 'shall take such reasonable measures as may be available to them to ensure compliance' with the duty to use international standard. *Ibid.*, Art. 3.1 (emphasis added).
50. *Ibid.*, Art. 2.4 (emphasis added).

Article 2.4 can be interpreted to mean that where a relevant international standard exists and the standard would be effective in accomplishing a government's legitimate regulatory objectives, the international standard *must* be used when the government acts by implementing a technical regulation to achieve its ends. For standards, the Code of Good Practice imposes a similar obligation on member nations as to their central government standardizing bodies.[51] As an incentive for compliance, the Agreement creates the rebuttable presumption that technical regulations prepared in accordance with relevant international standards will not be considered to create an unnecessary obstacle to trade.[52]

In a dispute over non-compliance with the TBT Agreement's obligations to use international standards as a basis for either technical regulations or standards, the burden of proof would fall on the complaining party. That party would bear the burden of proving that: (i) a relevant international standard had not been used as a basis for the national technical regulation or standard; and (ii) the international standard is effective and appropriate to fulfil the legitimate objectives pursued by the national authorities through the regulation or standard.[53] This two-part inquiry can actually be broken into five detailed elements to be considered. First, as a threshold question, is the point of whether the specification in dispute should be considered a 'technical regulation' or 'standard' under the TBT Agreement.[54] Second, a determination must be made as to whether the specification should be considered an 'international standard', a point to which I will return below. Third, an assessment should be made of whether the international standard is 'relevant'. Fourth, is the question of whether the national government involved used the international standard 'as a basis' for the national standard or technical regulation?

---

51. *Ibid.*, Art. 4 & Annex 3, para. F. WTO Members shall ensure that their central government standardizing bodies comply with the Code of Good Practice, which provides in para. F, similarly, that where relevant international standards exist (or their completion is imminent), standardizing bodies shall use them as a basis for the standards it develops, subject to terms virtually identical to those in Art. 2.4 above. *Ibid.* (emphasis added). As noted with technical regulations, WTO members are also required to 'take reasonable measures as may be available to them to ensure' that their local government and non-governmental bodies comply with these terms, including accepting and complying with the Code of Good Practice, Arts 3.1 & 4.1.
52. TBT Agreement, fn. 2 *supra*, Art. 2.5. Art. 2.5 provides in relevant part that '[w]henever a technical regulation is prepared, adopted or applied for one of the legitimate objectives explicitly mentioned in paragraph 2, and is in accordance with relevant international standards, it shall be rebuttably presumed not to create an unnecessary obstacle to international trade'. (Emphasis added).
53. Report of the Appellate Body, European Communities – Trade Description of Sardines, WT/DS231/AB/R, paras. 275 & 282 (26 Sep. 2001) ('EC-Sardines').
54. The Appellate Body interpreted the definition of 'technical regulation' in the EC-Asbestos case, and then applied it in the EC-Sardines case. There are three criteria to be satisfied: (i) the regulation must apply to an identifiable product or group of products; (ii) the regulation must lay down one or more characteristics of the product; and (iii) compliance with the product characteristics must be mandatory. Appellate Body Report, European Communities – Measures Affecting Asbestos and Asbestos-Containing Products, paras 80 WT/DS135/AB/R (adopted 5 Apr. 2001) ('EC-Asbestos'); EC-Sardines, fn. 53 *supra*.

And fifth, we can consider whether the international standard is effective and appropriate to fulfil the legitimate objectives pursued by the national authorities, and the related question of whether there are any 'fundamental technological problems' that would prevent adoption of the international standard.

In sum, as a means of harmonization, the TBT Agreement requires that, where relevant international standards exist (and subject to certain exceptions), governments who choose to adopt a technical regulation must use the international standard as a basis for any mandatory national standards.[55] In view of the responsibilities concerning use of 'international standards', it is noteworthy that this term is not defined in the TBT Agreement, although as discussed below, the WTO's Committee on Technical Barriers to Trade (CTBT)[56] has established a set of principles that improve on this situation by defining characteristics that should be integral to the process of developing an international standard. Moreover, if and when this becomes a potential point of contention, the WTO dispute settlement mechanism, as indicated below, should be in position to make a competent decision.

This TBT Agreement framework, coupled with the commercial reality that particular standards can become locked-in and therefore practicably indispensable in certain technology areas in order to achieve compatibility or interoperability, can limit Member countries' choices. The claim shared by certain WTO member countries, consequently, is that the mandatory adoption of international standards may come at a potentially significant cost.[57] Companies from developing and least developed countries may sometimes find it burdensome to meet the royalty payments charged by patent owners whose technologies are embedded in important technology standards. Some also complain of unfair treatment when seeking to participate in the international standards system, suggesting that IP rights create obstacles by hindering access to new technologies and imposing significant costs in the form of royalty payments. For example, in a formal communication to the WTO submitted 23 May 2005, the Chinese government stated:

> China is of the view that, IPR issues in preparing and adopting international standards have become an obstacle for Members to adopt international standards and facilitate international trade. It is necessary for the WTO to consider

55. *Ibid.*, Art. 2.4.
56. The CTBT oversees the TBT Agreement with a membership comprised of WTO member country representatives. The CTBT has responsibility for an on-going triennial review process and reviews accession commitments under the TBT Agreement for new WTO members.
57. For instance, in an April 2007 speech by Yi Xiaozhun, the vice-minister of China's Ministry of Commerce, he stated that '[d]elayed or inadequate IPR . . . disclosure, stringent IPR licensing conditions and expensive licensing fees run counter to fair competition, hinder the promotion and application of new technologies, obstruct the normal operation of international trade and impede the harmonious development of global economy and society.' (Reported by William New, 'China Leads Developing Country Push for Balance in IP and Standards', IPWATCH 1 (24 Apr. 2007), <www.ip-watch.org/weblog/2007/04/24/china-leads-developing-country-push-for-balance-in-ip-and-standards/>.

> negative impacts of this issue on multilateral trade and explore appropriate trade policies to resolve difficulties arising from this issue.[58]

It is clear that frictions between countries can develop, particularly at the international level, given the important role of intellectual property rights in technology standard-setting.[59] Despite the potential concerns, however, the TBT Agreement is silent on the issue of IP rights in international standard-setting, just as the TRIPS Agreement is silent on the issue of standards. For example, the qualified obligation under the TBT Agreement to use relevant international standards as a basis for national standards is not conditioned on whether a particular standard is encumbered by IP rights and therefore could be more costly to implement. Without a thoughtful and integrated approach on this issue, developing countries may build up national resentment towards the need to pay royalties to foreign patent owners in order to manufacture products complying with international standards set elsewhere.

Indeed, 'an uneasy tension exists between trade law obligations to use relevant international standards and the possibility that such adoption will entail payment of royalties to foreign IP owners'.[60] How do we balance the legitimate rights of IP owners to receive compensation, on the one hand, and the interests of those countries (and their constituents) pursuing technology transfer who are encouraged to pursue harmonization by implementing international standards, on the other hand? If the objective of the TBT Agreement is to use international standards as a means to overcome national trade barriers and facilitate increased technology transfer and trade, then additional consideration needs to be given to the role of intellectual property as it relates to international standards and, in particular, as an

---

58. WTO Committee on Technical Barriers to Trade, Communication from the People's Republic of China dated 23 May 2005 Communication from the People's Republic of China, G/TBT/W/251 (25 May 2005) para. 6.
59. This was the case when, in an unprecedented high-level letter of 15 Mar. 2004 to China's Vice Premier, Zeng Peiyan, three cabinet-level United States representatives – US Trade Representative Robert B. Zoellick, Secretary of Commerce Donald I. Evans, and Secretary of State Colin L. Powell – asserted the United States' objection to China's proposed national wireless technology standard, which differed from the existing international standard:

    > China would be the only country in the world mandating a specific encryption standard for general consumer use. Since this standard is unique to China, implementation will impose a significant new burden on both foreign and Chinese domestic suppliers. Furthermore, implementation of this standard will make Chinese products incompatible with internationally-accepted standards, isolating China from the larger world market.... We are particularly concerned that the new rules would require foreign suppliers to enter into joint ventures with Chinese companies and transfer technology to them. Such compelled investment and technology transfer would appear to be inconsistent with China's WTO commitments.

    The full text of the 15 Mar. 2004 letter was published by BUSINESS WEEK <www.businessweek.com/magazine/content/04_11/b3874018.htm>. The United States was concerned about the precedent that might be set if China was permitted to enforce mandatory national standards differing from the international standard in this high technology sector.
60. See Globalization and the Technology Standards Game, fn. 3 *supra*, 1404–1405.

element to be expressly factored into the legal framework provided by the TBT Agreement.[61]

My proposal is that to balance and achieve these objectives, the approach for legal recognition of international standards *within* the WTO regime should be *revised* to integrate an express policy toward intellectual property rights. The two areas, standards and intellectual property, are closely related and thus enhanced coordination is now required.[62] As discussed above, thoughtful rules governing IP rights have been developed by many of the diverse standard-setting bodies. I propose that it is time to integrate these well-established rules formally into the TBT Agreement framework.[63] The TBT Agreement addresses itself to the details of how standards and technical regulations should be 'prepared, adopted and applied.'[64] In this regard, how the standards development process deals with intellectual property rights should inform the assessment of whether a given standard may achieve recognition as an 'international standard', such that the TBT Agreement disciplines for international standards to apply.

The TBT Agreement, however, does not address the issue of IP rights in standard-setting. In this sense, there is an unhelpful 'disconnect' between the TBT Agreement's qualified mandate to use international standards and the reality that IP rights may be embedded in those standards and generate the concerns discussed above. In particular, while the obligation to use relevant international standards as a basis for standards arises under the TBT Agreement, the Agreement does not consider whether a particular standard is encumbered by IP rights and therefore could be more costly to implement. I discussed above the draconian remedy of a compulsory license that could be exercised in exceptional cases where there is a clear and anti-competitive abuse of IP rights. However, my position and proposal herein does not seek to diminish the recognition and enforcement of IP rights that may be embedded in technology standards that are developed through robust and transparent standard-setting processes; rather, I recommend that the rules established under the TBT Agreement should encourage a process of standards development that addresses relevant intellectual property issues in an open and transparent manner, so that resulting international standards, even if encumbered by intellectual property rights, have greater legitimacy. This can only happen by encouraging a process that requires disclosure of IP rights during the development of a standard in order to facilitate informed choice, and greater participation by all stakeholders in the process, even those from developing and least developed countries.

The WTO's CTBT adopted, as a central feature of its second triennial review of the TBT Agreement, the *Decision of the Committee on Principles for the*

61. *Ibid.*, 1405.
62. Indeed, there should be more coordination within the WTO at the working level between the TRIPS Council and the Committee on Technical Barriers to Trade.
63. I first made this proposal in Globalization and the Technology Standards Game, fn. 3 *supra*, 1479.
64. Reference to the preparation, adoption and application of technical regulations and standards is made throughout the TBT Agreement.

Development of International Standards, Guides, and Recommendations with Relation to Articles 2, 5, and Annex 3 of the Agreement (CTBT Principles).[65] The CTBT Principles address a comprehensive array of issues underpinning *fairness* in international standard-setting, including transparency, openness, impartiality and consensus, effectiveness and relevance, coherence, and measures to foster developing country participation. The CTBT provided in its Decision adopting the Principles that they 'should be observed' when international standards, guides, and recommendations are developed.[66] The Principles were established – in the absence of a definition – to clarify and strengthen the concept of the international standard. They are intended 'to improve the quality of international standards and to ensure effective application of the Agreement'.[67] Jan McDonald has put it another way: the CTBT has 'formulated criteria by which to determine whether an international standard can be used for [TBT Agreement] compliance'.[68] The logical extension of his view would appear to be that if a standard does not comply with the CTBT Principles, it should not be considered an *international standard* for purposes of Article 2.4 or Annex 3.[69]

---

65. CTBT, Decision of the Committee on Principles for the Development of International Standards, Guides, and Recommendations with Relation to Arts 2, 5 and Annex 3 of the Agreement, Annex 4, Second Triennial Review of the Operation and Implementation of the Agreement on Technical Barriers to Trade, G/TBT/9 (13 Nov. 2000).
66. *Ibid.*, para. 1.
67. *Ibid.* WTO members were concerned that difficulties might be encountered in relation to the use of certain international standards, and related trade problems could arise. In particular, '[a]dverse trade effects might arise from standards emanating from international bodies as defined in the Agreement which had no procedures for soliciting input from a wide range of interests.' *Ibid.*, para. 20. The Committee was also aware of the diversity of standards bodies and different approaches for standard-setting, yet observed that 'the obligation under the Agreement for Members to use international standards was the same.' *Ibid.*, para. 19.
68. Domestic Regulation, International Standards, and Technical Barriers to Trade, fn. 47 *supra*, 268.
69. There is debate about what constitutes an international standard. While the term is not defined in the TBT Agreement, an international standard is defined by the International Organization for Standardization (ISO) as 'a standard that is adopted by an international standardizing/standards organization', and an 'international standards organization' is 'one whose membership is open to the relevant national body from every country.' See ISO Guide 2, ch. 3.2.1.1.

    The President of the American Society for Testing and Materials (ASTM), an influential American standards body, offered a different definition of international standard, focusing on the importance of the process, as well as global acceptance and use by industry. He considered:

    > . . . a true international standard [to be] a standard that . . . [is] developed by a standardisation body . . . that is open to all interested parties regardless of national origin . . . has a demonstrated track record of global acceptance and use by the affected industries or regulatory bodies of various nations . . . is non-discriminatory by providing equal treatment . . . without favouring one company or nation over another.

    OECD, fn. 13 *supra*, 35 (quoting ASTM Standardization News (December 1997)). This statement tracks some of the principles in the CTBT Principles. My position, like the OECD, is to avoid adopting a narrow definition, and instead focus on the process for standard-setting as an important element, a process that includes recognized principles for dealing with IP rights.

For all of the wide-ranging principles touched upon relating to the standard-setting process, the CTBT Principles, like the TBT Agreement itself, are silent on the issue of intellectual property. My proposal is that the CTBT Principles should be amended to address expressly the treatment of IP rights (in particular, patents) in international standard-setting. Stated most broadly, the balance to be achieved is that standard-setting processes should respect the rights of IP owners to receive reasonable compensation, while ensuring that those seeking to implement standards can make an informed choice during the standards development process about whether they want to approve a standard encumbered by IP rights. This choice, in turn, may depend upon whether there is an assurance of access – on fair terms – to the IP rights incorporated in standards.[70]

My recommendation is that two basic and well-established principles – (i) early disclosure of IP rights (e.g., patent rights) in the standard-setting process, and (ii) declaration of position concerning willingness to license those rights – should form a part of the CTBT Principles governing the standards development cycle. Integrating these two principles into the rules established under the TBT Agreement will bring together the subjects of intellectual property and standard-setting within the WTO system, establishing an influential benchmark for Member countries. It would also reflect the recently announced unified IP policy of three prominent international standard-setting bodies, ISO, the International Electrotechnical Committee (IEC), and the International Telecommunication Union (ITU). These organizations joined together to form the World Standards Cooperation in 2001, with aims of promoting 'international consensus-based standardization worldwide' and undertaking initiatives to resolve issues regarding cooperation in the technical work of the organizations.[71] In March 2007, they announced the adoption of a harmonized approach to address the inclusion of IP rights (patents) in standards.[72] Their unified policy, the *Common Patent Policy for ITU-T/ITUR/ISO/IEC* (Common Patent Policy),[73] draws on the same two elements listed above, which have been present in their IP policies and the policies of many other standards organizations.

The Common Patent Policy first encourages the *disclosure* of patented technology necessary for the implementation of a standard before the standardization

70. This balanced objective is similar to one stated among the 'imperatives for action' in the American National Standards Institute's (ANSI) United States Standards Strategy: 'The standardization process must respect the rights of intellectual property owners while ensuring users have access to the intellectual property rights (IPR) incorporated into standards.' See Am Nat'l Standards Inst, United States Standards Strategy (2005), 5, <www.ansi.org/standards_activities/nss/usss.aspx?menuid=3#overview>.
71. ISO, World Standards Cooperation for the Information Society, ISO BULL., Comment (October 2003).
72. See Press Release, Int'l Telecom. Union, IEC, ISO, ITU Agree on Common Patent Policy (23 Mar. 2007), <www.itu.int/>.
73. The IEC, ISO, and ITU also jointly adopted Guidelines for the Implementation of the Common Patent Policy and a Patent Statement and License Declaration Form, <www.itu.int/dms_pub/itu-t/oth/04/04/T04040000010002PDFE.pdf>. (Which are intended to clarify and facilitate implementation of the Common Patent Policy.)

process is completed. In particular, 'any party participating in the work of ITU, ISO, or IEC should, from the outset, draw [their] attention . . . to any known patent or any known pending application, either their own or of other organizations'.[74] The Guidelines to the Policy explain that the words 'from the outset' imply that the disclosure should occur as early as possible during the development cycle for the relevant standard, taking into account that early drafts of a particular standard might be too vague to make this possible.[75] As to any doubts about the duty of care involved, the Guidelines provide that the 'information should be provided in good faith and on a best efforts basis, but there is no requirement for patent searches'.[76] Early disclosure facilitates an informed choice by organization members, based on technical and commercial considerations, of the technologies to be used in the standard.

The second aspect of the Common Patent Policy requires a holder of patents relevant to the standard to *make a licensing declaration* in relation to its disclosure of patent rights, choosing from among three basic alternatives. The patent holder can declare that it is:

(i) willing to negotiate licenses free of charge with other parties on a non-discriminatory basis on reasonable terms and conditions;
(ii) willing to negotiate licenses with other parties on a non-discriminatory basis on reasonable terms and conditions; or
(iii) unwilling to comply with the terms of paragraphs (i) or (ii), and in such case, the standard in question shall not include provisions dependent on the patent.[77]

The statement of licensing position should be made as early as possible in the standards development cycle. The Common Patent Policy has no requirement of compulsory licensing; such a stance would undermine incentives to participate in standard-setting activities.[78] The Policy makes it clear that if a patent holder chooses not to make its patent available for licensing, then the standard in question must not include specifications dependent on that patent.

74. *Ibid.*, para. 1.
75. Guidelines, fn. 73 *supra*, 3.
76. *Ibid.*
77. Common Patent Policy, fn. 72 *supra*, para. 2.
78. It is notable that in addition to Art. 31, Art. 40 of the TRIPS Agreement addresses licensing practices and therefore could have some relevance for standard-setting where IP issues arise. It provides in relevant part as follows:

(i) Members agree that some licensing practices or conditions pertaining to intellectual property rights which restrain competition may have adverse effects on trade and may impede the transfer and dissemination of technology.
(ii) Nothing in this Agreement shall prevent Members from specifying in their legislation licensing practices or conditions that may in particular cases constitute an abuse of intellectual property rights having an adverse effect on competition in the relevant market. As provided above, a Member may adopt, consistently with other provisions of this Agreement, appropriate measure to prevent or control such practices. . . .

TRIPS Agreement, Art. 40. <www.wto.org/english/tratop_e/trips_e/t_agm3d_e.htm>.

Making these changes will not undermine the interests of those seeking reasonable compensation for their IP rights. IP rights will be properly disclosed along with any attached conditions pertinent to licensing. Those involved in standard-setting will be able to evaluate all relevant commercial and technical considerations. The relative simplicity of focusing on these two important elements makes this approach suitable in the context of internationally diverse standard-setting systems, avoiding concerns of over-constraining regulation. This approach advances the accessibility and harmonization of standards. A draft text of my proposal is included below.

2.1 Proposal for Amendment of CTBT Principles

The CTBT Principles should be amended to add a new paragraph G, such as the following draft text:

G. Intellectual Property Rights

14. Intellectual property rights can be implicated in international standardization activity, particularly in the area of high technology. To encourage the accessibility and harmonization of international standards without undue constraints, the incentives for innovation provided by intellectual property rights should work together with making intellectual property resources available for broad implementation. In order to achieve this balance, standardizing bodies of WTO members should adopt policies to address intellectual property rights (in particular, patents) to encourage the early disclosure of any patent that may relate to standard(s) under development and a statement by any concerned patent holder as to its licensing position.

15. Accordingly, the intellectual property policy should, at a minimum, include requirements that:

- Any party participating in the work of an international standardizing body should, from the outset, draw attention to any known patent or any known pending application, whether its own or belonging to other organizations, that may relate to the standard(s) under development. This disclosure should occur as early as possible during the development cycle for the relevant standard, taking into account that early drafts for particular standards might be too vague to make this possible.
- A patent holder who has disclosed a patent under the terms above should make a licensing declaration, choosing from among the following three alternatives. The patent holder is:
  - (i) willing to negotiate licenses free of charge with other parties on a non-discriminatory basis on reasonable terms and conditions;
  - (ii) willing to negotiate licenses with other parties on a non-discriminatory basis on reasonable terms and conditions; or
  - (iii) unwilling to comply with the terms of paragraphs (i) or (ii), and in such case, the standard in question shall not include provisions dependent on the patent.

16. Information should be provided in good faith and on a best efforts basis.

Early disclosure of IP rights and a declaration concerning licensing position will facilitate informed choices by standard-setting bodies, so that the impact of any IP rights on technology selection can be evaluated in advance, thereby advancing the accessibility and harmonization of standards. The standards body can make a choice to use a standard that may be encumbered by IP rights, or seek to move the standard in a direction where such IP rights will not be implicated. The new policy incorporated into the WTO framework would thus send a strong signal to SSOs that do not already have such rules in place to address IP rights. At the same time, it would also send a message to developed and developing countries, and countries such as China, reinforcing their need to participate more fully in global standard-setting activities, rather than object regarding the burden of IP rights in existing international standards. In due time, a policy such as that outlined above, along with the other CTBT Principles including transparency, openness, impartiality and consensus, effectiveness and relevance, coherence, and new measures to foster developing country participation, should be incorporated into an amended text of the TBT Agreement itself. Bringing these CTBT revisions into the Agreement's terms will ensure that they are given the serious and full attention they deserve, thereby upholding Member countries to high standards of conduct in standard-setting activities, rather than the inadequate consideration that was given to such principles by the Appellate Body as described in the *EC-Sardines* case below.

A system that requires the adoption of existing and relevant international standards rests upon the legitimacy of the international standards themselves. A robust policy for dealing with IP rights in the standards development process is an important element to support this legitimacy. If the standards process is carried out properly – that is, in accordance with those relevant principles in the TBT Agreement, the Code of Good Practice and the CTBT Principles (amended to include a policy for intellectual property rights) – it becomes much more difficult to challenge the legitimacy of an international standard on the basis of IP rights or other grounds.

These changes will also bring progress for developing and least developed countries. As they grow increasingly sophisticated in their understanding of the standards-setting system, they will have an opportunity along with all other WTO Members to participate in standard-setting processes and make better informed choices about standards that may eventually be recognized as international standards. With participation in the standard-setting process and an understanding of any relevant IP issues, the standards that might apply in relevant technology sectors will be those that have grown out of an informed, transparent and inclusive decision-making process.

2.4 Dispute Settlement

Finally, one can ask what would be the consequences if a standard was not adopted through a transparent, open and impartial process, or if IP rights were somehow 'gamed' during the standard-setting process as in the *Dell* case discussed above. The qualified obligation in the TBT Agreement to adopt the relevant international standard could become more problematic. A WTO Member country could potentially claim exclusion from the standard-setting process as a basis initially for voting against international recognition, or later possibly as a reason for refusing to follow an international standard. Similarly, failure to follow certain rules concerning intellectual property rights, such as timely disclosure of patents, might provide a basis for questioning the legitimacy of a proposed international standard.

In disputes between private parties concerning alleged abuse in the standard-setting context, recourse can be made to the national courts or to an appropriate administrative agency (such as the United States Federal Trade Commission), with remedies such as damages in breach of contract cases or even compulsory licensing in extreme unfair competition cases finding justification under WTO rules. However, governmental actions concerning standards implementation will need to be assessed against WTO disciplines, with disputes heard in the WTO's dispute settlement system. If the question arises, for example, whether a given standard encumbered by IP rights should be considered as a 'relevant international standard' in light of the way in which IP issues were handled during the standards development process, this question might go before a WTO panel.

The *EC-Sardines* case focused on the status of the standard in dispute. In that case, the European Union argued that the standard in issue, Codex Stan 94,[79] should not be considered a 'relevant international standard' because it was not adopted by consensus.[80] The CTBT Principles include 'consensus' as one of the principles to be observed in international standard setting.[81] However, the Appellate Body, without reference to the CTBT Principles, (the CTBT Principles had just been issued), found that 'in our view, the European Communities' contention is essentially related to whether Codex Stan 94 meets the definition of a "standard" in Annex 1.2 of the TBT Agreement.'[82] As noted above, the TBT Agreement defines 'standard' as a '[d]ocument approved by a recognized body that provides ... rules, guidelines or characteristics for products or related processes and production methods.'[83] Reviewing this definition, the Appellate Body concluded that '[t]he issue before us ... is one of approval',[84] determining that the definition does not require approval by consensus, but rather a standard must only be approved by a 'recognized body'.[85] The Appellate Body in *EC-Sardines* also

79. In 1978 (and in 1995) the Codex Alimentarius Commission adopted a worldwide standard for canned sardines and sardine-type products: Codex Stan 94-1981, Rev. 1-1995 ('Codex Stan 94').
80. EC-Sardines, fn. 53 *supra*, para. 219.
81. Paragraph 8 of the CTBT Principles provides in relevant part that '[c]onsensus procedures should be established that seek to take into account the views of all parties concerned and to reconcile any conflicting arguments.' CTBT fn. 56 *supra*, para. 8.
82. EC-Sardines, fn. 53 *supra*, para. 220.
83. TBT Agreement, Annex 1 para. 2.
84. EC-Sardines, fn. 53 *supra*, para. 221.
85. *Ibid.*, para. 227.

discussed what it means to be a 'relevant' international standard: to be relevant, the international standard must 'bear upon, relate to, or be pertinent to' the national standard in question.[86]

The decision of the Appellate Body in the *EC-Sardines* case takes a narrow reading of the requirements to qualify as an 'international standard', without considering the CTBT Principles or other criteria that focus on the standards development process. While the Appellate Body's reading is arguably correct on the definition of the term 'standard', as set forth in Annex 1.2 of the TBT Agreement, it does appear to given any separate weight to the complete phrase, 'international standard'. Moreover, the Appellate Body does not appear to be eager to look inside the standards development process used by a particular 'recognized body'. However, given the purposes of the TBT Agreement, as discussed above, including encouraging international standards to promote technology transfer and trade while ensuring that technical regulations and standards do not create unnecessary obstacles, I would argue that a more broadly grounded interpretation is needed in order to consider whether any particular standard, when challenged, should qualify as achieving the status of an international standard. When a term such as 'international standard' is not defined in the TBT Agreement itself, the CTBT Principles should provide relevant and helpful guidance by directing attention to the *process for the development of the standard* as a key element to be considered in qualification. The process of standards development, whether in dimensions relating to IP issues or some other potential concern (e.g., inclusiveness), is especially important when applicable trade rules contain duties (albeit qualified) for WTO Members, including developed and developing countries, to use international standards as a means of harmonization. In view of all of the above considerations, my call for reform is that the CTBT Principles eventually be incorporated into the text of the TBT Agreement.

3 ADVISORY FACILITY ON STANDARD-SETTING FOR DEVELOPING AND LEAST DEVELOPED COUNTRIES

Finally, focusing on the specialized concerns of developing and least developed countries, technology standard-setting provides an example that requires understanding not only in the national context, but also in the international context where collaboration and balancing of rights is most needed. The players in technology standard-setting are largely from developed countries; however, there is a reason for this. Like the field of intellectual property in general, one of the realities that prevents

86. *Ibid.*, paras 228–232. The meaning of the phrase 'as a basis for' was also addressed in EC-Sardines. The Appellate Body, upon reviewing several dictionary definitions and its prior decision in EC-Hormones, reasoned that while 'based on' does not mean the same thing as 'conform to', nevertheless, 'there must be a very strong and very close relationship between two things' such that the international standard is the 'principal constituent', 'fundamental principle', or 'determining principle' of the national measure. *Ibid.*, paras 242, 245.

more widespread participation from developing or least developed countries is the lack of expertise or capacity to engage more fully in the process. They end up being marginalized. If developing countries were brought into standards-setting activities at an earlier stage, there would be obvious benefits, including training, development of expertise, early access, and eventually increased technical and strategic influence.[87]

Just as Article 66.2 of the TRIPS Agreement calls for special measures by WTO Members to encourage technology transfer to least developed country, so too the TBT Agreement contains provisions addressing the concerns of developing and least developed countries in the context of standards. Article 11.1 provides that WTO Members are required, if requested, to provide advice to developing country Members on the preparation of standards, while Article 11.2 similarly provides that Members, if requested, should provide developing country Members with technical assistance regarding the establishment of national standards bodies and participation in international standards bodies.[88] Article 12 of the Agreement recognizes that the WTO will provide differential and more favourable treatment to developing country Members, while taking into account their special development, financial and trade needs.[89] Moreover, reasonable measures are to be taken to ensure that international standardizing bodies are organized and operated in a way that facilitates active and representative participation by developing country Members, including making efforts to prepare international standards concerning products of special interest to developing country Members.[90] While the TBT Agreement creates a duty on WTO members to participate in international standard-setting activities (within the limits of their resources),[91] Article 12 further recognizes that the institutional and infrastructural problems facing developing country Members, as well as their stage of technological development, may hinder their ability to fully discharge their obligations under the TBT Agreement. In this respect, the CTBT is permitted to grant exceptions from obligations of Agreement.[92] Under both Articles 11 and 12, WTO Members are required to give priority especially to the needs of least-developed countries.[93] All of these measures, by increasing the capacities of developing and least developed country Members, will serve to facilitate technology transfer by promoting participation in the standard-setting system.

---

87. See Andrew Updegrove, 'Participating In Standard Setting Organizations: Value Propositions, Roles and Strategies', in The Essential Guide to Standards, Ch. 2. <www.consortiuminfo.org/essentialguide/>.
88. TBT Agreement, Art. 11.1 and 11.2.
89. *Ibid.*, Art. 12.1.
90. *Ibid.*, Art. 12.2 and 12.5.
91. *Ibid.*, Art. 2.6 & Annex 3, para. G.
92. *Ibid.*, Art. 12.8. The CTBT Principles refer to the 'development dimension', recognizing that constraints on developing countries to participate effectively should be taken into account in the standards development process. More specifically, '[t]angible ways of facilitating developing countries' participation in international standards development should be sought'. The focus under the Principles is on improving developing countries' participation in international activities, not granting them grounds for an exception to use discriminatory standards development processes at home.
93. *Ibid.*, Arts 11.8 and 12.8.

My proposal is intended to supplement these provisions in the TBT Agreement – it is modest yet concrete: to establish an Advisory Facility on Standard-Setting for Developing and Least Developed Countries. An Advisory Facility that is adequately funded can work with developing and least developed countries as they participate in specific standard-setting activities. The Advisory Facility should be available to provided case-specific advice as well as general training on both technical and legal issues. Similar advisory facilities have been established in other areas such as international trade,[94] while proposals have been launched for such facilities in the related area of international investment law and investor-State investment arbitration.[95] Moreover, a general mandate under the TBT Agreement already exists: Articles 11 and 12, as noted above, provide that WTO Members must provide technical assistance to developing countries and grant them differential and more favourable treatment. In addition, the International Organization for Standardization provides, in its Code of Ethics, that 'ISO parties shall contribute to actions to help its members from developing countries improve their capacity and their participation in international standardization'.[96]

An Advisory Facility would provide a focal point for funnelling the energies of WTO Members to provide valuable assistance to developing and least developed countries in the area of technology standard-setting. For all of the reasons discussed above, such a concrete step will ultimately yield results that support sustainable technology transfer.

4 CONCLUSION

In sum, returning to a theme of the conference in Ho Chi Minh City, we can ask what does sustainability mean for technology transfer. Do we, from a legal perspective, have the right legal mechanisms and institutions to attain the soft obligations set out in the TRIPS Agreement and other international conventions? In this paper, I argue that the standard-setting field is an area where both the relevant rules and institutional capacities are in need of significant improvement to assist developing and least developed countries reach sustainable technology transfer. I have shifted the focus from the TRIPS Agreement, more commonly the subject of attention when considering technology transfer, to the TBT Agreement and its support for standards as a means of facilitating technology transfer. With a

94. The Advisory Centre on WTO Law (ACWL) is located in Geneva and advises its developing country Members and least developed countries on all issues relating to WTO law. See ACWL website. <www.acwl.ch>. The ACWL gives free legal advice and training on WTO law and provides support in WTO dispute settlement proceedings at discounted rates. These services are available to the developing country Members of the ACWL (30 at present) and to least developed countries that are Members of the WTO or are in the process of acceding to the WTO (43 at present).
95. Inter-American Development Bank, Advisory Facility on Investor – State Dispute Settlement: Plan of Operations (RG-T1515). <www.iadb.org/document.cfm?id=1741656>.
96. See International Organization for Standardization, ISO Code of Ethics (2004). <www.iso.org/iso/codeethics_2004.pdf>.

system that enhances fairness and participation in the development of international standards, particularly where intellectual property rights are at stake, legitimacy in international standards will increase while steps can be taken to enhance the participation of developing and least developed countries. Once these initial steps are taken, the standard-setting system can mature under the discipline of the WTO system, including the dispute settlement mechanism which will hold Member countries to their commitments.

Chapter 8
Technology Transfer and Climate Change

Hari M. Osofsky

Widespread agreement exists that 'environmentally sound technologies'[1] should be transferred from developed to developing countries as part of a strategy for addressing climate change. This transfer, in theory, allows developing countries to focus on economic growth and poverty eradication goals, but in a more environmentally sound fashion that prevents their emissions trajectories from being as greenhouse gas intensive as developed country ones have been. This concept is included in general terms in core international environmental law documents, such as Agenda 21,[2] and with respect to climate change, in the United Nations Framework Convention on Climate Change (UNFCCC) and the agreements negotiated under it.[3] The Copenhagen Accord, for example, addresses technology transfer as a goal with respect to both mitigation and adaptation and establishes a Technology Transfer mechanism.[4]

1. For an in-depth exploration of this term's meaning, see s. 2 *infra*.
2. Agenda 21, s. IV, Ch. 34, U.N. Doc. A/ CONF.151/26 (12 Aug. 1992). <www.un.org/esa/dsd/agenda21/res_agenda21_34.shtml>.
3. United Nations Framework Convention on Climate Change, Art. 4.5, 9 May 1992, S. Treaty Doc. No. 102-38, 1771 U.N.T.S. 164, 166, 170, (UNFCC). <http://untreaty.un.org/English/notpubl/unfccc_eng.pdf>.
4. See Copenhagen Accord of 18 Dec. 2009, Decision -/CP.15. <http://unfccc.int/files/meetings/cop_15/application/pdf/cop15_cph_auv.pdf>.

Hans Henrik Lidgard, Jeffery Atik & Tu Thanh Nguyen, *Sustainable Technology Transfer*, pp. 177–194.

The more recent Cancun Agreements develop that technology mechanism further by establishing a Technology Executive Committee and a Climate Technology Centre and Network as its component parts.[5]

Although many efforts have been made to interpret and implement the language on technology transfer in the UNFCCC, disagreement continues to take place over crucial details of how such technology transfer should take place and the role of law in facilitating those transfers. For example, under the Clean Development Mechanism (CDM), developed countries can get credit for greenhouse gas emissions reductions projects in developing countries, and much controversy has ensued about whether that mechanism is achieving its goals or allowing developed countries to avoid emissions reductions. A 2008 article by Professor Michael Wara examined the nature of CDM projects and found that a substantial percentage of them were not focused on core sustainable energy technology development.[6] Similarly, numerous scholars have examined the interaction between the climate change and intellectual property treaty regimes, and raised concerns over the ways in which the intellectual property regime might hinder needed technology transfer.[7]

This chapter contributes to the dialogue over appropriate legal approaches to climate change technology transfer by providing a law and geography analysis of the technology itself and the way in which it is transferred. It argues that effective technology transfer requires a nuanced analysis of what should be considered environmentally sound in the context of climate change, and how that transfer can be accomplished most effectively in the particular places that receive it. Using the example of technology needs assessment, the chapter argues that more fluid governance models, which are less centred on the nation-state, have the potential to help assist in framing appropriate transfers.

Part II begins by considering definitional difficulties at the heart of the climate change technology transfer regime. While some core technologies are clearly environmentally sound, definitional issues abound at the margins. Especially as new technology with uncertain consequences emerges, key questions arise over what technology should actually be transferred. The part further situates these

5. Outcome of the work of the Ad Hoc Working Group on Long-Term Cooperative Action under the Convention, Draft decision -/CP.16, December 2010 (hereinafter 'Long-Term Cooperative Action Outcome'). (Advanced version). <http://unfccc.int/files/meetings/cop_16/application/pdf/cop16_lca.pdf>. For a broader summary of the outcomes of the Cancun negotiations, see Cesare Romano & Elizabeth Burleson, 'The Cancún Climate Conference', 15 *ASIL Insight*, 1 (2011).
6. Michael Wara, 'Measuring the Clean Development Mechanism's Performance and Potential', 55 *UCLA L. Rev.* 1759 (2008).
7. See e.g., Alexander Adam, 'Technology Transfer to Combat Climate Change: Opportunities and Obligations', 9 *J. High Tech. L.* 1 (2009); Elizabeth Burleson, '*Energy Policy*, Intellectual Property, and Technology Transfer to Address Climate Change', 18 *Transnat'l L. & Contemp. Probs.* 69 (2009); Lorelyn Hall, 'Technology Transfers under the United Nations Framework Convention on Climate Change', 2005 *Colo. J. Int'l. Envtl. L. & Pol'y* 59 (2005); Cameron Hutchison, 'Does TRIPS Facilitate or Impeded Climate Change Technology Transfer into Developing Countries', 3 *U. Ottawa L. & Tech. J.* 517 (2006).

technology-transfer issues amid a myriad of broader issues around defining environmentally sound technology in the context of climate change. For instance, increasing interest in climate change geoengineering, which involves interventions into the climate system intended to reverse climate change or limit its impacts, poses particularly difficult questions of when technology should be considered environmentally sound.[8]

Part III explains that these definitional dilemmas are magnified by the nation-state centred regulatory framework addressing the problem of climate change and the varying needs of places within developing countries. Although the climate change debate has traditionally focused on the international level, an emerging scholarly literature recognizes the crucial role that subnational governments and nongovernmental entities play in climate change regulation.[9] Moreover, the most

---

8. For instance, in their proposal for an IPCC expert meeting on geoengineering, the Co-Chairs of Working Groups I, II, and III indicate that 'geoengineering itself may constitute "dangerous anthropogenic interferences with the climate system"' under Art. 2 of the United Nations Framework Convention on Climate Change. IPCC Secretariat, Thirty-Second Session of the IPCC, Busan, 11–14 Oct. 2010, IPCC-XXXII/Doc. 5 (3.IX.2010), Agenda Item: 4.2, The IPCC Fifth Assessment Report (AR5), Proposal for an IPCC Expert Meeting on Geoengineering (Submitted by the Co-Chairs of Working Group I, II and III) IPCC-XXXII/Doc.5, 1 (Proposal for IPCC Expert Meeting).
9. For examples of this burgeoning literature, see Carolyn Kousky & Stephen H. Schneider, 'Global Climate Policy: Will Cities Lead the Way?', 3 *Climate Pol'y* 1, 11 (2003); Janet Koven Levit, 'Bottom-Up International Lawmaking: Reflections on the New Haven School of International Law', 32 *Yale J. Int'l L.* 383, 402–405 (2007); Judith Resnik, 'Law's Migration: American Exceptionalism, Silent Dialogues, and Federalism's Multiple Ports of Entry', 115 *Yale L. J.* 1564, 1627–1633 (2006); Judith Resnik, Joshua Civin & Joseph Frueh, 'Ratifying Kyoto at the Local Level: Sovereigntism, Federalism, and Translocal Organizations of Government Actors (TOGAS)', 50 *Ariz. L. Rev.* 709, 727–728 (2008); see also Randall S. Abate, 'Kyoto or Not, Here We Come: The Promise and Perils of the Piecemeal Approach to Climate Change Regulation in the United States', 15 *Cornell J.L. & Pub. Pol'y* 369 (2006); Donald A. Brown, 'Thinking Globally and Acting Locally: The Emergence of Global Environmental Problems and the Critical Need to Develop Sustainable Development Programs at State and Local Levels in the United States', 5 *Dick. J. Envtl. L. & Pol'y* 175 (1996); Ann E. Carlson, 'Implementing Greenhouse Gas Emissions Caps: A Case Study of the Los Angeles Department of Water and Power', 55 *UCLA. L. Rev.* 1479 (2008); Kirsten Engel, 'AALS Symposium: State and Local Climate Change Initiatives: What Is Motivating State and Local Governments to Address a Global Problem and What Does This Say About Federalism and Environmental Law?', 38 *Urb. L.* 1015 (2006); Robert B. McKinstry, Jr., 'Laboratories for Local Solutions for Global Problems: State, Local and Private Leadership in Developing Strategies to Mitigate the Causes and Effects of Climate Change', 12 *Penn. St. Envtl. L. Rev.*, 15 (2004); Hari M. Osofsky, 'Climate Change Litigation as Pluralist Legal Dialogue?', 26 *Stan. Envtl. L.J.* & 43 *Stan. J. Int'l L.* 181 (2007) (Joint Issue); Hari M. Osofsky, 'Is Climate Change "International"?: Litigation's Diagonal Regulatory Role', 49 *Va. J. Int'l L.* 585 (2009); Hari M. Osofsky, 'Local Approaches to Transnational Corporate Responsibility: Mapping the Role of Sub-National Climate Change Litigation', 20 *Pac. McGeorge GLOBAL Bus. & Dev. L.J.* 143 (2007); Hari M. Osofsky & Janet Koven Levit, 'The Scale of Networks: Local Climate Change Coalitions', 8 *Chi. J. Int'l L.* 409 (2008); Richard B. Stewart, 'States and Cities as Actors in Global Climate Regulation: Unitary vs. Plural Architectures', 50 *Ariz. L. Rev.* 681 (2008); Katherine Trisolini & Jonathan Zasloff, 'Cities, Land Use, and the Global Commons: Genesis and the Urban Politics of Climate Change', in *Adjudicating Climate Change: Sub-National, National, and Supra-National*

effective technological strategies vary widely across places, even within a particular developing country. Current approaches to needs assessment in the context of technology transfer treat each nation-state as an individual unit, and in the process, may miss the transnational interconnections and synergies among subnational entities.[10]

The chapter concludes in Part IV by exploring multidimensional strategies for regulating technology transfer more effectively. It draws from a variety of theoretical literatures exploring possibilities for more fluid, multiscalar regulatory regimes to propose ways in which current processes could address on-going problems of technology definition and place-based needs. It argues that greater involvement by smaller-scale actors in international-level assessments of what technology should be transferred is critical.

1 ALL CLIMATE CHANGE TECHNOLOGY WAS NOT CREATED EQUAL

The goal in climate change technology transfer is not, of course, to transfer any possible technology, but rather to transfer technology that is environmentally sound. However, determining what technology should qualify poses important legal and ethical concerns. What should be the standards for determining whether a technology is environmentally sound generally and with respect to climate change in particular? To what extent should the developing country recipient, and the entities within the developing country which are actually receiving it, influence the determination of soundness? This Part explores these questions by considering the definition of environmentally sound technology in the context of technology transfer and of climate change technology more broadly, with a focus on emerging geoengineering technologies.

Agenda 21 defines 'environmentally sound technologies' as those which 'protect the environment, are less polluting, use all resources in a more sustainable manner, recycle more of their wastes and products, and handle residual wastes in a more acceptable manner than the technologies for which they were substitutes'.[11]

Approaches, ed. William C.G. Burns & Hari M. Osofsky (New York: Cambridge University Press, 2009). <http://papers.ssrn.com/sol3/papers.cfm?abstract_id=1267314>; Laura Kosloff & Mark Trexler, 'State Climate Change Initiatives: Think Locally, Act Globally' 18 *Wtr. Nat. Resources & Env't* 46 (2004); William Andreen et al., 'Cooperative Federalism and Climate Change: Why Federal, State, and Local Governments Must Continue to Partner', Center for Progressive Reform, 29 May 2008. <http://progressiveregulation.org/articles/Cooperative_Federalism_and_Climate_Change.pdf>; Alice Kaswan, 'Climate Change, Consumption, and Cities', 36 *Fordham Urb. L.J.* 253 (2009).

10. For an in-depth discussion of this issue, see s. 3.
11. Agenda 21, fn. 2 *supra*, at s. IV, Ch. 34, 34.1.

It further explains the importance of technology transfer in this context and proposes to:

> promote, facilitate, and finance, as appropriate, the access to and the transfer of environmentally sound technologies and corresponding know-how, in particular to developing countries, on favourable terms, including on concessional and preferential terms, as mutually agreed, taking into account the need to protect intellectual property rights as well as the special needs of developing countries for the implementation of Agenda 21.[12]

Although these Agenda 21 goals are relatively uncontroversial in their general terms, definitional questions abound: How do we decide which technologies meet these criteria and who should be the decision maker? What are appropriate ways to transfer the technology, how favourable should the terms be, and what does it mean to protect intellectual property rights and the special needs of developing countries? As a 2003 UNEP report notes in the context of phytotechnology:

> Defining environmentally sound technologies in an absolute sense is difficult since the environmental performance of a technology depends upon its impacts on specific human populations, biota and ecosystems, and the availability of supporting infrastructure and human resources for the management, monitoring and maintenance of the technology, as well as the sustainability of natural systems. The soundness of environmental technology is also influenced by temporal and geographical factors, to the extent that some technologies may be environmentally sound now but may be replaced in the future by even cleaner technologies.[13]

A core challenge for technology transfer is to navigate the complexities of technological development and scientific uncertainty, whether with respect to climate change or other environmental concerns, in evaluating what technologies are appropriate, particularly at the margins.

In the climate change law context, these types of issues have arisen during years of negotiation over the meaning of and appropriate action under Article 4.5 of the UNFCCC, which reads:

> The developed country Parties and other developed Parties included in Annex II shall take all practicable steps to promote, facilitate and finance, as appropriate, the transfer of, or access to, environmentally sound technologies and know-how to other Parties, particularly developing country Parties, to enable them to implement the provisions of the Convention. In this process, the developed country Parties shall support the development and enhancement of endogenous capacities and technologies of developing country Parties.

12. *Ibid.*, at s. IV, Ch. 34, 34.14 (b).
13. UNEP, Phytotechnologies: 'A Technical Approach in Environmental Management' (2003). <www.unep.or.jp/ietc/publications/freshwater/fms7/2.asp>.

Other Parties and organizations in a position to do so may also assist in facilitating the transfer of such technologies.[14]

A number of agreements reached through UNFCCC Conferences of the Parties have reinforced the importance of creating effective mechanisms for technology transfer and established and extended an expert group on technology transfer.[15] The expert group has focused on a country-by-country technology needs assessment and an exploration of how to make implementation more effective.[16] The Cancun Agreements maintain this approach even while establishing new structures, stating that 'technology needs must be nationally determined, based on national circumstances and priorities'.[17]

While the needs assessment to some extent answers the question of which environmentally sound technologies are appropriate for particular countries, ambiguities remain about the term more generally and climate change agreements continue to focus primarily on implementation issues rather than core definitional issues. For example, the Copenhagen Accord uses language such as: 'We agree that developed countries shall provide adequate, predictable and sustainable financial resources, technology and capacity-building to support the implementation of adaptation action in developing countries'[18] and 'In order to enhance action on development and transfer of technology we decide to establish a Technology Mechanism to accelerate technology development and transfer in support of action on adaptation and mitigation that will be guided by a country-driven approach and be based on national circumstances and priorities'.[19]

Specific mechanisms attempt to define appropriate technology more clearly, but they do not eliminate all ambiguities. For example, the Cancun Agreements contain more guidance for what constitutes a qualifying CDM project. They resolve that carbon capture and storage in geological formations qualifies, and specify qualifying carbon dioxide emissions reductions or removal enhancements to supplement existing requirements.[20] However, these additions seem unlikely to resolve the bigger picture problems highlighted in Wara's piece.[21]

Much of the time, definitional ambiguity will not matter greatly. Most people agree on the value of energy efficient cars or supporting renewable energy in ways

14. UNFCCC, fn. 3 *supra*, at 4.5.
15. UNFCCC, Development and Transfer of Technologies. <http://unfccc.int/cooperation_and_support/technology/items/1126.php>.
16. UNFCCC, Expert Group on Technology Transfer: Five Years of Work (2007).
17. Long-Term Cooperative Action Outcome, fn. 5 *supra*, at Art. 114.
18. Copenhagen Accord, fn. 4 *supra*, at Art. 3.
19. *Ibid.*, at Art. 11.
20. See Carbon Dioxide Capture and Storage in Geological Formations as Clean Development Mechanism Project Activities, Decision -/CMP.6, Dec. 2010 (advance version). <http://unfccc.int/files/meetings/cop_16/application/pdf/cop16_cmp_ccs.pdf>; Further Guidance Relating to the Clean Development Mechanism, Decision -/CMP.6, Dec. 2010 (advanced version). <http://unfccc.int/files/meetings/cop_16/conference_documents/application/pdf/20101204_cop16_cmp_guidance_cdm.pdf>.
21. Wara, fn. 6 *supra*.

that will work in local contexts. But other technologies raise difficult ethical and legal questions in the technology transfer context. For example, should hydropower or nuclear power be viewed as environmentally sound technologies moving forward? They both have a very low greenhouse gas emissions profile, but raise other environmental concerns which become worse when improper implementation takes place.[22] Moreover, even the environmental soundness of core renewable energy technologies often is disputed at the implementation phase. Wind power is loud and kills birds, which has led to locally-based objections.[23] Similarly, some types of solar power technology pose concerns in the desert climates that are very sunny because of their water use.[24] These general debates over what technologies should be viewed as environmentally sound in the climate change context spill over into how people view what should be transferred.

While many disputes take place over well-established but evolving technologies, other approaches raise serious concerns because they are at the limits of scientific understanding and technological development. The recent BP *Deepwater Horizon* oil spill in the United States reinforced the risks that can occur at the boundaries of technological knowledge.[25] Complex risk assessments must take place in assessing the appropriate stage for transferring (or even implementing in developed countries) emergent technology. Beyond questions about whether the technology itself is safe enough, new and complex technologies require major on-the-ground support to function effectively.

These hard questions regarding the meaning of environmental soundness do not just arise in the technology transfer context but with respect to the broader dilemmas of how to address the problem of climate change. Policymakers and civil society increasingly despair over the possibilities for mitigation adequate to prevent major climate change,[26] which has led to more serious consideration of

---

22. For an interesting analysis of the complexities of using nuclear power to address climate change in the United States context, see Tanya Mortensen, 'An Unattainable Wedge: Four Limiting Effects on the Expansion of Nuclear Power', 5 *Envt'l & Energy L. & Pol'y J.* 60 (2010). For a thoughtful exploration of ethical issues surrounding hydropower, see John McPhee, *Encounters with the archdruid* (New York: Farrar, Straus and Giroux 1971).
23. For an analysis of some of the complexities of siting wind turbines and a cooperative federalism proposal, see Patricia E. Salkin & Ashira Pelman Ostrow, 'Cooperative Federalism and Wind: A New Framework for Achieving Sustainability', 7 *Hofstra L. Rev.* 1049 (2009).
24. See Osha Gray Davidson, 'Solar Farms That Sip Desert Water', *Phoenix Sun*, 17 Jul. 2009, <http://thephoenixsun.com/archives/4030>. For a discussion of solar rights allocation, see Sarah C. Bronin, 'Solar Rights', 89 *B.U. L. Rev.* 1217 (2009).
25. For a discussion of issues arising from the BP *Deepwater Horizon* Oil spill, see Curry L. Hagerty & Jonathan L. Ramseur, Deepwater Horizon Oil Spill: Selected Issues for Congress, CSR Report for Congress, 30 Jul. 2010. <www.fas.org/sgp/crs/misc/R41262.pdf>. I have discussed the governance concerns arising from these technology dilemmas in the context of the BP Deepwater Horizon Oil Spill in Hari M. Osofsky, 'Multidimensional Governance and the BP Deepwater Horizon Oil Spill', 63 Fla L Rev 1077 (2011).
26. The thresholds for concentration levels of greenhouse gases likely to result in major climate change are somewhat uncertain and controversial. However, there is a high level of consensus that the risks increase as the concentrations of greenhouse gases do. For a proposal that 'that human changes to atmospheric CO_2 concentrations should not exceed 350 parts per million by

emerging geoengineering technologies.[27] As high-level scientific meetings increasingly take place on geoengineering, including the IPCC deciding to include

---

volume, and that radiative forcing should not exceed 1 watt per square metre above pre-industrial levels', see Johan Rockström et al., 'A Safe Operating Space for Humanity', 461 *Nature* 472 (2009). In August 2011, the monthly mean carbon dioxide measured at Mauna Loa Observatory, Hawaii was 390.02 parts per million. See U.S. Department of Commerce, National Oceanic & Atmospheric Administration, NOAA Research, Earth System Research Laboratory: Global Monitoring Division, Trends in Atmospheric Carbon Dioxide, <www.esrl.noaa.gov/gmd/ccgg/trends/>. Our failure to keep greenhouse gases below these thresholds has uncertain, but potentially devastating, consequences. For one exposition of the potential human and biological consequences of exceeding these thresholds, see Mary Christina Wood, '"You Can't Negotiate with a Beetle": Environmental Law for a New Ecological Age', 50 *Nat Resources J* 167 (2010). For analyses of the inequalities regarding impacts and adaptations, see *Intergovernmental Panel on Climate Change, Fourth Assessment Report: Climate Change 2007* (AR4), ed. Rajendra K. Pachuari & Andy Reisinger (Geneva, Switzerland: 2007), Foreword and Preface. <www.ipcc.ch/publications_and_data/ar4/syr/en/contents.html>; Ruth Gordon, 'Climate Change and the Poorest Nations: Further Reflections on Global Inequality', 78 *U Colo L Rev* 1559 (2007). For discussions of the complexities of addressing the refugee concerns that climate change appears likely to cause, see Bonnie Docherty & Tyler Giannini, 'Confronting a Rising Tide: A Proposal for a Convention on Climate Change Refugees', 33 *Harv Envtl L Rev* 349 (2009); Gaim Kibreab, 'Climate Change and Human Migration: A Tenuous Relationship?', 20 *Fordham Envtl L Rev* 357 (2009); Elizabeth Burleson, 'Climate Change Displacement to Refugee', 5 *J Envtl L & Litig* 19 (2010). For explorations of the national security implications of climate change, see CNA Corporation, National Security and the Threat of Climate Change (2007). <www.cna.org/nationalsecurity/climate/report/SecurityandClimate_Final.pdf>; Department of Public Information, News and Media Division, New York, Security Council, 5663rd Meeting (AM & PM), SC/9000, 'Security Council Holds First-Ever Debate on the Impact of Climate Change on Peace and Security, Hearing over 50 Speakers', 17 Apr. 2007. <www.un.org/News/Press/docs/2007/sc9000.doc.htm>.

27. For examples of this dialogue, see Steven E. Koonin et al., 'Climate Engineering Responses To Climate Emergencies V', *Novim* 28 Jul. 2009, <www.novim.org/attachments/031_037_Novim%20Report%20Final%2007.28.09.pdf>; Prof. John Shepherd FRS, 'Geoengineering the Climate: Science, Governance and Uncertainty', The Royal Society 1 Sep. 2009. <http://royalsociety.org/geoengineering-the-climate/>; Bjørnar Egede-Nissen, Henry David Venema, IISD, 'Desperate Times, Desperate Measures: Advancing the Geoengineering Debate at the Arctic Council' (2009). <www.arcticgovernance.org/desperate-times-desperate-measures-advancing-the-geoengineering-debate-at-the-arctic-council.4782242-142902.html>; Steven E. Koonin et al., 'Climate Engineering Responses to Climate Emergencies', Novim 28 Jul. 2009; *Committee on Geological and Geotechnical Engineering in the New Millennium; Opportunities for Research and Technological Innovation*, Committee on Geological and Geotechnical Engineering, National Research Council (Washington, DC: National Academies Press, 2006); Albert C. Lin, 'Geoengineering Governance', 8 Issues In Legal Scholarship Art. 2 (2009); David W. Schnare, 'Climate Change and the Uncomfortable Middle Ground: "The Geoengineering" and "No Regrets" Policy Alternative', Paper delivered at the Thomas Jefferson Institute for Public Policy 4 March 2008 <www.heartland.org/events/NewYork08/newyork2008-ppt.html>; 2009 ESI/MITEI/CGCS Symposium: 'Engineering a Cooler Earth: Can We Do It? Should We Try?', 30 Oct. 2009; Ecological Society of America, Symposium 21: 'The Environmental Effects of Geoengineering', 6 Aug. 2009; The National Academies: 'America's Climate Choices, Geoengineering Options to Respond to Climate Change: Steps to Establish a Research Agenda', 15–16 Jun. 2009; American Enterprise Institute for Public Policy Research, Geoengineering: 'A Revolutionary Approach to Climate Change', 3 Jun. 2008.

it throughout its Fifth Assessment Report, major disagreement exists about the level of uncertainty, risks, and benefits posed by the two leading approaches to geoengineering, solar radiation management (SRM) and carbon dioxide removal (CDR). SRM approaches do not decrease greenhouse gas concentrations, but attempt to offset their effects through techniques that increase surface or marine cloud reflectivity or create blocks to incoming solar radiation. CDR techniques, on the other hand, remove greenhouse gases from the atmosphere through land and ocean management techniques.[28] SRM approaches, because they address effects, have the potential to work in shorter timescales, whereas CDR approaches actually engage the underlying problem.[29]

The geography of these techniques and of the problem they attempt to address paired with scientific uncertainty raise serious concerns regarding what it would mean to implement geoengineering in an environmentally sound manner. First, an underlying geographic mismatch exists between the scale of implementation and impact. Many of these approaches are not technically difficult, which means that a country, nongovernmental organization, or company might be able to experiment with them unilaterally. However, these experiments initiated in particular places might have local, regional, or global effects depending on the technique used and the scale at which it is implemented. As the IPCC Working Group co-chairs note, '[u]nilateral action may have environmental side effects on other countries and regions, and may not appropriately address the global scale of the issue'.[30]

Second, there is scientific uncertainty (and disagreement) about the geography of the impacts of various techniques. For instance, Michael MacCraken suggests the Arctic sea ice melt may be able to be countered with aerosol injections timed during sunlit months to avoid impacting the ozone layer,[31] while other scientists argue that aerosols in the quantities needed to address climate change would have serious ozone implications and that aerosols may move around in

---

28. See Proposal for IPCC Expert Meeting, fn. 8 *supra*, at 1. For an analysis of carbon storage and sequestration, which is generally treated as distinct from geoengineering proposals, but is arguably part of CDR, see Alexandra B. Klass & Elizabeth J. Wilson, 'Climate Change, Carbon Sequestration, and Property Rights', 2010 *U. Ill L Rev* 363 (2010).
29. See Proposal for IPCC Expert Meeting, fn. 8 *supra*, at 1. Bayden Russell and Sean Connell raise concerns about geoengineering strategies that do not reduce the concentration of atmospheric carbon dioxide. See Bayden D. Russell & Sean D. Connell, 'Letters: Honing the Geoengineering Strategy', 327 *Science* 144 (2010).
30. Proposal for IPCC Expert Meeting, fn. 8 *supra*, at 1. Joshua Horton argues that 'fear of unilateralism is largely misplaced, grounded more in unexamined policy assumptions than in reasoned analysis of the strategic situation faced by states'. Joshua B. Horton, 'Geoengineering and the Myth of Unilateralism: Pressures and Prospects for International Cooperation' 4 *Stan J L Sci & Pol'y* 56 (published online 2011).
31. See Michael MacCracken, 'Beyond Mitigation: Potential Options for Counter-Balancing the Climatic and Environmental Consequences of the Rising Concentrations of Greenhouse Gases' (World Bank Pol'y Res., Working Paper No. 4938, 2009). <www.climate.org/PDF/WorldBank_Beyond-Mitigation_MacCracken.pdf>. For an example of other scholarship supportive of SRM, see T.M.L. Wigley, 'A Combined Mitigation/Geoengineering Approach to Climate Stablization', 314 *Science* 452 (2006).

unexpected ways.[32] Similarly, regional scale models for SRM's effectiveness come up with radically different results depending on the way in which sea-ice dynamism, dynamic-ocean feedback, ocean-atmosphere dynamics, cloud dynamics, and El Niño are integrated into them.[33]

Third, scientists struggle with questions about the geographic scale needed for effective testing and implementation and about how risks match up with scale and location. For example, Alan Robock, Martin Bunz, Ben Kravitz, and Georgiz Stenchikov argue that 'geoengineering [referring to SRM] cannot be tested without full-scale implementation'.[34] If they are correct, testing would essentially be the same as implementation, with all of its attendant risks. With respect to a CDR approach like ocean iron fertilization, location may matter more than scale from a risk perspective; a smaller-scale experiment in an ecologically sensitive area may bring more attendant harms than a larger-scale experiment in a less sensitive location.[35]

Finally, many techniques involve diverse geographic effects and risk trade-offs. SRM might reduce temperatures but not address ocean acidification at a global scale and might impact regional precipitation patterns, such as altering the Asian monsoons, in ways that harm people.[36] Some SRM approaches pose risks of sudden, massive climate change if they are stopped, which means that a commitment to them now may limit the options of future generations.[37] CDR approaches

32. For an example of scholarship arguing that proposed sulphur aerosol geoengineering schemes would accelerate ozone depletion, see Simone Tilmes, Rolf Müller & Ross Salawitch, 'The Senstivity of Polar Ozone Depletion to Proposed Geoengineering Schemes', 320 *Science* 1201 (2008) (published online 4 Apr. 2008). For further discussion of these issues, see Bidisha Banerjee, 'The Limitations of Geoengineering Governance in a World of Uncertainty', 4 *Stan J L Sci & Pol'y* 15 (published online 2011); Alan Robock, 'Whither Geoengineering?', 320 *Science* 1166 (2008).
33. See Banerjee, fn. 32 *supra*, at 25–26 (comparing Video: David Battisti, Climate Engineering with Aerosols – Predictable Consequences?, MIT Earth System Initiative Symposium-Climate Engineering Symposium, (2 Dec. 2009). <http://techtv.mit.edu/genres/32-science/videos/4636-climate-engineering-with-aerosolspredictable-consequences> with Philip J. Rasch et al., 'Exploring the Geoengineerings of Climate Using Stratospheric Sulfate Aerosols: The Role of Particle Size', 33 *Geophysical Res Letters* 1 (2008).
34. Alan Robock et al., 'A Test for Geoengineering?', 327 *Science* 530 (January 2010).
35. Randall S. Abate & Andrew B. Greenlee, 'Sowing Seeds Uncertain: Ocean Iron Fertilization, Climate Change, and the International Environmental Law Framework', 27 *Pace Envtl L Rev* 555 (2010).
36. See Proposal for IPCC Expert Meeting, note 8 supra, at 1; Gabriele C. Hegerl & Susan Solomon, *'Risks of Climate Engineering'*, 325 *Science* 955 (2009).
37. For an analysis of this issue, see William C.G. Burns, 'Climate Geoengineering: Solar Radiation Management and its Implications for Intergenerational Equity' 4 *Stan J L Sci & Pol'y*, 37 (published online 2011) (citing Sci. & Tech. Comm., The Regulation of Geoengineering, 2009-10, H.C. 221, at 16; 59, H. Damon Matthews & Ken Caldeira, 'Transient Climate-Carbon', 104 *PNAS* 9951 (2007); B. Govindasamy et al., 'Impact of Geoengineering Schemes on the Terrestrial Biosphere', 29 *Geophysical Res Letters* 18-1, 18-3 (2002); Peter G. Brewer, 'Evaluating a Technological Fix for Climate', 104 *PNAS* 9915, 9915 (2007); John C. Moore, Svetlana Jevrejeva & Aslak Grinstad, 'Efficacy of Geoengineering to Limit 21st Century Sea-Level Rise', 107 *PNAS* 15,699 (2010) >; Victor Brovkin et al., 'Geoengineering Climate by

like large-scale bio-char and afforestation, paired with aggressive mitigation, might be able to play a minor supportive role in keeping temperatures down, but critics argue that implementation at this scale could use too much of the arable land needed for food production.[38] The difficulty of determining how these concerns should translate into an assessment of environmental soundness is exacerbated by the existence of separate treaty regimes for the ocean, atmosphere, climate change, outer space, as well as cross-cutting ones on environmental modification and biological diversity.[39] This division means that treating geoengineering holistically is difficult; SRM will interact with a different set of existing treaties than CDR (with some overlap around climate change and environmental modification and some differences within each of the two categories) due to the varying ways that they will impact the planet and many of these treaties make their own determination about when technology is environmentally sound or dangerous. This complexity is compounded by the fact that some of the potentially relevant treaty party or bodies have addressed geoengineering and others have not.[40] In addition, broader questions remain about which types/implementation approaches should be covered and whether some approaches may be small enough in scale or impact that particular treaties would not apply to them.[41]

Stratospheric Sulfur Injections: Earth System Vulnerability to Technological Failure', 92 *Climatic Change* 243, 254 (2009); Eli Kintisch, 'Scientists Say Continued Warming Warrants Closer Look at Drastic Fixes', 318 *Science* 1054, 1055 (2007); Robock, fn. 32 *supra*.

38. See Banerjee, fn. 32 *supra*, at 26; T. M. Lenton & N. E. Vaughan, 'The Radiative Forcing Potential of Different Climate Geoengineering Options', 9 *Atmos Chem Phys Discuss* 2559 (2009).
39. See UNFCCC, fn. 3 *supra* (climate change); Convention on the Prohibition of Military or Any Other Hostile Use of Environmental Modification Techniques, 18 May 1977, 31 U.S.T. 333, T.I.A.S. 9614 (environmental modification); Convention on Biological Diversity, 5 Jun. 1992, 1760 U.N.T.S. 143 (biological diversity); Convention on the Prevention of Marine Pollution by Dumping of Wastes and Other Matter, 29 Dec. 1972, 1046 U.N.T.S. 120 (oceans); 1996 Protocol to the Convention on the Prevention of Marine Pollution by Dumping of Wastes and Other Matter, 7 Nov. 1996), 36 I.L.M. 1 (oceans); United Nations Convention on the Law of the Sea, 10 Dec. 1982, 1833 U.N.T.S. 397 (oceans); Montreal Protocol on Substances that Deplete the Ozone Layer, 16 Sep. 1987, S. Treaty Doc. No. 100-10, 1522 U.N.T.S. 29 (atmosphere); Convention on Long-Range Transboundary Air Pollution, 13 Nov. 1979, 18 I.L.M. 1442 (atmosphere); Protocol to the 1979 Convention on Long-Range Transboundary Air Pollution on the Reduction of Sulphur Emissions or Their Transboundary Fluxes by at Least 30 Per Cent, 8 Jul. 1985, 27 I.L.M. 707, 1480 U.N.T.S. 217 (atmosphere); Protocol to the 1979 Convention on Long-Range Transboundary Air Pollution on Further Reduction of Sulphur Emissions, 14 Jun. 1994, 33 I.L.M. 1542 (atmosphere); Treaty on Principles Governing the Activities of States in the Exploration and Use of Outer Space, Including the Moon and Other Celestial Bodies, 27 Jan. 1967, 18 U.S.T. 2410, 610 U.N.T.S. 205, *reprinted in* 6 I.L.M. 386 (1967) (outer space). For an analysis of these conventions and how their various provisions might apply and have been applied to geoengineering, see Albert C. Lin, 'International Legal Regimes & Principles Relevant to Geoengineering' (27 May 2011 draft manuscript on file with author).
40. For an analysis of the differences in how relevant treaties have addressed ocean iron fertilization, see Abate & Greenlee, fn. 35 *supra*; see also Karen N. Scott, 'The Day After Tomorrow: Ocean CO_2 Sequestration and the Future of Climate Change', 18 *Geo Int'l Envtl L Rev* 57 (2005) (exploring the compatibility of carbon dioxide at sea with applicable international and regional law).
41. See Abate & Greenlee, fn. 35 *supra*.

None of these questions about either technology transfer or geoengineering can be answered simply by putting a group of scientific and legal experts in a room because they involve core value choices. While it is easy to support technology transfer that represents a win-win, the marginal cases are very troubling. For the purposes of this chapter's more narrow focus on transfers, when widespread disagreement exists about the value of a technology or significant uncertainty exists about its effectiveness or safety, those engaged in technology transfer must make difficult judgments about whether these technologies should be brought to developing countries and how the input from various interested parties should influence that judgment call through official mechanisms.

2 DILEMMAS OF SCALE IN CLIMATE CHANGE TECHNOLOGY TRANSFER

These complex questions are made harder by problems of scale, which are the focus of this Part. As I have explored in depth in previous scholarship, climate change is a fundamentally multiscalar problem which requires multiscalar regulatory solutions. Specifically, emissions in particular places interact with the global climate system and cause varying impacts in a wide range of places which differ in their capacity to adapt. With respect to mitigation, decisions at individual, local, state, national, and international levels, as well as regional levels in between, help to shape the extent and type of greenhouse gas emissions. These multi-level structures similarly intersect with efforts to adapt to climate change's impacts, which vary greatly from place to place.[42]

These issues of scale matter in a climate change technology transfer context because most of the international agreements on climate change focus on the nation-state as the critical governmental unit. To the extent that smaller-scale units make international agreements—such as the cities, states, and provinces doing so at the 2009 Copenhagen negotiations—they are not explicitly incorporated into the UNFCCC framework and are not viewed as legally binding under international law.[43] The approach to technology transfer under the UNFCCC reflects this bias towards the national scale, as developing country needs are assessed on a national basis, even though the national assessments themselves

42. Osofsky, 'Is Climate Change "International"?', fn. 9 *supra*.

43. Hari M. Osofsky, 'Multiscalar Governance and Climate Change: Reflections on the Role of States and Cities at Copenhagen', 25 *Maryland J. Int'l L.* 64 (2010). For examples of the subnational agreements of Copenhagen, see Press Release, Office of the Governor of California, Gov. Schwarzenegger Announces New Coalition of Subnational Leaders to Combat Climate Change (14 Dec. 2009). <http://gov.ca.gov/press-release/14032/>. (Press Release); The Copenhagen Climate Summit for Mayors, Copenhagen, Denmark 14–17 Dec. 2009, 'Cities Act: Copenhagen Climate Communiqué', 16 Dec. 2009, (Communiqué).<www.kk.dk/Nyheder/2009/December/~/media/B5A397DC695C409983462723E31C995E.ashx>; The City Climate Catalogue, The City Climate Catalogue, Copenhagen City Climate Catalogue – List of Commitments <www.climate-catalogue.org/index.php?id=6870>.

focus on multiple scales and a multiscalar network consolidated in the Climate Technology Centre provides support.[44]

While this treatment of nation-states as core units comports with international law, which views nation-states as its primary subjects and objects, it potentially misses critical interconnections. Each nation-state under the current approach does a multiscalar needs assessment that includes small and large scales and short and long timeframes; smaller-scale governmental entities often are included as key stakeholders in the nation-state needs assessment.[45] However, such an assessment may not fully address transnational relationships among cities and states, as well as synergies within regions that cut across national borders. It assumes that technology transfer should focus on the nation-state and its component parts.

These potential gaps pose issues for the most efficient and effective technology transfer. If subnational entities in different countries have similar needs, they ideally would collaborate in the technology transfer process and learn from one another's experiences. That might happen under the current system, but the nation-state-based structure of information-gathering and transfer makes those kinds of interchanges less likely. Although nation-states are the parties under the UNFCCC and play an important organizing role in technology transfer, too narrow a focus on them as the primary organizing scale of government is limiting.

Moreover, the ethical questions posed in the previous Part are made harder by national-level organization. What is most environmentally sound will vary from place to place, often at relatively small physical scales which overlap with multiple subnational governmental entities. The current focus on the nation-state as primary organizer, even if it includes key subnational stakeholders within that country, may not always include the full range of internal subnational needs or experiences of subnational entities in other countries that might be relevant.

3 TOWARDS A MULTIDIMENSIONAL APPROACH TO CLIMATE CHANGE TECHNOLOGY TRANSFER

This Part draws from new governance, legal pluralism, and geographic network theory to propose alternative ways of addressing the dilemmas of appropriate and effective technology transfer raised in the previous parts. It argues that inclusion of cross-border coalitions of subnational governmental entities in assessments of needs and environmental soundness would address the shortcomings of current approaches outlined above.

Although treaty-making in the climate change context largely follows a traditional international law model of sovereign state consent, possibilities for greater integration of subnational entities abound. For example, as I have explored in previous scholarship, the recent Copenhagen Accord and nation-state implementation

44. Long-Term Cooperative Action Outcome, fn. 5 *supra*, at Arts 114, 120 & 123; UNFCCC & UNDP, Handbook for Conducting Technology Needs Assessment for Climate Change (2010).
45. *Ibid.*

pledges have focused at the national-level despite the simultaneous meetings and agreements among subnational entities at Copenhagen. That disconnect raises broader questions about whether the UNFCCC process should be changed, either formally or informally, to include subnational networks more fully.[46] This chapter has a more narrow focus on technology transfer and so does not tackle those larger questions except within this context. Specifically, this Part looks at ways in which technology needs assessment could proceed differently, and in so doing, engage the nuances of local contexts more effectively.

Much of international legal theory has moved away from strict Westphalian models that are entirely centred on the nation-state.[47] As I have analysed previously,[48] these alternatives fall into three main, partially overlapping categories. Modified Westphalians still view the nation-state as the primary subject and object of international law, but recognize the nuanced interactions between the nation-state and other actors that help create those international norms.[49] Pluralists de-centre the nation-state further, arguing that the wide range of interactions among multiple actors comprise international law. They consider many processes beyond ones based on nation-state consent as serving in a law-creating role.[50]

---

46. See Osofsky, fn. 43 *supra*.
47. This chapter uses the term 'Westphalian' to mean theories of international law that view nation-states as its primary subjects and objects and the consent of sovereign and equal nation-states as driving the process. For discussions of Westphalian approaches, see Ian Brownlie, *Principles of Public International Law*, 6th edn (New York: Oxford University Press, 2003), 287–288; Michael J. Kelly, 'Pulling at the Threads of Westphalia: '"Involuntary Sovereignty Waiver" – Revolutionary International Legal Theory or Return to Rule by the Great Powers', 10 *UCLA J. Int'l L. & Foreign Aff.* 361, 383 (2005).
48. Hari M Osofsky, 'The Geography of Climate Change Litigation Part II: Narratives of Massachusetts v. EPA', 8 *Chi. J. Int'l L.* 573 (2008); Osofsky, fn. 43 *supra*.
49. For examples of the range of viewpoints contained within modified Westphalian approaches, compare Jack L. Goldsmith & Eric A. Posner, *The Limits of International Law* (New York: Oxford University Press, 2005), with Harold Hongju Koh, 'Jefferson Memorial Lecture: Transnational Legal Process after September 11th', 22 *Berkeley J. Int'l L.* 337, 339 (2004).
50. For examples of the global legal pluralism literature, see Robert B. Ahdieh, 'Dialectical Regulation' 38 *Conn. L. Rev.* 863 (2006); Diane Marie Amann, 'Abu Graib', 153 *U. Pa. L. Rev.* 2085 (2005); Diane Marie Amann, 'Calling Children to Account: The Proposal for a Juvenile Chamber in the Special Court for Sierra Leone', 29 *Pepp. L. Rev.* 167 (2001); Elena A. Baylis, 'Parallel Courts in Post-Conflict Kosovo', 32 *Yale J. Int'l L.* 1 (2007); Paul Schiff Berman, 'Global Legal Pluralism', 80 *S. Cal. L. Rev.* 1155 (2007); William W. Burke-White, 'International Legal Pluralism', 25 *Mich. J. Int'l L.* 963 (2004); Janet Koven Levit, 'A Bottom-Up Approach to International Law Making: The Tale of Three Trade Finance Instruments', 30 *Yale J. Int'l L.* 125 (2005); Ralf Michaels, 'The Re-Statement of Non-State Law: The State, Choice of Law, and the Challenge from Global Legal Pluralism', 51 *Wayne L. Rev.* 1209 (2005). Global legal pluralism often builds upon the New Haven School approach. For examples of that approach, see Harold D. Lasswell & Myres S. McDougal, *Jurisprudence for a Free Society: Studies in Law, Science and Policy*, vol. I (New Haven Press/Kluwer Law, 1992), xxi; Richard A. Falk, 'Casting the Spell: The New Haven School of International Law', 104 *Yale L.J.* 1991 (1995); Myres S. McDougal & Harold D. Lasswell, 'The Identification and Appraisal of Diverse Systems of Public Order', 53 *Am. J. Int'l L.* 1 (1959); Myres S. McDougal et al., 'The World Constitutive Process of Authoritative Decisions', 19 *J. Legal Educ.* 253 (1967); W. Michael Reisman, 'International Lawmaking: A Process of Communication', 75 *Am. Soc'y Int'l L. Proc.* 101 (1981).

Finally, critical scholars question the legitimacy of the international legal system. They tend to divide among the three previous categories in their view of how international law is currently made, but offer distinct arguments for rethinking and/or reconstituting the international legal system.[51]

The current approach to needs assessment arguably follows either a strict or a modified Westphalian model. It focuses country by country, but acknowledges the importance of key stakeholders within countries as helping to determine national needs and a broad multiscalar network as helping to meet them. To the extent that one thinks of only the ultimate nation-state perspective as mattering in the implementation of climate change technology transfer, then the current approach is strict Westphalian. If the input of multiple stakeholders is viewed as a more direct part of international law implementation and norm development, then the needs assessment approach is modified Westphalian. Either way, the recipient nation-state is the core actor and its consent serves as a critical driver in the technology transfer.

However, a more inclusive, pluralist alternative remains. Namely, if the current needs assessment were done on a country-by-country basis, but also expanded to explore subnational linkages, international law implementation would somewhat decentred. While the nation-state would remain a crucial actor, perhaps even the most important one given the treaty structure under which the needs assessment is taking place,[52] other groupings would also guide how technology transfer proceeds.

Both geographic network theory and new governance theory could help undergird such a modification. Geographer Kevin Cox, for example, has argued that each scale is made up of both core entities and interactions at that level and cross-level interactions. Due to this multiscalar quality of each scale, a particular level, such as the local, is actually comprised through a multiscalar network.[53]

---

51. For examples of the critical international law literature, see Balakrishnan Rajagopal, *International Law from Below: Development, Social Movement and Third World Resistance* (Cambridge: Cambridge University Press, 2003); Symposium: Critical Race Theory and International Law: Convergence and Divergence, 45 Vill. L. Rev. 827, 827–1220 (2000); David Kennedy, 'When Renewal Repeats: Thinking Against the Box', 32 *NYU J. Int'l L. & Pol.* 335, 476–500 (2000); Obiora Chinedu Okafor, 'Newness, Imperialism, and International Legal Reform in Our Time: A TWAIL Perspective', 43 *Osgoode Hall L.J.* 171 (2005).
52. Many accounts that could be characterized as pluralist still recognize the nation-state as a critical international legal actor, even as they acknowledge other actors as important. For example, Anne-Marie Slaughter states in *A New World Order* that 'The state is not the only actor in the international system, but it is still the most important actor.' Anne-Marie Slaughter, *A New World Order* (Princeton, New Jersey: Princeton University Press, 2004).
53. See Kevin R. Cox, 'Spaces of Dependence, Spaces of Engagement and the Politics of Scale, or Looking for Local Politics', 17 *Pol. Geography* 1 (1998). For analyses of Cox's approach, see Kevin R. Cox, 'Representation and Power in the Politics of Scale', 17 *Pol. Geography* 41 (1998); Katherine T. Jones, 'Scale as Epistemology', 17 *Pol. Geography* 25 (1998); Dennis R. Judd, 'The Case of the Missing Scales: A Commentary on Cox', 17 *Pol. Geography* 29 (1998); Michael Peter Smith, 'Looking for the Global Spaces in Local Politics', 17 *Pol. Geography* 35 (1998); Lynn A. Staeheli, 'Globalization and the Scales of Citizenship', 19 *Geography Res. F.* 60 (1999). I have previously applied Cox's work to climate change governance issues. See, e.g., Osofsky, 'Is Climate Change "International"?', fn. 9 *supra*.

The current approach to climate change technology needs assessment recognizes the national as comprised of its component parts but does not fully capture the scale of those component parts as they interact transnationally.

New governance approaches similarly explore how a wide range of stakeholders could be incorporated more meaningfully into decision-making processes, particularly those regarding multiscalar problems.[54] The current needs assessment guide arguably has elements of new governance through its recognition that more entities than just the national government should assess national needs. However, to the extent that an intra-national assessment of stakeholders is incomplete due to subnational entities interacting transnationally and having needs groupings, a new governance analysis might suggest a stakeholder approach that is more inclusive of transnational groupings.

The current UNFCCC process has room for a more inclusive approach without major treaty modification. The technology needs assessment process already includes a wide range of stakeholders in its nation-state based process.[55] Moreover, the network approach being used on the assistance side (consolidated in the Climate Technology Centre) could be expanded to contribute to the needs assessment process as well. While such an expansion would need to be careful to respect the sovereignty of the technology-receiving state party, it could provide helpful input to the national process.[56] This part's proposal is thus merely an extension of how things are done, but one that has the potential to make progress on both definitional and scalar challenges. Namely, if the stakeholders assessing needs and appropriate technology to address them are grouped both intra-nationally and transnationally, the assessment of what is environmentally sound for a place can be grounded in experiences of like places, even if they are within other countries. Similarly, to the extent that commonalities exist across transnational groupings of subnational entities, scalar groupings that go beyond the concentric circles of local, state, national, international become more possible. Regional collaborations beyond just regional groupings of nation-states may emerge and help to guide effective transfer that is targeted at needs in a more nuanced fashion.[57]

---

54. For examples of the new governance literature, see Gráinne de Búrca & Joanne Scott (eds) *Law and New Governance in The EU and US* (Oxford and Portland, OR: Hart Publishing, 2006); Bradley C. Karkkainen, '"New Governance" in Legal Thought and in the World: Some Splitting as Antidote to Overzealous Lumping', 89 *Minn. L. Rev.* 471 (2004); Orly Lobel, 'Setting the Agenda for New Governance Research', 89 *Minn. L. Rev.* 498 (2004); and Orly Lobel, 'The Renew Deal: The Fall of Regulation and the Rise of Governance in Contemporary Legal Thought', 89 *Minn. L. Rev.* 342 (2004); J.B. Ruhl & James E. Salzman, 'Climate Change, Dead Zones, and Massive Problems in the Administrative State: A Guide for Whittling Away', 98 *Cal. L. Rev.* 59 (2010).
55. See UNFCCC & UNDP, Handbook for Conducting Technology Needs Assessment for Climate Change, fn. 44 *supra*.
56. See Long-Term Cooperative Action Outcome, fn. 5 *supra*, at Art. 123.
57. For examples of analyses of the role of the region in the geography literature, see William B. Meyer et al., 'The Local-Global Continuum', in *Geography's Inner Worlds: Pervasive Themes in Contemporary American Geography*, ed. Ronald F. Abler, Melvin G. Marcus & Judy M. Olson (New Brunswick, New Jersey: Rutgers University Press 1992), 255; Anssi Paasi, 'Place

Such an approach would raise important implementation issues which are beyond the scope of this chapter. For example, how should these additional processes for needs assessment be integrated with national needs assessments? Would they be separate processes or part of a larger process? What would it mean in practice to integrate these analyses and still respect the sovereignty of the developing country party? These relationships would need to be well-defined through the UNFCCC's continuing process of developing the technology mechanism for such hybridity to work.

Thinking in these more fluid terms about governance, then, does not change the fact that technology transfer involves hard questions that often must be addressed in context-specific ways, but it gives additional tools for answering them may help to fit the scale of needs more effectively. Even within this treaty regime governed through nation-state consent, the expert process for assessing technology transfer needs creates room for more pluralist approaches that allow non-traditional orderings into international law processes. Such approaches do not change the formal international legal interactions directly, but indirectly influence national and international understandings and more importantly, help to reconstitute the nature of the international law at issue.[58] If climate change technology transfer can better engage the complex questions of scale and definition through innovative interactions, it can model a better way ahead for international efforts to address climate change.

More broadly, bringing more pluralist, integrative approaches to the multiscalar legal regime addressing climate change technology creates governance structures that can help to deal with very difficult technology issues like the disputes surrounding geoengineering discussed above. To the extent that a wide range of entities are dialoguing in a variety of formations about when technology is environmentally appropriate and needed, multipolar engagement in regulating the use of potentially dangerous technologies with global impacts become more likely. These dynamics, layered together with whatever international legal regime emerges to harmonize the myriad of treaties applicable to geoengineering, can help to minimize unilateral action that fails to appropriately analyse the risks involved.

4 CONCLUSION

As the international legal community struggles to address challenging problems like climate change, technology transfer is an important part of the solution. Making sure that developing countries have access to technology that they need to limit

---

and Region: Looking through the Prism of Scale', 28 *Progress in Human Geography* 536 (2004). For a discussion of regional groupings among nation-states in the climate change technology transfer context, see Ivan Vera, 'Climate Change and Technology Transfer: The Need for a Regional Perspective', UN-DESA Policy Brief No. 18, <www.un.org/esa/policy/policybriefs/policybrief18.pdf>. For a discussion of the role that universities can play in technology transfer, see Elizabeth Burleson & Winslow Burleson, 'Innovation Cooperation: Energy Biosciences and Law', 2011 *U. Ill. L. Rev.* 651 (2011).

58. See Osofsky, 'Climate Change Litigation as Pluralist Legal Dialogue?', fn. 9 *supra*.

emissions while addressing their economic needs is critical to any long-term emissions reduction strategy. This chapter does not dispute this fact, but grapples with how difficult achieving these goals are in actuality and the need for creative approaches that embrace complexities of governance. It uses the example of expert needs assessment to engage what such conceptual creativity might actually mean in practice.

In the end, more pluralist approaches to international law cannot make the problems themselves less hard. Technology transfer will continue to involve difficult value questions, scientific uncertainty, and fixed governance structures of varying functionality. Moreover, these transfers occur in a broader context in which mitigation is failing and policymakers increasingly contemplate unpalatable options like adapting to major climate change or risking large-scale geoengineering, both of which involve serious equity concerns.

But these strategies can change the dialogue in needed ways by including interactions and relations that actually matter in international law creation and implementation, but are not captured well by a strict Westphalian narrative. That nuanced conversation is needed to make sure that we make the best choices possible in addressing the challenges of climate change and the emerging technologies that attempt to mitigate or even reverse the atmospheric processes.

Chapter 9

Technology Transfer and Benefit Sharing under the Biodiversity Convention

Peter Gottschalk

What is technology transfer? Despite several decades of international negotiations there still seem to be conceptual confusion about the meaning of technology transfer. The need for a definition is often pointed out in both academic and policy-oriented debates but there seem to be no consensus in sight. To understand the debates it is important to acknowledge that technology transfer is not an objective phenomenon to be found 'out there'; it is a holistic construct made up by political actors. Therefore, I argue, the conceptual confusion rather reflects that there are continuing and on-going discursive struggles over the meaning of technology transfer.

Not since the 1970s has technology transfer been subject to so much debate as currently. Most of the attention is directed to the climate negotiations but technology transfer has also been addressed in the recently concluded negotiations of the Nagoya Protocol under the Convention on Biological Diversity.[1] After two years of intense negotiations under the 'Nagoya Roadmap', the Tenth Conference of the

1. 1992 Convention on Biological Diversity, Adopted Rio de Janeiro 5 Jun. 1992, entered into force 29 Dec. 1993, 1760 U.N.T.S. 79 (Convention or CBD).

Hans Henrik Lidgard, Jeffery Atik & Tu Thanh Nguyen, *Sustainable Technology Transfer*, pp. 195–220.

Parties to the Convention (COP 10) adopted the Nagoya Protocol on 29 October 2010 in Nagoya, Japan.[2] At the ceremony in New York on 11 February 2011, when the Protocol was opened for signature, it was advertised as 'a new legal instrument at the service of sustainable development'.[3] Two days later, on 4 February 2011, it was announced that the Global Environmental Facility approved a USD 1 million project to facilitate rapid ratification. The Protocol consists of 30 articles and an Annex. The objective of the Nagoya Protocol is the fair and equitable sharing of the benefits arising from the utilization of genetic resources. Three basic issues form the core of the Protocol, often called 'the ABC of ABS'; Access, Benefits-sharing, and Compliance. The Nagoya Protocol also contains an operative provision on technology transfer, Article 23, which will be set out and analysed in detail below.[4]

The preamble to the Nagoya Protocol contains (p. 5) the following statement:

> Recognizing the important contribution to sustainable development made by technology transfer and cooperation to build research and innovation capacities for adding value to genetic resources in developing countries, in accordance with Articles 16 and 19 of the Convention.

This statement clearly shows that biodiversity policy is not only concerned with environmental problems but also the control of the handling of new technologies and the societal regulation of global distributional conflicts, which are directly connected with the use of biodiversity.[5] Yet this recognition also begs the question 'in what way technology transfer actually contributes to sustainable development?'

While the Annex to the Protocol contains many options for benefit-sharing, the focus of this chapter is on technology transfer as a means to achieve it. As such, the analysis is of a particular aspect of the Protocol where issues concerning benefit-sharing, sustainable development and technology transfer meet and interlink. Although this point may, at first seem narrow, it has strong future implications not just for the Protocol and implementation of the CBD but also for broader notions of sustainable development.

The aim of this chapter then, is to assess the state of law in relation to Article 23. A 'traditional' assessment of the state of law by use of established sources of law (treaty and customary law) will, however, not takes us far. In order to make a deeper analysis of the state of law, I will use a constructivist approach. In doing so, I will place technology transfer in a wider context – not just within the framework of the CBD but also (at least to some extent) within global debates about technology

2. Nagoya Protocol on Access to Genetic Resources and the Fair and Equitable Sharing of Benefits Arising from their Utilization to the Convention on Biological Diversity, Nagoya, 29 Oct. 2010, Annex I to Depositary Notification, Reference C.N.782.2010.TREATIES-1, To be open for signature in New York from 2 Feb. 2011 until 1 Feb. 2012 (Nagoya Protocol). <http://treaties.un.org/doc/source/signature/2010/CN782E.pdf>.
3. Press release on 2 Feb. 2011. <www.cbd.int/doc/press/2011/pr-2011-02-02-abs-en.pdf>.
4. See further below.
5. Ulrich Brand, & Görg Christoph, 'Post-Fordist Governance of Nature: The Internationalization of the State and the Case of Genetic Resources – A Neo-Poulantzian Perspective', 15 *Rev Int'l Political Econ*, 567, 589 (2008).

transfer. Therefore, we need to ask different questions than we usually do and must search for answers in other 'places' than in the established sources of law.

Despite failed attempts to define technology transfer, there seems to be wide agreement that technology transfer has been unsuccessful.[6] For example, Sands states that it is widely recognized that the treaty provisions on technology transfer are inadequate.[7] I accept these observations as correct although they do not necessarily say anything about the state of law – ineffectiveness of technology transfer is not necessarily due to lack of legal obligations. As will be made clear, technology transfer obligations – as treaty obligations – are often formally binding. The problem is rather that these obligations are articulated in such open-ended and vague language that it is difficult to identify precisely what the obligation is. As such, these obligations are better described as soft law.[8] There is an absence of legal rules obligating developed countries to transfer technologies to developing countries despite persistent attempts and determined norm entrepreneurship on behalf of developing countries to create binding technology transfer obligations.

What I will attempt here is an assessment of the state of law, not just in terms of positivist black or white law but also in terms of constructivist shades of grey. Drawing from Brunnée and Toope's interactional account of international law,[9] I view law as particular kinds of social norms. Their theory enables us to assess the state of law in different shades of grey and to prescribe what is needed for law to further develop in desirable directions. I argue that what is needed for technology transfer concerns both conceptual and normative development in ways that I will set out and develop below. It is, of course, beyond the scope of this chapter to do a more comprehensive assessment of the state of the law.

In trying to identify shared understandings (norms), I also want to explore the relationships between open-ended concepts such as sustainable development, technology transfer and benefit-sharing. Although the kind of functions they may perform may be questioned, they at least help organize the discursive spaces. Relationships between, for example, sustainable development and technology transfer are not necessary but rather strategically linked by actors.[10]

An important tool used by negotiators – as by human beings in general – is concepts. 'Concepts are born of the necessity that individuals have to organise the information they learn or to systematise the experiences they go through.'[11] In fact, negotiators need concepts both as tools and as weapons in discursive struggles to

---

6. A problem with such statements is that the bases of such statements may be questioned, i.e. if there are no generally agreed definitions, how can we know?
7. Phillippe Sands, *Principles of International Environmental Law* (Cambridge: Cambridge University Press, 2003), 1038.
8. Alan Boyle. 'Soft Law in International Law-Making', in *International Law*, ed. Malcolm D. Evans (Oxford: Oxford University Press, 2010), 122.
9. Jutta Brunnée & Stephen Toope, *Legitimacy and Legality in International Law. An Interactional Account* (Cambridge: Cambridge University Press, 2010).
10. David W. Leebron, 'Linkages', 96 *Am J Int'l L* 5 (2002).
11. Claude Bélanger, 'Concepts in Social Science and History', <www2.marianopolis.edu/quebeckhistory/events/concepts.htm>.

get what they want. As such, and on a micro-sociological level, concepts can be regarded as one of many techniques negotiators use to arrive at gradually more stringent obligations to secure technology transfer.[12]

I emphasize that, methodologically this is a sketch, a first draft or an experiment employing ideas drawn from conceptual history[13] to inquire into the emergence and development of the concept of technology transfer. Although it follows immediately from the social approach underlying this chapter, it should nevertheless be made explicit that I rely on an instrumental (pragmatist) view of concepts. From such a view,'[c]entral concepts function as sites of action around which vocabularies, rhetoric and discourses of political language turn'.[14] I therefore suggest that conceptual history is a way to study legal change. Conceptual change reflects socio-legal change. This relationship is not unidirectional; concepts also help constitute the world. Conceptual change then becomes a way to study (understand and explain) legal change. Conceptual history also has the potential to reveal the main contentions around technology transfer and point to possible ways forward to overcome these contentions.

The structure is as follows. The first part of this chapter provides a brief account of the historical 'career' of the concept of technology transfer. It starts by addressing the question 'What is technology transfer?' and considers how the concept historically was formed. It then moves on to explore what can be done with the concept of technology transfer. The second part turns to technology transfer in the context of the CBD. In the third part, the recent adoption of the Nagoya Protocol is analysed. Possible conceptual interaction and effects of the institutional context for norm making are also discussed. The chapter concludes with some prospects for future developments.

1 WHAT IS TECHNOLOGY TRANSFER?

Before taking a closer look at technology transfer in the ABS negotiations, we need to know some of the history and the global debates shaping the concept of technology transfer. On the surface, the legacy from the negotiations for a New International Economic Order (NIEO) may not be immediately visible in the debates on technology transfer in the context of the CBD and the Nagoya Protocol, but they are certainly there.

---

12. Since concepts are used by human beings to communicate and learn they may seem too fundamental for human existence to be regarded as 'tools' used in negotiations. However, a focus on concepts makes more sense if we acknowledge that international negotiations in many respects are processes of learning.
13. Inspiration is drawn from Reinhart Koselleck, especially his *The Practice of Conceptual History. Timing History, Spacing Concepts* (Stanford: Stanford University Press, 2002) and from George Lakoff & Mark Johnson's *Metaphors We Live By* (Chicago: University of Chicago Press, 2003).
14. 'What Is Conceptual History?' <ww.concepta-net.org/conceptual_history>.

What, then, is technology transfer? The question has been repeated over and over again in policy-oriented debates over technology transfer, but the meaning of technology transfer still seems to cause confusion. Should this really be such a mystery? Perhaps we are asking the wrong question. In my view, the more appropriate and productive question in this context is – what can be done with the concept of technology transfer?[15] So, let us start by taking a brief look of the historical 'career' of the concept of technology transfer, and what has been done with it.

Historically, there have been several attempts to create international rules on technology transfer and provisions on technology transfer can be found in most Multilateral Environmental Agreements (MEA's). The origins of the concept are probably found in national law, especially in the laws of Latin American countries. At the international level, however, the issue of transfer of technology was first voiced in the context of the claims for a NIEO. Susan Sell notes that it was when developing countries linked technology transfer to economic development that they began to see the advantages of a multilateral approach.[16] Paragraph 4(p) of the Declaration on the Establishment of a New International Economic Order[17] UNGA Res. 3201 (S-VI) provides:

> The new international economic order should be founded on full respect for the following principles: . . . (p) To give to the developing countries access to the achievements of modern science and technology, to promote the transfer of technology and the creation of indigenous technology for the benefit of developing countries in forms and in accordance with procedures that are suited to their economies.

NIEO also established a Programme of Action on the Establishment of a New International Economic Order (UNGA Res. 3202 (S-VI)). An important part of this program was a demand for International Code of Conduct on the Transfer of Technology.

The technology transfer concept under NIEO had strong political underpinnings. One political idea was that countries should be transformed by technology. Ernst Haas argues:

> Experts in developing countries have come to think of technology transfer as an overarching concept, instead of worrying separately about such matters as foreign capital inflows, obtaining patents for specific products, finding markets for their products, or building a certain type of factory. . . . Instead of being ends in themselves, they become means toward a more complicated end – the achievement of technological self-reliance. Instead of being effects, they are

---

15. Bartelson Jens, 'Making Sense of Global Civil Society', 12 *Eur J Int'l Rel* 371–395 (2006).
16. Susan K Sell, *Power and Ideas: North-South Politics of Intellectual Property and Antitrust* (New York: State University of New York Press, 1998), 19.
17. See UN General Assembly resolutions A/Res/3201 (S-VI) and A/Res/3202 (S-VI), 1 May 1974. U.N. G.A.O.R. Supp. (No.76), 29 U.N. Doc. A 9559 (1974).

reconceived as causes leading to more basic effects such as wealth, prestige, status, autonomy.[18]

The preamble to the draft Code of Conduct echoed the same message: '[B]elieving that technology is the key to progress of mankind and that all peoples have the right to benefit from the advances and development in science and technology in order to improve their standards of living'.[19] The idea behind this statement was that there is a 'universal heritage' of welfare based upon technology common to all states in the world. From this the idea of a right to 'access' to technology is derived. The imbalance caused by Western colonialism should in that way be readjusted.[20]

A Programme of Action was also established under the NIEO. Section IV entitled 'Transfer of Technology' stated that '[a]ll efforts should be made: a. to formulate an international code of conduct for the transfer of technology corresponding to needs and conditions prevalent in developing countries'. But most of the efforts of the international community to elaborate rules on technology transfer have been unsuccessful. NIEO did not manage to produce any (binding) law with regard to technology transfer. At best, it produced a number of drafts which may be considered soft law. The reasons why the negotiations for an International Code of Conduct on the Transfer of Technology collapsed are complex but it is clear that one important reason was that the negotiators could not agree on the legal nature of the Code. Developing countries pressed hard for a legally binding Code while the US resisted this.[21]

One of the first treaties to include more detailed rules on technology transfer is the United Nations Convention on the Law of the Sea[22] (UNCLOS). Part XIV of UNCLOS contains 13 articles on the development and transfer of marine technology. The Convention adopts language that is relied upon in the following instruments adopted at the United Nations Conference on the Environment and Development (UNCED) in Rio.[23] In addition, many of the techniques adopted in subsequent environmental agreements can be traced directly to UNCLOS.[24] Further developments followed in the negotiations leading up to the Montreal Protocol on Substances that Deplete the Ozone Layer.[25] One of the most contentious issues in those negotiations was technology transfer and assistance to developing countries. The technology at issue was environmentally safe chemical

18. Ernst B. Haas, *When Knowledge is Power: Three Models of Change in International Organizations* (Berkeley: University of California Press, 1990), 77.
19. UNCTAD Draft International Code of Conduct for the Transfer of Technology. UN Doc. TD/CODE TOT/14 (28 Mar. 1979).
20. Wolfgang Fikentscher, *The Draft International Code of Conduct on the Transfer of Technology – A Study in Third World Development*, IIC Studies in Industrial Property and Copyright Law, vol. 4 (Basel: Verlag Chemie, 1980), 7–8.
21. See Sell, fn. 16 *supra* for a comprehensive analysis.
22. United Nations Convention on the Law of the Sea, 16 Nov. 1994, 21 ILM 1245 (1982).
23. Sands, fn. 7 *supra*, 1039–1040.
24. *Ibid.*
25. Montreal Protocol on Substances that Deplete the Ozone Layer to the Vienna Convention for Protection of the Ozone Layer, 1 Jan. 1989, 26 ILM 1541 (1987).

substitutes to chlorofluorocarbons (CFC's). These discussions raised concerns on whether intellectual property rights would make necessary technology transfers burdensome for developing countries.[26] The language first adopted in UNCLOS also appears in the Convention and the UNFCCC (United Nations Framework Convention on Climate Change). Agenda 21, which has been described as a 'non-binding blueprint and action plan for a global partnership for sustainable development',[27] devoted a whole chapter, Chapter 34 entitled 'Transfer of Environmentally Sound Technology', to the issue of technology transfer. History thus tells us that technology transfer provisions seem to be good examples of legal transplants. Wiener has noted that 'in practice, busy diplomats tend to use what they know: having negotiated a treaty on one topic, they apply its concepts and design to the next treaty on a new issue'.[28] The negative side of this is that opposition to a concept formed in one treaty setting also transfers to the next. Although most of the norms on technology transfer have legally binding form they are examples of soft formulations which demonstrate considerable uncertainty with regard to their specific content and leave much room for interpretation and elaboration.

We can conclude that the technology transfer concept produced by the NIEO negotiations has a strong programmatic character. Obviously, this is a normative version of the technology transfer concept denoting a political project towards greater multilateralism and international cooperation. As such, technology transfer should rather be described as a political concept which captures visions or imaginations of what technology transfer can and should do, rather than what it does currently. The main contention about technology transfer was, and still is, about the conditions for the transfer of the technology, i.e. whether technology should be transferred at concessional or commercial rates. This contention still divides the technology transfer concept into two distinct versions – one interventionist and regulatory, and another market-based.

2 TECHNOLOGY TRANSFER IN THE CONTEXT OF THE CONVENTION

Any discussion on the CBD and technology transfer must start with acknowledging that technology transfer is given an important role in the convention. Article 1 reads as follows:

> The objectives of this Convention, the be pursued in accordance with its relevant provisions, are the conservation of biological diversity, the sustainable use of its components, and the fair and equitable sharing of the benefits arising out of the utilization of genetic resources, including by appropriate access to

26. Karen T. Litfin, *Ozone Discourses. Science and Politics in Global Environmental Cooperation* (New York: Columbia University Press, 1994).
27. UNCED Report, A/CONF.151/26/Rev.1, 1993.
28. Jonathan B. Wiener, 'Something Borrowed for Something Blue: Legal Transplants and the Evolution of Global Environmental Law', 27 *Ecology L Q* 1300, 1301 (2001).

> genetic resources and by appropriate transfer of relevant technologies, taking into account all rights over those resources and to technologies, and by appropriate funding.

Moreover, the function of technology transfer in the context of the CBD is most explicitly revealed in Article 20. Article 20(4) of the CBD reads:

> [T]he extent to which developing country Parties will effectively implement their commitments under this Convention will depend on the effective implementation by developed country Parties of their commitments under this Convention related to financial resources and transfer of technology and will take fully into account the fact that economic and social development and eradication of poverty are the first and overriding priorities of the developing country Parties.

Technology transfer thus primarily functions as an enabler of meaningful developing country participation in the Convention and as a legal concept; technology transfer gives expression to the principle of common but differential treatment.[29] Thus, differentiation of responsibilities is justified by differences among countries in financial and technological capacity.

2.1 Provisions on Technology Transfer in the CBD

In addition to Article 1 and Article 20(4) mentioned above, issues relating to technology transfer and cooperation are addressed in Articles 16 to 19 of the CBD. Other provisions, such as Article 15, contain important elements regarding technology transfer in the context of access to genetic resources and benefit sharing, and it should be noted that the Bonn Guidelines[30] identify technology transfer as a benefit-sharing option. Article 22 contains a provision on intellectual property rights, which is also of relevance to technology transfer issues.

As with most other technology transfer provisions, many of the provisions in the CBD – such as Articles 16–19 – are 'soft' in the sense that they are open-textured or general in their content and wording. The CBD is a result of a complex bargaining process achieved through a network of North-South compromises. The most relevant trade-off in this context was concessions on the protection of intellectual property rights (Article 16(2)–(3)) made by the South on one hand, and technology transfer (Article 16(3)–(5)) and the rights of indigenous peoples' and local communities' rights (Article 8(j)) made by the North on the other hand.[31]

Moreover, the treatment of technology transfer was one of the issues that the USA explicitly objected to at the Rio Summit. The USA, in a declaration made in

29. Alastair Iles, 'Rethinking Differential Obligations: Equity under the Biodiversity Convention', 16 *Leiden Journal of International Law* 249 (2003).
30. Bonn Guidelines on Access to Genetic Resources and Fair and Equitable Sharing of the Benefits Arising Out of their Utilization, COP 6 Decision VI/24.
31. Désirée M. McGraw, 'The CBD – Key Characteristics and Implications for Implementation', 11 *Rev Eur Community and Int'l Envtl L (RECIEL)*, 24, 6–27 (2002).

Rio, stated the following: 'As a matter of substance, we find particularly unsatisfactory the text's treatment of intellectual property rights; finances, including, importantly, the role of the Global Environmental Facility (GEF); technology transfer and biotechnology'.[32]

The two elements which make up technology transfer are concepts in their own rights, and we need to take a closer look at them. Both the concept of technology and the concept of transfer are concepts close to our personal experiences. Let us first take a closer look at the concept of 'technology'.

2.1.1 Technology

The word technology comes from the Greek *technologia*. Definitions of technology abound and are often related to the use of knowledge, the kind of techniques, processes, tools, crafts, resources, systems or methods of organization in order to solve problems or serve some purpose. The term technology has, however, changed significantly over the last 200 years, but it should be noted that there is disagreement over the definition of technology between on the one hand scientists and engineers and on the other hand, social scientists. Scientists and engineers tend to view technology as applied natural science while social scientists usually emphasize technology as a cultural activity.

Relevant technologies under the CBD are those that contribute to meeting the three objectives of the CBD, i.e. technologies that are relevant to the conservation and sustainable use of biodiversity or make use of genetic resources and do not cause significant damage to the environment. Technologies relevant to the conservation and sustainable use of biodiversity are broad and often considered not readily identifiable due to scientific uncertainty (which also complicates the management of natural resources in terms of biodiversity), while technologies relevant to Access and Benefit Sharing ABS (genetic resources) are perceived as more readily identifiable. Although technology is not defined in the CBD,[33] technology explicitly includes biotechnology ('"technology" includes biotechnology', Article 2). '"Biotechnology" means the variability among living organisms from all sources including, *inter alia*, terrestrial, marine and other aquatic ecosystems and the ecological complexes of which they are part; this includes diversity within species, between species and of ecosystems' (Article 2). The rapid developments of modern biotechnology and biochemistry including genomics, proteomics and synthetic biology makes also biotechnology a rapidly changing concept. Does, for example, synthetic biology fall within the scope of the CBD?[34] Also technologies that were developed

32. US Declaration made at the UNEP, 22 May 22 1993; 31 I.L.M. 848.
33. Sands, for example, states that '[t]echnology transfer is a term which is frequently used, with little consideration given to what it actually means'. But he does not try to tell us 'what it [*technology transfer*] actually means' or could mean. Sands, fn. 7 *supra*, 1039.
34. '40 Shades of Technology Transfer. A Compilation of Concepts' (Institute for Biodiversity – Network (IBN), Germany, 2004). <ww.cbd.int/doc/case/-studies/tttc/tttc-00038-en.pdf>, illustrates the diversity in technology well.

and are used by indigenous and local communities embodying traditional lifestyles are included into the concept of technology (compare Article 8(j) of the CBD). Technology is generally understood as including both 'hard' and 'soft' technology. Coughlin claims that the CBD uses the term 'technology' too broadly with the result that 'the negotiating parties end up in an all-or-nothing struggle for rights over the entire universe of technology and fail to see the possibilities for compromise that inhere in an approach which divides up the pie'. He uses the Merck-INBio Agreement to analyse the CBD provisions on technology transfer and argue that three different kinds of technology can be distinguished in the Agreement. First, there are conservation technologies which include scientific understandings of ecology and biology. Second, there is technology necessary for biotechnological research which can be used to identify and produce a marketable pharmaceutical. Third, there is technology which enables imitation of specific existing pharmaceuticals or agricultural products. According to Coughlin, it would be possible to clearly distinguish between different classes of technology similar to the Merck-INBio Agreement.[35]

2.1.2 Transfer and Cooperation

What should be done with technologies and how, is contained in the concept of 'transfer'. The concept of 'cooperation' reaches further still. But understandings about the meaning of 'transfer and cooperation' are diverse. This is also why calls for conceptual transformation into other concepts such as technology governance have been made.

The concept of 'transfer' has a more outstanding legal character than 'technology' but it is also the term 'transfer' that often tends to be conceptually stretched far from the meaning it has in ordinary or legal language. Only in political language can meaning be allowed to stretch so far; the concept of 'transfer', then, acquires a more metaphorical character. The IPCC has however used a broad and inclusive understanding of 'transfer'.[36] Their definition 'encompasses diffusion of technologies and technology cooperation across and within countries. It covers technology transfer processes between developed countries, developing countries and countries with economies in transition, amongst developing countries, amongst developing countries and amongst countries with economies in transition'. Therefore there are reasons to think carefully about what is transferred and the nature of these transfer processes. The IPCC further states that the concept of 'transfer' comprises learning processes to understand, utilize and replicate technology, including the capacity to choose it and adapt it to local conditions and integrate it with indigenous technologies. However, when we speak about 'technology cooperation', as developing countries often prefer to do, the concept of transfer is exchanged for a more appropriate concept. There are also other concepts which are sometimes linked to technology in order to stretch the meaning of technology

35. Michael D. Coughlin, 'Using the Merck-INBio Agreement to Clarify the Convention on Biological Diversity', 31 *Colum J Transnat'l L* 31 358, 358–360 (1993–1994).
36. Footnote 45 *infra*.

transfer. One such concept is 'capacity building'. Capacity building relates more to the conditions prevailing in the country receiving the technology than to the conditions for the transfer. According to Roffe and Tesfachew:

> The assumption was that once technology was transferred, technological upgrading would take place which, in turn, would enhance local productive capacity. Consequently, much of the analysis focused on the imperfections of the technology transfer process and on the rôle played by transnational corporations. Little attention was paid to domestic absorptive and adaptation capacity.[37]

Andersen, Sarma and Taddino[38] also identify lack of attention to local conditions and inattention to market incentives as obstacles to technology transfer.

2.2 Technology Transfer as a Cross-Cutting Issue

The CBD process has produced, among other things, a *Programme of Work on Technology Transfer and Technological and Scientific Cooperation.*[39] Further, the Conference of Parties (COP) to the Convention has established an Ad Hoc Technical Experts Group on Technology Transfer and Scientific and Technological Cooperation (AHTEG), and in decision VIII/12, COP requested the Executive Secretary to explore the possibilities of developing a 'Biodiversity Technology Initiative'. At the meeting of the AHTEG, in Geneva (10–12 September 2007), the experts noted, among other things, the importance of clearly conceptualizing or defining technology transfer. Further, the Group of Experts highlighted the importance of linking the technology transfer activities of the CBD to other processes and fora focused on technology transfer, such as the World Trade Organization (WTO). The COP has repeatedly emphasized that further work is required to help develop a common appreciation of the relationship between intellectual property rights and the relevant provisions of the TRIPS Agreement,[40] and the CBD, in particular on issues relating to technology transfer.[41] Recently, a report on the role of intellectual property rights in technology transfer was published by the Secretariat to the CBD.[42] Although I do not address intellectual property issues in this

37. Pedro Roffe & Taffere Tesfachew, 'Revisiting the Technology Transfer Debate: Lessons for the New WTO Working Group', *Bridges*, 2 Feb. 2002 <www.iprsoneline.org/ictsd/docs/RoffeTesfachewBridgesYear6N2February2002.pdf>.
38. Stephen O. Andersen, K. Madhava Sarma & Kristenten N. Taddonio, *Technology Transfer for the Ozone Layer: Lessons for Climate Change* (London: Earthscan, 2007), 8.
39. Adopted by COP 7 in Decision VII/29.
40. Agreement on Trade-Related Aspects of Intellectual Property Rights (TRIPS). The TRIPS Agreement is Annex 1C to the Marrakesh Agreement Establishing the World Trade Organization, done at Marrakesh 15 Apr. 1994, 33 I.L.M. 1125, 869 U.N.T.S. 299.
41. See e.g., COP 3 Decision III/17.
42. UNEP/CBD/AHTEG-TTSTC/1/OTH/TECHSTUDY.

text, it should be noted that the role of intellectual property rights in the development and transfer of climate change adaptation and mitigation technologies has emerged as a particularly contentious issue.[43]

An appropriate starting point, in my view, is the interdisciplinary nature of technology transfer. There is still an absence of any coherent comprehensive theory of technology transfer.[44] There are numerous institutions working with technology transfer issues. In the contemporary political language this is a 'governance' problem, i.e. it concerns relations between different regimes (institutions and rules), for example the relationship between WTO rules (TRIPS Articles 66-67) and the CBD. A focus on legal aspects on technology transfer might comprise a number of different aspects such as, for example, institutional rules or human rights. A move towards 'technology governance' could also include issues concerning institutional design (governance structures).

2.3 Negotiating Technology Transfer

As can be seen there is still considerable uncertainty about the nature of and kinds of technologies relevant to technology transfer under the CBD. The situation is similar to the climate regime. Even though there is no established definition of technology and technology transfer in the UNFCCC, at least with reference to climate technologies a broader definition suggested by the Intergovernmental Panel on Climate Change (IPCC)[45] is increasingly being implied defining technology not only in terms of equipment but also covers relevant flow of hardware, software, information and knowledge.[46] The IPCC has defined technology transfer as:

a broad set of processes covering the flows of know-how, experience and equipment for mitigating and adapting to climate change amongst different stakeholders such as governments, private sector entities, financial institutions, non-governmental organizations (NGOs) and research/education institutions.[47]

Note that the IPCC definition of technology transfer goes beyond what can be found in any treaty provision.

Before going further, it is necessary to make clear that Article 23 of the Nagoya Protocol aims to achieve a fairly narrow purpose – the sharing of benefits – while technology transfer often is used in an instrumental way for a number of

43. See for example the joint study 'Patents and clean energy: bridging the gap between evidence and policy' (Munich: EPO/UNEP/ICTSD, 2010).
44. 'Trade and Climate Change. A report by the United Nations Environment Programme and the World Trade Organization' (Switzerland: WTO Secretariat, 2009), 42.
45. The Intergovernmental Panel on Climate Change (IPCC) was jointly established by the United Nations Environment Programme (UNEP) and the World Meteorological Organization (WMO); confirmed by the UN General Assembly through resolution 43/53 on 6 Dec. 1988.
46. Alina Averchenkova, 'The Outcomes of Copenhagen: The Negotiations & the Accord', UNDP Environment & Energy Group Climate Policy Series, 2010, 24.
47. 'Methodological and Technological Issues in Technology Transfer. A Special Report of the Working Group III of the IPCC. Summary for Policymakers', at 3.

different purposes. Sands, for example, holds the opinion that '[o]n technology transfer, Agenda 21 aims to accommodate so many differing views that it is difficult to see how its provisions could be translated into meaningful, practical action'.[48] Here, the programmatic character of the technology transfer concept becomes obvious.

The EU, in its thematic report on technology transfer and cooperation under the CBD, enclosed a report to the WTO on 'Implementation of Article 66.2' and a communication to the WTO Working Group on Trade and Transfer of Technology.[49] In the communication, the EU states: 'The [EU] intends to contribute to this work programme, which should address the issue of technology transfer in a comprehensive and ambitious manner, taking due account of the objective of sustainable development'.[50] While this statement is good evidence of emerging shared understandings of strong linkages between technology transfer and sustainable development it is not difficult to find examples of a lack of shared understandings about the meaning of technology transfer. The EU for example suggests that '[i]n order to move towards a common understanding of technology transfer, the Working Group should seek to agree on a clear, simple and workable definition of its two components [technology and transfer]'.[51]

Finally, the original mandate of the Ad Hoc Open-ended Working Group on Access and Benefit-sharing (WGABS) was to negotiate an international regime. I would argue that this has consequences for the assessment and the future workings of the Nagoya Protocol. As pointed out by Lang, '[l]egal regimes not only regulate behaviour within their domain of operation, but also help constitute the social domains on which they act'.[52] A focus on the constitutive function of norms would direct our attention to other parts of the process. The discourse on technology transfer is rooted in controversies in the NIEO debates which still create considerable division between the North and the South.[53]

Despite its important function in the CBD, in particular for the third objective, technology transfer has not had a central place in the ABS negotiations. However, as discussed above, technology transfer is at the core of North-South relations and therefore fundamental issues of equity (justice) arise. For the South technology transfer has always been about justice.[54] Whether shared meanings on technology

48. Phillipe Sands, 'International Law in the Field of Sustainable Development', 65 *British Yearbook of International Law* 375 (1994).
49. The EU thematic report on technology transfer and cooperation thematic can be found on the Convention webpage. <www.cbd.int/doc/world/eur/eur-nr-stc-en.pdf>.
50. EU thematic report on technology transfer and cooperation, 66.
51. *Ibid.*
52. Andrew Lang, 'Legal Regimes and Regimes of Knowledge: Governing Global Services Trade', LSE Law, Society and Economy Working Papers 15/2009, at 1.
53. Sell, fn. 16 *supra*.
54. Historically, 'justice' in this context mostly referred to 'redistributive justice'. In contemporary political and policy debates reference is more often made to 'global justice'. Meanings of 'global justice' are however under construction. One of the processes through which meanings of 'global justice' are becoming, I argue, is in the making of the concept of benefit-sharing. See Thomas Nagel, 'The Problem of Global Justice', 33 *Philosophy & Public Affairs* 113

transfer as a means of benefit-sharing (justice) has developed among all actors (also the North) remains to be seen. In any case, it might be too early to make any meaningful assessment. The most important function of Article 23 (or the shared understandings flowing from the process) is probably to enable local, contextually adapted solutions 'rather than the roll out of one big blueprint'.[55] Only the future can tell whether this will happen. It is clear though that Article 23 contains several important elements, such as sensitivity to ecological vulnerability and attention to vertical normative relationships, which require further development. So what remains to be done in practice? One thing that could be done is to differentiate states' reporting obligations so that technology transfer becomes more visible. If reporting obligations would clearly define the kind of activities that count as technology transfer, states could not 'hide' behind pragmatism or benevolence.[56] As long as technology transfer activities are not acknowledged for being synonymous with technology transfer and not foreign aid or something else, prospects for future legal developments remain weak. Another significant step toward legal recognition would be to acknowledge that technology transfer is required by justice. Such recognition could catalyse the emergence of shared understandings and social norms.

2.4 Sustainable Technology Transfer?

Technology transfer and sustainable development might in many ways look like strange bedfellows. They could both be considered 'essentially contested concepts'[57] which analytically may seem more or less incapable of interacting in meaningful ways. However, it is policymakers who have linked them and it is the processes (co-production) that link and give meaning to them in different contexts which is my main concern, not the elaboration of static operational definitions. More important, sustainable development forms part of the context in which the technology transfer norms at issue here (Article 23 of the Nagoya Protocol) have been negotiated and normatively developed. In this context technology transfer and sustainable development are mutually constitutive, and viewed as a pragmatic undertaking, the relative vagueness of the concepts probably enhances conceptual interaction. In fact, the institutional context in which these two concepts meet must also be taken into account.

(2005); Gunther Teubner, 'Self-subversive Justice: Contingency or Transcendence Formula of Law?', 72 *Mod L Rev* 1 (2009), for interesting discussions.

55. Maja Goepel, 'Formulating Future Just Policies: Applying the Delhi Sustainable Development Law Principles', 2 *Sustainability* 1695 (2010).
56. Brunnée and Toope note that many industrialized countries prefer to see financial and technical assistance as a matter of pragmatism or benevolence, rather than something required by law; see fn. 9 *supra*, 161.
57. Perhaps I should add 'at best'. For example Dryzek argue that '[s]ustainable development should be treated as a discourse, not a concept capable of precise definition'; John S. Dryzek, *Deliberative Global Politics: Discourse and Democracy in a Divided World* (Oxford: John Wiley & Sons, 2006), 17.

2.4.1 The Institutional Context for Norm Making

The CBD is – together with the UNFCCC and the United Nations Convention to Combat Desertification (UNCCD)[58] – often regarded as a 'new' kind of environmental agreement which incorporates a number of innovative elements as part of sustainable development discourse: global character, strong North-South dimension, role of participatory process, long-term horizon, and innovative implementation schemes.[59] One distinguishing feature is the dynamic development of norms taking place particularly through adoption of protocols to the conventions. The Nagoya Protocol for example finds its legal basis in Article 28 of the CBD, which states the following: 'The Contracting Parties shall cooperate in the formulation and adoption of protocols to this Convention'. According to Article 1 of the CBD it has three objectives: (1) conservation of biological diversity, (2) sustainable use of its components, and (3) the fair and equitable sharing of the benefits derived from the use of genetic resources, including by appropriate access to genetic resources and by appropriate transfer of relevant technologies. The CBD recognizes both the intrinsic value of biodiversity and the conservation of biodiversity as a 'common concern of humankind'. The CBD reaffirms that states have sovereign rights over their own biological resources and specifically sets out that principle in the operative part of the convention, in Article 3. According to McGraw, the CBD's key innovation is the interrelationship between conservation, sustainable use and benefit sharing.[60] In other words, the CBD is not just about conservation; it is also an economic agreement. The economic nature of issues is apparent in the ABS issue. It is also apparent in the context of, for example, technology transfer and scientific cooperation. In fact, some commentators have remarked that the treaty might just as well have been denominated the 'Convention on Biotechnology Transfer'.[61] As mentioned above, the adoption of Articles 15 to 19 (providing for the sharing of biotechnology research results) was the ultimate reason the US did not ratify the CBD. But it is the post-agreement negotiations that have been most challenging, largely due to the fact that several of the most contentious issues were left unresolved at the time of the adoption of the CBD.[62] Both the negotiation process that led to the adoption of the Cartagena Protocol on Biosafety[63] and the ABS negotiations provide good evidence.

---

58. In those Countries Experiencing Serious Drought and/or Desertification, Particularly in Africa, Paris, 14 Oct. 1994, Entry into force: 26 Dec. 1996.1954 U.N.T.S. 3/ [2000] A.T.S. 18/ 33. (UNCCD).
59. Hans Bruyninckx, 'Sustainable Development: The Institutionalization of a Contested Policy Concept', in *Palgrave Advances in International Environmental Politics*, ed. Michele Betsill, Kathryn Hochstetler & Dimitris Stevis (Basingstoke: Palgrave Macmillan, 2006), 287.
60. 'The CBD – Key Characteristics and Implications for Implementation', fn. 8 *supra*, 24.
61. Dan L. Burk, Kenneth Barovsky & Gladys H. Monroy, 'Biodiversity and Biotechnology', 260 *Science* 1900 (25 Jun. 1993).
62. 'The CBD – Key Characteristics and Implications for Implementation', fn. 8 *supra*, 28.
63. Cartahena Protocol on Biosafety to the Convention on Biological Diversity, 39 I.L.M. 1027 (2000); UN Doc. UNEP/CBD/ExCOP/1/3/ (2000).

According to Footer it is clear that COP Decision VII/19 – the decision that set the terms of reference for the WGABS for the elaboration of an international ABS regime – breaks new ground in terms of institutional governance by promoting an institutionalized process that is both deliberative and participatory.[64] As such, the process is an example of more innovative policy-making discourses – an example of a sustainable development policy process.[65] The CBD is thus best viewed as a 'framework sustainable development convention' which overlaps with agreements beyond the environmental realm.[66] Indeed, framework conventions are a particular kind of institutional arrangement. The COP, assisted by, and interact with, various sub-bodies like the WGABS, provides a forum for on-going dialogue. COP's 'represent hybrids between issue-specific diplomatic conferences and the permanent plenary bodies of international organizations'.[67] The framework approach is 'designed to postpone difficult negotiating issues, but to keep at them'.[68] As such the framework approach is thought to be able, over time, to produce shared understandings about common problems.[69] Sustainable development discourse also calls for the recognition and promotion of strong linkages between biodiversity and a number of other issues; poverty eradication, food security, provision of fresh water, soil conservation and human health.[70]

2.4.2 The Principle of Common but Differentiated Responsibilities

In view of the important function technology transfer performs in the CBD, not just for meaningful developing country participation but also for further normative development, we will take a closer look at the principle of common, but differentiated, responsibilities. According to Birnie et al. the most significant function of the principle of common, but differentiated, responsibilities is to establish a framework for cooperation and compromise. They characterize the principle as a 'framework principle' and point out that it is principally an obligation to cooperate in developing the law.[71] Iles, however, explores how differential obligations can be

---

64. Mary E. Footer, 'Agricultural Biotechnology, Food Security and Human Rights', in *Biotechnology and International Law*, ed. Francesco Francioni, & Tullio Scovazzi (Oxford: Hart Publishing, 2006), 269.
65. 'Sustainable Development: The Institutionalization of a Contested Policy Concept', fn. 59 *supra*, 268.
66. 'The CBD – Key Characteristics and Implications for Implementation', fn. 8 *supra*.
67. Brunnèe Jutta, 'COPing with Consent: Law-Making under International Environmental Agreements' 15 *Leiden J Int'l L* 1, 16 (2002).
68. *Ibid.*, 8.
69. Jutta Brunnée & Stephen Toope, 'Environmental Security and Freshwater Resources: Ecosystem Regime Building', 91 *Am J Int'l L* 26, 29–34 (1997).
70. Catherine Redgwell, 'Biotechnology, Biodiversity and Sustainable Development', in *Biotechnology and International Law*, ed. Francesco Francioni & Tullio Scovazzi (Oxford: Hart Publishing, 2006), 66.
71. Birnie Patricia, Alan Boyle & Catherine Redgwell, *International Law & the Environment* (Oxford: Oxford University Press, 2009), 133.

reworked in terms of environmental justice in the context of biological diversity and suggests a focus on their implementation rather than on negotiation.[72]

The principle of common but differentiated responsibility evolved from the notion of the common heritage of mankind and is a manifestation of general principles of equity in international law. From a functional point of view justice legitimizes asymmetry of commitment. In view of the principle of sovereign equality, differential treatment obviously needs both political and legal justification. The principle of common but differentiated responsibility first appeared in Principle 7 of the Rio Declaration, which states: 'In view of the different contributions to global environmental degradation, states have common but differentiated responsibilities'.[73] Common responsibility is necessary due to economic interdependence and the global nature of environmental problems. Common responsibility is rooted in the principle of cooperation while differentiated responsibilities are an articulation of the acknowledgement of industrial country contributions to the global environmental crisis.[74] Differentiated responsibility also reflects differing capacities of states to take remedial measures. It could, however, be argued that asymmetric commitments have been cast narrowly in the sense that these commitments remain static and focused on treaty-making instead of treaty implementation.[75] Further, actual decision-making and everyday practice under the CBD demonstrate that differential obligations have been greatly limited in their scope and effect.[76]

A significant problem with the principle of common but differentiated responsibility is that it is state-centric. It reflects the global political organization of sovereign states and addresses justice between states, not between individuals.[77] Notwithstanding the fact that environmental justice is sometimes about intra-societal relations, global environmental justice is about the global distribution of burdens and benefits. In order to move beyond state interests towards more cosmopolitan conceptions of justice, a focus on other processes than negotiation processes are necessary.

Although the distinction between developed and developing countries in the CBD is not as clear as in the UNFCCC, I would argue that there are opportunities for further differentiation. Yet while debates on further differentiation continue in the UNFCCC, this can hardly be said to be the case with regard to the CBD.

---

72. 'Rethinking Differential Obligations: Equity under the Biodiversity Convention', fn. 29 *supra*, 219.
73. Rio Declaration on Environment and Development (Annex 2); Report of the UN Conference on Environment and Development, UN Doc A/CONF.151/26, 31 I.L.M. 874.
74. Lavanya Rajamani, 'The Principle of Common but Differentiated Responsibility and the Balance of Commitments under the Climate Regime', 9 *Rev Eur Community & Int'l Env'l L (RECIEL)* 120, 121–122 (2000).
75. Rethinking Differential Obligations: Equity under the Biodiversity Convention', fn. 29 *supra*, 219.
76. *Ibid.*, 229–232.
77. Bradley C. Parks & J. Timmons Roberts, 'Environmental and Ecological Justice', in *Palgrave Advances in International Environmental Politics*, ed. Michele Betsill, Kathryn Hochstetler & Dimitris Stevis (Basingstoke: Palgrave Macmillan, 2006), 338.

3 ASSESSING ARTICLE 23

Article 23 of the Protocol reads as follows:

> In accordance with Articles 15, 16, 18 and 19 of the Convention, the Parties shall collaborate and cooperate in technical and scientific research and development programmes, including biotechnological research activities, as a means to achieve the objective of this Protocol. The Parties undertake to promote and encourage access to technology by, and transfer of technology to, developing country Parties, including the least developed and small island developing States among them, and Parties with economies in transition, in order to enable the development and strengthening of a sound and viable technological and scientific base for the attainment of the objectives of the Convention and this Protocol. Where possible and appropriate such collaborative activities shall take place in and with a Party or the Parties providing genetic resources that is the country or are the countries of origin of such resources or a Party or Parties that have acquired the genetic resources in accordance with the Convention.

Furthermore, Article 5 ('Fair and Equitable Benefit-Sharing') of the Protocol, to which Article 23 refers, develops Articles 1 and 15 of the CBD. Article 5.1 states:

> In accordance with Article 15, paragraphs 3 and 7 of the Convention, benefits arising from the utilization of genetic resources as well as subsequent applications and commercialization shall be shared in a fair and equitable way with the Party providing such resources that is the country of origin of such resources or a Party that has acquired the genetic resources in accordance with the Convention. Such sharing shall be upon mutually agreed terms.

The objective of the Protocol is set out in Article 1:

> The objective of this Protocol is the fair and equitable sharing of the benefits arising from the utilization of genetic resources, including by appropriate access to genetic resources and by appropriate transfer of relevant technologies, taking into account all rights over those resources and to technologies, and by appropriate funding, thereby contributing to the conservation of biological diversity and the sustainable use of its components.

The concept of benefit-sharing is open-ended. It still has not acquired any stable meaning. What can safely be said, however, is that what is fundamentally at stake in this elaboration of the concept of benefit-sharing is justice. However, concepts and theories of global justice are in the early stages of formation.[78] As will become evident below, considerations of justice in the context of the CBD are expressed mainly through the principle of common but differentiated responsibilities. Equity considerations, of course, have a central place in the third objective of the CBD; it is still the 'fair and equitable' benefit-sharing that concerns us.

---

78. Thomas Nagel, 'The Problem of Global Justice', 33 *Phil & Pub Aff* 113 (2005).

Before looking further into the Article I will reproduce the latest bracketed draft version of Article 18*bis* (before the finally adopted version) which reads:

> In accordance with Articles 15, 16, 18 and 19 of the Convention, Parties shall collaborate and cooperate in scientific research and development programmes, including biotechnological research activities, as a means to achieve the objective of this Protocol. [This collaboration and cooperation [shall] [should] include, inter alia, measures by developed country Parties that provide incentives, to companies and institutions within their jurisdiction,][Parties undertake] to promote and encourage access to technology by, and transfer of technology to, developing country Parties, including the least developed and small island developing states among them and Parties with economies in transition, in order to enable the development and strengthening of a sound and viable technological and scientific base for the attainment of the objectives of the Convention and of this Protocol. Where possible, [such] collaborative activities [shall][should] take place [with][in] the [[countries] [country][providing] country of origin of] genetic resources.[79]

A careful reading shows that there are several levels of commitment or obligation in the article. The main legal obligation flowing from the Article, to 'collaborate and cooperate . . . ', simply reiterates cross-referenced Articles 15, 16, 18 and 19 of the CBD and thus does not go beyond what is already in the CBD. The obligation to 'promote and encourage . . . ' takes a step further. However, while earlier versions contained an obligation to 'provide incentives to companies and institutions within their jurisdiction'[80] the adopted version solely contains an obligation to 'promote and encourage . . . ' Somehow this makes the earlier draft version more explicit concerning one of the fundamental problems confronting the negotiators; the fact that the targets for regulation are non-State actors. The next level of obligation concerns where collaborative activities geographically shall take place. The inclusion of the last sentence in Article 23 – 'Where possible and appropriate such collaborative activities shall take place in and with a Party or the Parties providing genetic resources . . . ' – was important to developing countries, and from a cultural view of technology which see technology as a result of needs arising in a particular societal context, rather than being driven by some kind of universal inner logic, this makes sense.

79. The draft text is as it was after the resumed meeting of the ING in Montreal from 18 to 21 Sep. 2010. Note that square brackets indicate texts lacking consensus. See the Annex to the Report of the Second Part of the Ninth Meeting of the Ad Hoc Open-ended Working Group on Access and Benefit-Sharing, UNEP/CBD/COP/10/5/Add.4.

80. This language ('to provide incentives') can also be found in Art. 66.2 TRIPS. However, we must not look outside the CBD to find such language; Art. 11 of the CBD calls on parties to 'adopt economically and socially sound measures that act as incentives for the conservation and sustainable use of components of biological diversity'. This language rather reflects ongoing debates on regulatory techniques; whether traditional forms of 'command-and-control' regulation, or new techniques which make use of economic incentives should be used. See e.g., Sands, fn. 48 *supra*, 368.

As may be seen, the negotiators had several brackets to take care of in Nagoya. The remaining square brackets were taken care of by the negotiators through bargaining 'in a spirit of compromise' late at night the day before the Protocol was adopted. Still, there was not much debate about these provisions among the negotiators. As can be seen it is mostly the formal legal status (bindingness) which is contested. For example, how technology transfer can be used to share benefits was not much debated in the negotiations. Article 5.4 of the Protocol however states that 'Benefits may include monetary and non-monetary benefits, including but not limited to those listed in Annex 1'. Annex 1 is an extended version of language first articulated in the Bonn Guidelines and in that context more debate has taken place. Annex 1, for example, mentions 'Joint ownership of relevant intellectual property rights' as a monetary benefit-sharing option, and 'Sharing of research and developments results', 'Collaboration, cooperation and contribution in scientific research and development programmes, particularly biotechnological research activities, where possible in the country providing genetic resources' and 'Strengthening capacities for technology transfer' as non-monetary benefit-sharing options. All of these concepts can now be found in the Protocol.

Next we may ask what is not in the provision. The provision does, for example, not say anything about the nature of technology. The problem with defining technology transfer becomes evident. Note that Article 1 of the CBD says 'by appropriate transfer of relevant technologies' which makes the qualification 'relevant' decisive. However, the lack of definition risks defining technology transfer as anything from a single patent license to technological self-reliance (sovereignty). In fact, it does not seem that Article 23 does much to clarify or develop the technology transfer provisions in the CBD.

From a 'systemic' point of view, it must be pointed out that the Protocol is not a self-contained system. I should also point to the fact that Article 22 of the Protocol deals with Capacity and Article 24 deals with Non-Parties (read the USA).[81] Article 22.5 states:

> Measures in accordance with paragraphs 1 to 4 may include, inter alia; (g) 'technology transfer', and infrastructure and technical capacity to make such technology transfer sustainable.

Considering the strong links between biodiversity and issues such as food security and health that sustainability discourse calls for,[82] there are good reasons for broadening the meaning of 'relevant technologies' or even re-think the concept of technology.

---

81. Article 24 of the Nagoya Protocol on Non-Parties reads: 'The Parties shall encourage Non-Parties to adhere to this Protocol and to contribute appropriate information to the Access and Benefit-Sharing Clearing House.'
82. Catherine Redgwell, 'Biotechnology, Biodiversity and Sustainable Development', in *Biotechnology and International Law*, ed. Francesco Francioni & Tullio Scovazzi (Oxford: Hart Publishing, 2006), 122.

The references to sustainable development in the preamble and Article 22 are important. Authority to act and develop international environmental law can be found in the Rio Declaration – Principle 27 mandates the international community to 'further develop international law in the field of sustainable development'. Exactly what that means is not clear but it would not be too far reaching to say that sustainable development has become the overarching concept which is supposed to guide us into a future world. In reality, however, the situation should rather be described as 'stalled legalization'.[83] Environmental law has not 'hardened' but is still mostly of a soft nature. A question is whether the formal legal status of technology transfer provisions are of significance or if we rather should look to the effectiveness of technology transfer in more a more socio-legal fashion, i.e. rather than asking what the law is we should ask what it can do.

I will show that Article 23 can also be understood in terms of environmental justice, i.e. the obligations 'can be construed to require justice in everyday benefit-sharing and sustainable use instead of only in transfers of resources between nation-states'.[84] Such an interpretation is also consistent with recent thinking on the evolution of benefit-sharing obligations.[85] A perhaps too optimistic reading of the Article would be that it responds to some of the normative challenges confronting international environmental law in general. However, it is doubtful whether we can talk about a 'new legal ethic'. I would rather say that Article 23, as it is articulated, provides opportunities for going beyond State-to-State obligations in the implementation at a national level. In fact, implementation should also be regarded as part of the normative process. Law-making and application are not considered strictly separate in an interactional framework, and new normative understandings may arise through interpretative processes or processes designed to promote compliance.[86]

The provision further makes distinctions between developed and developing states, and among developing States two kinds of developing States; least developed and small island developing States. In addition to those, Parties with economies in transition are recited. It is here that the provision most explicitly gives expression to the principle of common but differentiated responsibilities.

In this context it should be noted that one of the most debated questions is whether the division between developed and developing countries is too simplified. While Article 23 responds to this complex reality by also including countries with 'economies in transition', the principle of common but differentiated responsibilities has been debated more in the context of climate change, especially the Kyoto Protocol,[87] where the appropriateness of the institutionalized North-South

83. Ken Conca, 'Environmental Governance after Johannesburg: From Stalled Legalization to Environmental Human Rights?', 1 *J Int'l L & Int'l Rel* 121 (2005).
84. Rethinking Differential Obligations: Equity Under the Biodiversity Convention, fn. 29 *supra*, 249.
85. Elisa Morgera & Elsa Tsioumani, 'The Evolution of Benefit Sharing: Linking Biodiversity and Community Livelihoods', 19 *Rev Eur Community & Int'l Env'l L (RECIEL)* 150 (2010).
86. COPing with Consent: Law-Making under International Environmental Agreements, fn. 67 *supra*, 36.
87. Kyoto Protocol to the United Nations Framework Convention on Climate Change, 10 Dec. 1997, 37 I.L.M. 22 (1998) (Kyoto Protocol).

division and corresponding legal obligations has been questioned.[88] In fact, this was the issue that made the climate meeting at COP 15 in Copenhagen a failure.

The International Law Association (ILA) 2002 New Delhi Declaration on the Principles of International Law Related to Sustainable Development[89] is regarded the 'most current benchmark of important principles of international law on sustainable development'.[90] ILA principle 3.2 reads: 'Differentiation of responsibilities, whilst primarily based on the contribution that a State made to the emergence of environmental problems, must also take into account the economic and developmental situation of the State, in accordance with paragraph 3.3.' Principle 3.3 reads: 'The special needs and interests of developing countries and of countries with economies in transition, with particular regard to least developed countries and those affected adversely by environmental, social and developmental considerations, should be recognized'. Moreover, Principle 3.4 declares:

> Developed countries bear a special burden of responsibility in reducing and eliminating unsustainable patterns of production and consumption and in contributing to capacity-building in developing countries, *inter alia* by providing financial assistance and access to environmentally sound technology. In particular, developed countries should play a leading role and assume primary responsibility in matters of relevance to sustainable development.

Thus, it is not unlikely that the inclusion in the Protocol text of 'economies in transition' finds its origin in the ILA's 2002 New Delhi Declaration, which encourages the recognition of countries with economies in transition. Moreover, the reference to 'small island developing States' implies that ecological differences such as greater vulnerability among small island developing States is acknowledged and taken into account, i.e. also pointing in the direction of ecological sustainability.[91]

Redgwell claims that 'one of the challenges facing international environmental law in the twenty-first century is achieving a holistic and integrated approach to environmental regulation which applies within as well as between and beyond states' – what Elliott has called 'a new legal ethic'[92] – going on to refer to the CBD as an example of a treaty instrument moving in this direction.[93] Should this be

88. *Legitimacy and Legality in International Law. An Interactional Account*, fn. 9 *supra*, 154.
89. ILA Resolution 3/2002: New Delhi Declaration on the Principles of International Law Related to Sustainable Development, in ILA, Report of the Seventieth Conference, New Delhi (London: ILA, 2002). <www.ila-hq.org>.
90. Goepel, fn. 55 *supra*, 1697. Note that the ILA is an NGO and that the authority of the Delhi Declaration is to be found in the content, not its form.
91. Compare also the Convention's Preamble; '*Acknowledging further* that special provision is required to meet the needs of developing countries, including the provision of new and additional financial resources and appropriate access to relevant technologies'; '*Noting* in this regard the special conditions of the least developed countries and small island states.'
92. Lorraine M. Elliott, *The Global Politics of the Environment*, 2nd edn (Oxford: Palgrave Macmillan, 2004).
93. Redgwell Catherine, 'International Environmental Law', in *International Law*, 3rd edn, ed. Malcolm D. Evans (Oxford: Oxford University Press, 2010), 688.

true for the making of Article 23, then the challenge ahead is to translate the provision into everyday practices.

So how might technology transfer contribute to the development of the concept of benefit-sharing? And how might they both be affected by sustainable development? In policy debates, technology transfer is becoming more and more linked to sustainable development. The recognition in the Preamble to the Nagoya Protocol is just one of many examples. In fact, '[i]t is the latter two objectives [sustainable use of biodiversity and the fair and equitable sharing of the benefits] which 'bring the CBD into the field of sustainable development'.[94]

According to the United Nation Environment Programme (UNEP). '[t]he Convention's provisions on technology transfer reflect the global consensus among countries that the development, transfer, adaptation and diffusion of technologies, as well as the building of related capacities are critical for achieving sustainable development'.[95] This statement is certainly correct but such consensus cannot be taken as reflective of a social norm. Further normative interaction and deliberation is required for a social norm to evolve. As a general matter, transfer of technology should, of course, be sustainable but since technology transfer can also be a way to realize sustainable development, it should also be supportive of wider development objectives such as the Millennium Development Goals (MDG's).

To sum-up, we can say that the concept of technology transfer is not so much an empirical or legal concept as it is a political (normative) concept capturing visions of what can be done with technology transfer. The kind of problems technology transfer is invoked to solve (or contribute to solving) is connected to broad issues such as sustainable development and global environmental justice. Putting the adjective 'sustainable' in front of technology transfer may mean more than we currently can imagine.

It may indeed be difficult to predict whether we will see any binding, precise international obligations to transfer technology in the future (unless perhaps under very well-defined circumstances). This begs the questions about the role of law in general in technology transfer, i.e. should technology transfer be governed by regulation (if so, what kind of regulation) or market mechanisms? In both cases law has a role to play, but a different one. Hoekman et al. have suggested that '[a] "soft law" approach that establishes broad guidelines and relies on ensuring transparency and accountability via regular multilateral monitoring of performance is likely to be a more effective mechanism to increase cooperation and compliance with commitments ... ' than binding and enforceable technology transfer commitments.[96]

94. International Law in the Field of Sustainable Development, fn. 48 *supra*, 335.
95. 'Technology Transfer and Cooperation under the Convention on Biological Diversity: Towards more Effective Implementation', United Nations Environment Programme (UNEP) (2010), 1, <www.unep.org/dec/PDF/TechnicalTransferCBD.pdf>.
96. Bernard M. Hoekman, Keith E. Maskus & Kamal Saggi, 'Transfer of Technology to Developing Countries: Unilateral and Multilateral Policy Options', *World Bank Policy Research Working Paper 3332*, (World Bank, 2004), 30.

As soft law, technology transfer may also perform important functions in the norm building process. Palmer points out that '[a]ll politicians know the value of ambiguity' and that soft law solutions change the political thinking on an issue.[97] According to Palmer '[s]oft law is where international law and international politics combine to build new norms'.[98]

4 CONCLUSIONS: TOWARDS TECHNOLOGY GOVERNANCE?

While we can safely conclude that the concept of technology transfer is mainly constituted by past experiences, contemporary debates on technology transfer – especially on climate change – show us that debates on access to and transfer of technology are alive and kicking. Although much remains to be done before we can say that technology transfer is a successful enterprise, a significant conceptual shift may be on the move. Gerstetter et al. note that 'the emergence of some developing countries as major players in the climate technology field is likely to result in changed political dynamics and may bring up new issues on who will provide technologies to whom and under what conditions'.[99] Whether the opinion that technology transfer is inefficient or even regarded as a failure, is due to lack of legal obligation is an open question. Nevertheless, both conceptual and normative development is required for technology transfer to become sustainable and successful. Conceptual and normative development, I argue, go hand in hand. More specifically, concepts need clarification and obligations need differentiation.

Several voices raised at the climate meeting in Copenhagen emphasized the need to see beyond technology transfer, i.e. that technology transfer is a narrower concept which has its place under the 'umbrella concept' technology governance. Among those who raised this issue where the Indian Minister of Environment and Forests, Jairam Ramesh and the Director General of the World Intellectual Property Organization (WIPO), Francis Gurry. In fact, one of the outcomes of COP 15 in Copenhagen was the creation of a technology mechanism. Although the institutional structures are still to be decided thinking in terms of institutional design implies thinking of technology transfer as legally providing norms in terms of framework/process rather than substantive obligations.

The design of the climate technology mechanism was however debated. An international group of more than 150 NGO's also released a joint declaration on

---

97. Geoffrey Palmer, 'New Ways to Make International Environmental Law', 86 *Am J Int'l L*, 259, 269 (1992).
98. *Ibid.*, 269.
99. Christiane Gerstetter, Dominic Marcellino & Elena von Sperber, 'Technology Transfer in the International Climate Negotiations – The State of Play and Suggestions for the Way Forward', 3 *Climate and Carbon L Rev* 3 (2010).

Friday 11 December 2009 – 'Let's look before we leap'[100] – calling upon governments in the climate negotiations in Copenhagen to revise the draft text on technology transfer (according to NGOs the climate negotiation text, as per 10 December 2009, included 468 references to technology transfer) to ensure that technology assessment is included and that the precautionary principle is taken into account. NGO's raised concerns that historically, the focus had been on the economic dimensions of technology transfer, while ignoring ecological and social dimensions. This, according to NGO's, is greatly due to private ownership (multinational corporations) of technology. NGO's call for ecologically sustainable, socially just and economically affordable technology transfers. Clearly, developing countries with the support of NGO's are determined to re-shape the technology transfer agenda by including ecological, social and justice issues. In other words, sustainable technology transfer.

Also in the context of the CBD similar issues are raised. Consider the following citation from the suggested strategy for the practical implementation of the programme of work on technology transfer and scientific and technological cooperation:

> Technology transfer, in particular in the context of the third objective of the Convention will not be effective as an on-off and one-way activity, but needs to embedded in a participatory decision-making process as well as in integrated, long-term scientific and technological cooperation, which, as based on reciprocity, would also provide a key mechanism for the effective building or enhancement of capacity in developing countries and countries with economies in transition.[101]

It is of course an immensely complex task – far beyond the scope of this article to also consider the norms of science, which would require a much more careful analysis of the existence of shared understandings and emerging social norms – to map the whole normative landscape under such conditions. In the context of the CBD there are however already institutional structures in place to address these issues. In fact, the Subsidiary Body on Scientific, Technical and Technological Advice (SBSTTA), created under the CBD, could fulfil the function of a technology governing body. According to Article 25(c) of the CBD the SBSTTA, among other things, shall:

> (c) Identify innovative, efficient and state-of-the-art technologies and know-how relating to the conservation and sustainable use of biological diversity and advice on the ways and means of promoting development and/or transferring technologies;. . . .'

Finally, it is often suggested that sustainable development can be structurally conceived as a pillared structure consisting of international environmental law, international human rights law and international economic law. Such a structure

100. ETC Group, Declaration. <ww.etcgroup.org/en/node/4956>.
101. UNEP/CBD/AHTEG-TTSTC/1/5, at 26 (para. 3).

presents sustainable development as an integrating and interactive effort.[102] However, as I have developed my argument in this chapter, the third leg of sustainable development i.e. human rights has been pushed into the background. I believe there are very good reasons for taking human rights more seriously in the field of sustainable development and for integrating ideas about the right to development into concepts of sustainable development. Human rights certainly have important contributions to make to the realization of sustainable development. One such, in my view obvious, contribution would be the right to participate in decision-making related to the environment.[103] There is no doubt significant policy and academic interest in the governance of science and technology for sustainable development,[104] and this is precisely the message in the preamble to the Nagoya Protocol discussed above. In fact, this is one of the most important tasks confronting biodiversity governance today and in the future.

102. Dominic McGoldrick, 'Sustainable Development and Human Rights: An Integrated Conception', 45 *Int'l Comp L Q* 796, 797 (1996).
103. *Ibid.*, 804–805.
104. Peter, Newell, 'Democratising Biotechnology? Deliberation, Participation and Social Regulation in a Neo-liberal World', 36 *Review of International Studies* 471 (2010).

Chapter 10

Sustainable Technology Transfer, Climate Control, and Renewable Power in Vietnam

Steven Ferrey

I have been privileged to work in Vietnam on energy-sector projects with and for the Government of Vietnam, during the past decade. This chapter focuses on the initiatives of, and opportunities for, the Government in the power sector, and how that integrates with other energy-sector efforts in developing countries in Asia. It will examine the new emphasis on control of climate change and global warming gas (GHG) emissions, and the important impact of sustainable technology transfer on future infrastructure development. It will also focus on the importance of sustainable technology transfer as part of collective efforts of 200 world nations to substantially reduce the emissions of global warming gases.

This question arises at an important time. The world stands at a crossroad because, in the next decade, there will be an unprecedented, massive investment in electrification in developing nations. Once installed, those generation facilities will remain in place, contributing to global warming – or not – for 40 years and in many cases longer.[1] According to Rajendra Pachauri, United Nations International Panel on Climate Change (IPCC) Chairman: 'What we do in the next two to three years will determine our future'. These choices in energy technology, made now, will

1. National Energy Foundation, 'Fuel Consumption Stats, Primary Energy Consumption by Fuel' (1999). <http://www.nef1.org/ea/eastats.html>.

Hans Henrik Lidgard, Jeffery Atik & Tu Thanh Nguyen, *Sustainable Technology Transfer*, pp. 221–249.

certainly be the signature of the world carbon footprint for the remainder of this century, during which we may pass the point of no return in terms of global warming impacts.[2]

There must be sustainable technology transfer of renewable energy and demand-management technologies to reduce atmospheric carbon concentrations. There is no Kyoto Protocol[3] requirement that developed economies make any shift to zero-carbon or low-carbon renewable power, and the Clean Development Mechanism (CDM)[4] under the Protocol has navigated another direction in that the majority of projects sanctioned are not renewable energy projects. This article sets forth the proven successful model in several Asian countries, including now in Vietnam that alters the transfer of energy production resources.

My first time in Vietnam was in December 2002. I received a phone call when I was working in Thailand and India, asking if I could come to Hanoi to talk with government officials and stakeholders. After those meetings in Hanoi, the Government of Vietnam proceeded to explore renewable energy options and I came back in 2004–2005 to work with the government and economist Robert Vernstrom, to identify systemic legal and regulatory barriers and incentives to renewable energy development within the Vietnam legal system. That completed, I was asked to come back to Hanoi again in 2007, working with the new energy regulatory authority of Vietnam, ERAV, creating the legal documents and regulatory changes for Vietnam to launch a renewable energy program to encourage the development of independent small power projects.

This was a major and significant step for the Republic of Vietnam. I was delighted to assist in a modest way with that effort. Vietnam joined with other developing nations in Asia moving to the forefront to encourage sustainable renewable energy development and control of carbon emissions. There are today additional ways to monetize the value of these initiatives in Vietnam, as the world moves forward with carbon control. These involve the CDM mechanism under the Kyoto Protocol.

Looking retrospectively, Vietnam has not enjoyed these carbon incentives to the degree of its Asian neighbours, as shown in later figures in this chapter. Yet, with growth in real gross domestic product at rates in excess of 8% annually during 2005–2007, with one of the fastest electricity growth rates in the world, and described by the World Bank as 'one of the best-performing developing economies in the world',[5] Vietnam deserves to be at the front of these carbon technology transfer and benefits.

---

2. James Hansen et al., 'Target Atmospheric CO_2: Where Should Humanity Aim?', 2 *Open Atmos. Sci. J.* 217, 225–226 (2008), (Hansen Where Should Humanity Aim?); Bill McKibben, 'Civilization's Last Chance: The Planet Is Nearing a Tipping Point on Climate Change, and It Gets Much Worse Fast', *L.A. Times*, 11 May 2008, <www.latimes.com/news/printedition/opinion/la-op-mckibben11-2008may11,0,2392815.story>.
3. Kyoto Protocol to the United Nations Framework Convention on Climate Change, 10 Dec. 1997, 37 I.L.M. 22 (1998) (Kyoto Protocol).
4. See <http://cdm.unfccc.int/index.html>.
5. World Bank, 'Vietnam: Laying the Foundation for Steady Growth', (February 2007) <siteresources.worldbank.org/intvietnam/overview/21594788/ida-vietnam.pdf>.

Let me start by analysing some of the other countries in Asia and what they have done. In this context, I note some points about the Vietnam experience, which becomes an important springboard for efforts to create and monetize tradable carbon 'offsets' to control CO_2 emissions, and develop the renewable infrastructure of Vietnam.

1 THE GLOBAL CLIMATE CHALLENGE AND ENERGY TECHNOLOGY

1.1 The Role of Power Technology

More than one-third of CO_2 emissions in the US are attributable to the electric power sector.[6] This is similar to other developed nations. Energy use, and the construction of fossil-fuel fired power generation facilities, is increasing as population growth and development continue, especially in developing nations.[7] The majority of energy and power generation expansion will occur just in Asia over the next decades.[8] Unabated, this exponential increase in power demand could tip the global environment thermostat to run-away global warming risk, regardless of what the US, the European Union and other developed nations do to reign in their carbon emissions.[9]

Ninety-eight per cent of anthropogenic CO_2 emissions are from combustion of fossil fuels.[10] Fossil fuel generation results in 64% of total atmospheric CO_2, and this amount has increased significantly since 1990.[11] Power derived from burning gaseous, liquid and solid fossil fuels used to create electric power release copious quantities of CO_2 into the environment.[12] Most countries are using fossil fuels, not renewable power resources, to satisfy this exponential increase in demand.

---

6. See U.S. Energy Information Administration (EIA), 'Emission of Greenhouse Gases in the United States 2005', (February 2007), (EIA US Greenhouse Gases 2005), <www.eia.doe.gov/oiaf/1605/ggrpt/summary/carbon.html>.
7. World Bank Statement, Ministerial Segment – COP11 – Montreal 4, <http://siteresources.worldbank.org/ESSDNETWORK/Resources/MINISTERIALSEGMENTCOP11Montreal.pdf>. International Energy Agency, 'World Energy Outlook 2004', (IEA World Energy Outlook 2004). <www.worldenergyoutlook.org/2010.asp>.
8. *Ibid.*
9. Climate Change 2007 – The Physical Science Basis Contribution of Working Group I to the Fourth Assessment Report of the Intergovernmental Panel on Climate Change *Summary for Policy Makers*. <www.ipcc.ch/ipccreports/ar4-wg1.htm>.
10. EIA, Emission of Greenhouse Gases in the United States, 1998 (1999).
11. EIAUS Greenhouse Gases 2005, Executive Summary, at 2–3, Frequently Asked Global Change Questions (30 Apr. 2008). <http://cdiac.ornl.gov/faq.html>.
12. The amount of carbon released per unit of usable energy decreased each time as human populations moved from wood to coal as the dominant CO_2-releasing fuel in the late nineteenth century, and again moved from coal to oil in the mid-twentieth century, and will move toward natural gas in the future. See Steven Ferrey, *The Law of Independent Power: Environmental Law Series*, § 2.1, 28th edn (Clark Boardman Callaghan, 2010) (Law of Independent Power).

GHGs increased about 70% during the 35 years between 1970–2004, with the combustion of fossil fuels accounting for 70% of GHG emissions, and electric power generation responsible for 40% of these CO_2 emissions – with coal-fired electric power generation accounting for about 70% of the emissions in this sector.[13] Global energy-related emissions are expected to increase 57% from 2005 to 2030.[14] From any perspective, this is a fast-growing phenomenon. Electric power demand is continuing to increase.[15] At current rates of energy development, energy-related CO_2 emissions in 2050 would be 250% of their current levels under the existent pattern.[16]

All forecasts of the US Department of Energy, the International Energy Agency, and independent forecasters, agree that greenhouse gas (GHG) emissions will increase exponentially, not decrease, during the foreseeable future.[17] GHG emissions are a function of modern society's traditional practice of using fire to manipulate the universe – particularly combusting fossil fuels for electric power production.[18]

The average annual growth rate in primary energy use in developing countries from 1990 to 2001 grew by 3.2% per year, compared in industrialized countries where growth over the same period was 1.5% annually.[19] The US Department of Energy forecasts that energy demand in developing Asia will double over the next 25 years.[20]

The International Energy Agency projected that it will require an investment of $16 trillion by 2030 to meet the world's energy requirements, with USD 5 trillion of that amount allocated to electric power production, primarily in Asia and Africa.[21] It is expected that global energy use will double by 2040 and triple by 2060, creating a tremendous demand on existing fuel sources.[22] The April 2008

13. Joëlle de Sepibus, 'The Liberalisation of the Power Industry in the European Union and Its Impact on Climate Change – A Legal Analysis of the Internal Market in Electricity', NCCR, Switzerland, Trade Regulation Working Paper No. 2008/10 (1 May 2008), 2–4. <http://papers.ssrn.com/sol3/papers.cfm?abstract_id=1137823>.
14. U.S. GAO, 'International Climate Change Programs: Lessons Learned from the European Union's Emissions Trading Scheme and the Kyoto Protocol's Clean Development Mechanism', GAO-09-151, 18 Nov. 2008, 48. <www.gao.gov/products/GAO-09-151>.
15. See e.g., IEA, World Energy Outlook 2004.
16. IEA, 'Energy-Technology Perspectives – Scenarios and Strategies to 2050, 2006' <www.iea.org/publications/free_new_Desc.asp?PUBS_ID=2078>.
17. EIA, 'International Energy Outlook 2007', Ch. 7 – 'Energy-Related Carbon Dioxide Emissions', <www.eia.doe.gov/oiaf/ieo/emissions.html> [this link no longer takes you to the 2007 version. The 2010 version is available at <www.eia.doe.gov/oiaf/ieo/world.html>.
18. About three-quarters of the anthropogenic sources of carbon in the atmosphere is the result of the combustion of fossil fuels, while 25% is the result of deforestation and resultant inability of the biosphere to assimilate and reprocess this chemical compound. McGraw-Hill, *World Energy Outlook*, June 1997.RI/McGraw-Hill.
19. 2004 World Energy Outlook, at fn. 15 *supra*.
20. See U.S. DoE, EIA, *International Energy Outlook*. <www.eia.doe.gov/oiaf/ieo/>.
21. IEA, 'World Energy Investment Outlook 2003' at 3, 343, <www.iea.org/textbase/nppdf/free/2003/weio.pd>.
22. IEA, World Energy Outlook 2004.

Bangkok talks of world representatives following the Bali conference, concluded that a post-2012 international carbon scheme should look much like the pre-2012 Kyoto regime, including trading of allowances and the creation of additional credits or 'offsets' through the existing CDM Kyoto mechanisms.[23]

1.2 The Sustainable Technology Future

The world is increasingly smaller and hotter. Global warming is proving to be the conspicuous environmental challenge of this century, to date and for the foreseeable future. It has emerged as a meta environmental issue, subsuming within its reach those conventional environmental issues of air pollution, water supply, ecosystem management and environmental equity.

The fundamental touchstone technology for development in all nations is electricity. As electricity is used in place of fossil fuels and human labour, less overall energy is used and more productive and efficient operations occur in certain segments of society.[24] The rule of sustainable technology becomes a critical variable.

There is a three-part algorithm with inputs of (1) population growth, (2) energy intensity of development, and (3) choice of technology, that determines the every larger carbon footprint of modern civilization. Only the last of these three inputs appears within the realistic control of political institutions in the current world order. There are now advanced real-time demonstrations of what will change the technology vector in developing nations. The majority of power generation expansion will occur just in Asia over the next decades.[25]

In a world where burning fossil fuels is the dominant electric energy signature; intensified use of electric power foretells a direct increase in carbon emissions. Some projections estimate that by 2030, Asia alone will emit 60% of the world's carbon emissions.[26] The future is directly dependent on whether high-carbon fossil fuels or zero-carbon or low-carbon renewable technologies are chosen now to generate power to meet this new, more intensive electricity demand. These choices in sustainable energy technology made now certainly will be the future signature of our carbon footprint during the crucial period of the next half century during which we may pass the point of no return in terms of global warming.

Choices for new power generation technology today translate directly to the size of tomorrow's carbon footprint. Global CO_2 emissions are rising at the rate of

---

23. Eric Lyman, '"Progress" of Bangkok Talks Shows Much Still to Be Done for 2009 Global Agreement', BNA Env't Rep, (11 Apr. 2008), 704. For discussion of Joint Implementation (JI). <http://unfccc.int/kyoto_protocol/mechanisms/joint_implementation/items/1674.php>, see *infra* for information on CDM.
24. See Clark Gellings & Richard Lordan, 'The Power Delivery System of the Future', *Electricity Daily* (January–February 2004) at Conclusion.
25. IEA World Energy Outlook, 2004.
26. See generally Deborah E. Cooper, Note, 'The Kyoto Protocol and China: Global Warming's Sleeping Giant', 11 *Geo. Int'l Envt'l L Rev* 401, 405 (1999).

approximately 10% per year.[27] The GHG mix of electric energy sources is within legal control by government policies and incentives. Whether or not governments will divert the fossil fuel vector toward renewable or other low-carbon electric energy deployment, and do so in time to avert a global warming tipping point is in part a function of whether there is a clear roadmap on how to navigate the carbon-concerned future of development.

Unprecedented deployment of renewable energy generation alternatives is required to alter this trend.[28] The technology exists to accomplish this. Renewable power is the long-term low-carbon application for developing nations. Renewable energy can provide opportunities for poverty alleviation, supply energy, and enhances energy security by relying on domestic resources.[29] Unlike fossil fuels, renewable resources are widely disseminated across the globe. While many nations – particularly developing nations – have no significant fossil fuel reserves of oil, coal or natural gas, every nation has significant renewable energy in some form – hydropower, sun light, wind, agricultural biomass waste, wood, ocean wave power, etc.

This requires implementation of new regulatory mechanisms for technology transfer and deployment. It is essential to get the infrastructure right, as it controls the form and function of long-term GHG emission mechanisms. 'The stakes, for all life on the planet, surpass those of any previous crisis.'[30]

Without this roadmap, the percentage of fossil fuels in the mix – and thus the potential sources of GHGs in the power sector – are forecast to remain relatively constant. The International Energy Agency in Paris forecasts that by 2030, world demand for energy will grow by 59% and fossil fuel sources will still supply 82% of the total, with non-carbon renewable energy sources supplying only 6%.[31] So, without a radical restructuring of the institutional framework in the power sectors of world nations, business-as-usual leads to a continuing significant increase, rather than the required decrease, in annual CO_2 emissions, compounding of GHG concentrations in the atmosphere, and the repercussions of rapid global warming.

By 2030, the position of developed and developing nations will have reversed, with developing countries emitting the dominant share of CO_2 emissions, and that share increasing over time into the future. GHGs in the twenty-first century predominately are about power generation. The single-point nature of power plants' emissions, the centralized nature of most power plant decisions in developing nations, and the rapidly increasing demand for electricity, make electricity generating plants the logical focus for an intelligent legal and policy assault on GHG

27. See Ray Purdy, 'The Legal Implications of Carbon Capture and Storage under the Sea', 7 *Sustainable Dev. L. & Pol'y* 22 (2006–2007).
28. Neal J. Cabral, 'The Role of Renewable Portfolio Standards in the Context of a National Carbon Cap-and-Trade Program', 8 *Sustainable Dev. L. & Pol'y* 13, 14–15 (2007–2008).
29. Mohamed T. El-Ashry, 'Framework for a Post-Kyoto Climate Change Agreement', 8 *Sustainable Dev. L. & Pol'y* 2, 2–3 (2007–2008).
30. Hansen Where Should Humanity Aim?
31. IEA World Energy Outlook 2004.

emissions. There are solutions with application and implementation of renewable electric power generation technologies.[32]

Let me note in the next section several key findings analysing what will and will not work to mitigate GHG emissions in the electric power appetite of the world.[33]

2 WHAT WORKS IN DEVELOPING NATIONS?

There are trustworthy models of how renewable power can be implemented even in previously non-competitive or monopolized electric power sectors in developing countries. It is in these nations where renewable power sustainable technology can be deployed *ab initio* and where it can be the structural backbone of energy development. This is primarily a legal challenge – to properly create the institutional framework, the laws and regulations, and the contractual relationships to facilitate renewable energy development. This is a prerequisite to the financial challenge that will not be met in the modern economy without an appropriate legal structure.

2.1 Leading Asian Countries

A handful of developing Asian nations have pioneered renewable electric energy programs to reduce the emission of greenhouse gases. Between 1993 and 2010, these nations in Asia have been among the first in developing small power producer (SPP) programs to promote renewable energy development in their countries. These programs create important models and lessons on how to harvest success on global warming policy in the energy sector.

They have achieved in just a few years a substantial contribution of new renewable small power projects to the national energy supply, achieving in just a few years almost 5% of total power supply in states in India, in Sri Lanka and in Thailand, from small renewable energy initiatives. Approximately 60% of all new power generation capacity financed in developing countries is in Asia.[34]

The five models analysed are drawn from countries with different forms of government and have different predominant fuel sources in their generation base (hydro, coal, gas, oil). Table 10.1 displays key comparative elements of program design and implementation in five of the programs surveyed.

---

32. See Law of Independent Power, § 2:11.
33. Steven Ferrey, 'Small Power Purchase Agreement Application for Renewable Energy Development: Lessons from Five Asian Countries', World Bank with support from United Nations Development Program, 2004. <http://siteresources.worldbank.org/INTEAPREGTOPENERGY/Resources/Power-Purchase.pdf>.
34. R. David Gray & John Schuster 'The East Asian Financial Crises-Fallout for Private Power Projects', *Viewpoints* Note No. 146 (World Bank, August 1998), 1. <http://rru.worldbank.org/documents/publicpolicyjournal/146gray.pdf>.

Table 10.1 Comparative Program Overview

Country Program	*Year Begun*	*Maximum Size (MW)*	*Premium for Renewable Energy*	*Primary Fuel*	*Eligible PPA Solicitation*
Thailand	1992	60 or <90	Yes, competitive bid	Gas	Controlled period
Indonesia	1993	<30 Java <15 other island grids	No	Renewable energy	Controlled Period
Sri Lanka	1998	<10	No	Hydro	Open offer
India: Andhra Pradesh	1995	<20 Prior <50	Yes, in tariff	Wind	Open offer
India: Tamil Nadu	1995	<50	No	Wind	Open offer

Table 10.2 displays salient comparative elements of legal design of the power purchase agreement (PPA) and contractual entitlement in five of the Asia-nation programs surveyed. To explain several concepts, third-party sales allow the renewable power generator to sell at retail to power consumers directly, bypassing the wholesale sale to the state utility. This provides alternative options to secure a revenue stream to such a project. Self-service wheeling allows use of the utility transmission system to put power into the power grid at, for example, the wind generation site, and withdraws an equivalent amount of power at one's factory or business at a distant location. This essentially allows a virtual geographic bridge between a power generation source and the owner's point of consumption of that power. Net metering is the ability to sell surplus self-generated power to the utility grid, receiving a credit or turning one's retail consumption meter in reverse to reflect such sale back to the utility.[35] Each of these regulatory embellishments benefits the independent small power seller.

Note the differing policies in different programs on direct retail third-party sales, self-wheeling, and net metering or energy banking.[36] Table 10.3 displays comparative elements of the PPA tariff in these same countries. The tariff sets the price that the country's utility pays to purchase wholesale renewable power produced under the renewable energy programs. Avoided cost is the cost at which the utility which purchases the power of the small power seller could either add generating capacity to generate that additional amount of power, or purchase

35. See S. Ferrey, 'Nothing But Net: Renewable Energy and the Environment, MidAmerican Legal Fictions, and Supremacy Doctrine', 14 *Duke Env L. & Pol F* 1, 52–65 (2003) (Nothing but Net).
36. Ferrey, fn. 33 *supra*, 14. For a discussion of these topics, See generally Law of Independent Power, §§ 10:1 and 4:26–4:27.

Table 10.2 Comparative PPA Elements

Country Program	*Standard PPA?*	*Maximum Years*	*Third-party Sales*	*Self-service Wheeling*	*Net meter-Banking*
Thailand	Yes	20–25 firm 5 non-firm	No, under consideration	No, under consideration	Yes, if <1 MW
Indonesia	Yes	20 firm 5 non-firm	No	Yes	No
Sri Lanka	Yes	15	No	No	No
India: Andhra Pradesh	Not formally, but a de facto standardized form	20	No, previously allowed	Yes, but very expensive	Yes
India: Tamil Nadu	In development	5–15	No, previously allowed	Yes	Yes

that amount of power from others in the wholesale market.[37] Avoided cost prices for the sale of power have been the cornerstone of the PURPA program in the US for 30 years, and is an internationally recognized equitable pricing principle for power sales.

Note that the avoided cost concept and a standardized PPA are generally utilized in most successful renewable energy programs in developing nations. This is consistent with the PURPA requirements in the United States' legal system.[38]

2.1.1 Sri Lanka

The experience in Sri Lanka is indicative that successful global warming renewable energy projects can be implemented even under the most difficult political circumstances. For two decades, Sri Lanka was engaged in a protracted civil war as the Tamil Tigers, an ethnic Tamil minority within a Sinhalese-dominated country, waged a bloody succession movement that claimed about 80,000 lives, including a former sitting Prime Minister of the nation.

As of 2007, the national utility grid in Sri Lanka had 1,800 MW of installed generation, double the amount from a decade earlier. A single standardized PPA and standardized power purchase tariff, designed to be fair to both private

37. 18 C.F.R. 292.101(6) (Definition 'Avoided costs'), see also Law of Independent Power, §§ 7:1–7:5.
38. 16 U.S.C. § 824a-2; 18 C.F.R. 292.101 et seq.

Table 10.3 Comparative Tariff Elements

Country Program	*Avoided Cost Basis*	*Indexed to Foreign Currency*	*Periodically Adjusted*	*Design Elements*
Thailand	Yes, energy and capacity payment for firm contracts only	No	Yes	Utility purchases 65% of off-peak power
Indonesia	Yes, both energy and capacity	Yes	Yes, for changes in avoided capacity cost	Steep on-peak incentives; differentiated for each island grid
Sri Lanka	Yes, energy only; non-dispatchable units receive less than full avoided energy cost	Not directly, but price linked to dollar-denominated imported oil price	Yes, and includes foreign fuel component	Calculated annually, based on three-year moving average imported oil price
Andhra Pradesh	Yes, not to exceed 90% of retail tariff	No	Yes	Reset every three years
Tamil Nadu	Exceeds voided cost	No	Yes	Higher tariff for biomass than wind

small renewable power producers and the purchasing utility, were the foundation for this renewable energy program.

Fifteen-year PPAs are available for projects up to 20 MW in size.[39] To attract wind and biomass projects, Sri Lanka has announced that it will move to a cost-based PPA tariff for SPPs differentiated for each renewable technology, so that wind and biomass will receive a higher tariff than small hydro projects.[40]

Small hydro and other renewable energy developers of facilities no larger than this threshold are allowed to sign a standardized PPA. All of the awarded Sri Lanka projects were small hydroelectric projects with the single exception of one small cogeneration facility. All were privately owned by independent developers. No one company dominated the successful projects. Sri Lanka had more than 50 operating SPP supplying more than 100 MW of power.[41] Another 25 SPPs were under

39. Ferrey, fn. 33 *supra*, 56.
40. See CEB website. <www.ceb.lk/PVT/PPP%20Home2.htmwww.ceblk>.
41. *Ibid.*

construction, plus another approximately 25 SPPs were under active development.[42] The term of the PPA is up to 20 years.[43]

2.1.2 Thailand

Thailand was one of the first countries in Asia to adopt a small power solicitation program. The Electricity Generation Authority of Thailand (EGAT) has installed about 22,000 MW of generation capacity.[44] The Thai program was modelled on elements of the PURPA small power program in the United States.

The Thai system employs competitive bidding by renewable energy SPPs to suppress and award subsidy payments. It has been successful in minimizing the cost of such subsidies and employing available subsidy funds to bring forth the maximum number of megawatts of new private power resources. However, such a competitive system requires that there be a controlled competitive solicitation process for SPPs. By contrast, India's states and Sri Lanka avoid a solicitation in favour of a continually open offer to sign PPAs and purchase power.

The regulations allow SPPs to deliver for sale to EGAT up to 60 MW, although up to 90 MW is within the discretion of EGAT to accept on a case-by-case basis. The program has not restricted participation to renewable sources.[45] As of the end of 2002, 71 SPPs had been accepted and obligated, with a total capacity of 2,330 MW.[46] The bulk of these projects were cogeneration projects, and most of these firm power projects are powered by natural gas. While still utilizing a lower polluting fossil fuel, gas-fired cogeneration plants, because they utilize the same fuel more efficiently for two purposes, are often grouped with other favoured energy projects. Contract terms of 20–25 years are the norm for these larger cogeneration projects under firm contracts.

2.1.3 India

India has become a major player in renewable generation and private sector power development. India is the tenth largest developer of small hydro facilities, and the fifth largest developer of wind power, as well as the fifth largest producer of photovoltaic (PV) systems, in the world. In India, state electricity boards provide electric power. A number of states have SPP programs.

---

42. *Ibid.*
43. Ferrey, fn. 33 *supra*, 56.
44. *Ibid.*, 20.
45. *Ibid.* Subsidies are available in the 2001–02 solicitation process for up to five years for renewable projects in the amount of not more than 0.36 baht per kWh (USD 0.01 per kWh). The subsidies are granted under the Energy Conservation Promotion Fund Committee (ENCON), established by the Energy Conservation Promotion Act, B.E. 2535 (1992). Two billion baht (USD 50 million) is allocated to such renewable project subsidies, in up to 300 MW of such projects contracted after June 2000. Selected projects must be in commercial operation by September 2004 or earlier.
46. Ferrey, fn. 33 *supra*, 22–23.

Of the more than 30 Indian states, the state of Andhra Pradesh is the most advanced in installing wind capacity, with an installed capacity of more than 7,000 MW. 189 MW of wind capacity are in operation. There is no formal standardized contract. Therefore, individual negotiation occurs with the state utility monopoly to determine the contract terms and prices. The state utility makes the determination of the purchase rate it will offer the SPP.

The Tamil Nadu system is more than 7,000 MW.[47] Tamil Nadu state has a significant fraction of India's wind turbine capacity and a significant percentage of biomass projects. A SPP maximum size limit of 50 MW is imposed.

In Tamil Nadu state, no formal standardized PPA is employed, although the utility has employed the same PPA design in every situation, thereby creating a de facto standardized PPA. Wheeling of power to an affiliated location—not to a third-party—is permitted. The tariff is higher for biomass projects than for wind, to reflect the former's non-intermittent, controllable power generation characteristics.[48]

Most of the SPP projects are wind, bagasse, cogeneration, biomass gassifier, and PV.

2.2 International Lessons in Renewable Development Technology Transfer

Even where developing nations feature different forms of government and have different predominant fuel sources in their generation base (hydro, coal, gas, oil), there are common principles that are present for successful small renewable energy programs for global warming mitigation. Several important lessons for legal and regulatory infrastructure of successful renewable program design and policy are revealed by this comparative analysis:

Transparent Regulatory Process. A transparent process is required to build investor, developer, and lender confidence.

Standardized PPA. All programs employ either *de jure* or *de facto* standardized PPAs, and most employ either an avoided cost-based tariff or avoided cost principles. All allow some form of long-term firm contract commitment.

Legal Dispute Resolution Mechanism. A legal framework for structured project development is necessary.

Allocation of Legal Risks. A variety of commercial, sovereign, currency, and regulatory risks are implicitly or expressly allocated in the power sector.[49] The Thai program reduces the SPP capacity payment where the SPP does not deliver. Tamil Nadu facilitates SPP power wheeling.

47. *Ibid.*, 49.
48. *Ibid.*, 53.
49. For a discussion of these topics, see Law of Independent Power, § 3:10.

Interconnection Requirements. Utilities must interconnect with renewable energy projects subject to a straightforward procedure to accomplish this without significant transaction costs or interconnection risk.

Legal Milestones and Bid Security. To eliminate the speculative non-development risk, the Thai program requires a bid security deposit of 500 baht per kW ($12 per kW) of capacity.[50] Sri Lanka in 2003 placed a six-month limit on the validity of LoIs granted for renewable projects and required bid security bonds of SL Rs 2,000 per kW (USD 20 per kW).[51] The Thai program also requires a deposit of 100 baht per kW (USD 2.50 per kW) of applicants, and more for larger sources.

Avoided Cost Principles. The State utility has a monopsony on the purchase of wholesale power in most of the electric sectors of nations of the world. Therefore, the power purchaser and transmission entity must be subject to objective PPA and tariff principles on avoided cost.

Renewable Set-Aside. The program in Thailand allocates entitlements and subsidies in order of the most preferred projects and the least required subsidy for renewable projects. A variant in 29 US states employs a renewable portfolio standard for a minimum percentage of power sold by each retail seller.[52]

Third-Party Sales. None of these Asian programs currently allows direct third-party retail sales of power by the SPP (except in limited industrial estate areas). However, other states in India do allow direct retail sales.[53]

Net Metering and Energy Banking. Energy banking is allowed in more than half the states in the United States in the form of 'net metering',[54] providing an exchange of renewable energy during a billing period. Several of the Asian countries adopted energy banking variants, and in 2009, Sri Lanka adopted net metering.

Every nation in the world has enough solar energy falling on either its roads or buildings to supply 100% of its electric requirements. Haphazardly designed or implemented renewable energy programs, as has happened with the Kyoto Protocol CDM program, will not succeed in the new environment where carbon emissions must be dramatically mitigated. Cumulatively the pioneer programs in

---

50. Ferrey, fn. 33 *supra*, 12, 16, 24.
51. *Ibid.*, 53, 58.
52. See S. Ferrey, 'Renewable Orphans: Adopting Legal Renewable Standards at the State Level', 19 *Electricity J* (March 2006), 52, 54. <www.sciencedirect.com/science?_ob=ArticleURL&_udi=B6VSS-4J624J6-2&_user=10&_coverDate=03%2F31%2F2006&_rdoc=1&_fmt=high&_orig=search&_origin=search&_sort=d&_docanchor=&view=c&_searchStrId=1548878851&_rerunOrigin=google&_acct=C000050221&_version=1&_urlVersion=0&_userid=10&md5= 2ff156d42721d836c65887eea142d292&searchtype=a>.
53. Ferrey, fn. 33 *supra*, 14.
54. Nothing but Net, 55–54.

developing countries provide an energy laboratory during the past decade leaving an inventory of successful program techniques.

Table 10.4 highlights the successful legal contours of programs which have been implemented in developing nations in fast-growing Asia. This table sets forth whether renewable projects come into the program through either a controlled bid/solicitation or through an open enrolment process, both of which have proved successful in different countries. Table 10.4 documents whether security deposits, project milestones, or neither is utilized to prevent projects in the program from stalling. How/if renewable projects are incentivized is set forth, as well as whether there are particular competitive components of the program. It is indicated whether a standardized long-term PPA is utilized to protect the renewable energy producer.

Elements of the tariff for the sale of power are highlighted, including whether it is based on acceptable avoided cost principles and whether the power seller is paid for sale of electrical capacity as well as energy supplied, and how that capacity payment is adjusted downward if there is a failure to supply by the power seller. It is also indicated whether or not the power buyer, the utility, has the ability to dispatch the generating unit, or tell it when it can operate on the system to sell power. Finally, the embellishments of power wheeling, energy banking, and net metering, each of which provides additional options to the power seller, are indicated in the table.

If one does not establish the correct legal structure, renewable energy programs in developing countries will fail. There is dramatic evidence of this.

3 VIETNAM INITIATIVES ON CARBON AND RENEWABLES

In this context of other Asia country renewable energy programs, let me offer a few comments about Vietnam's renewable program development. I have worked with the Government of Vietnam on what eventually became its new non-negotiable standardized small power program (NSSPP). This was a particularly different challenge for the country, and it has done well. I am pleased to have played a modest role in this program evolution, first working with the Ministry of Industry and Trade (MOIT) in 2002–2005, and later with the Energy Regulatory Authority of Vietnam (ERAV) in 2007–2008.

3.1 The Vietnam Sustainable Energy Potential

Vietnam has significant potential for the implementation of renewable energy and CDM projects.[55] Most of this potential lies in the energy sector: between 1996 and

55. Germany Trade & Invest, 'CDM Market Brief: Vietnam Case Study', 2009. For this portion of the paper, I draw on the excellent work of my colleagues Nhan Nguyen, Michael Mehling, and others.

Table 10.4 PPA Successful Management Design and Practices

Successful Design and Management Practice Features	*Thailand*	*Indonesia*	*Sri Lanka*	*India: Andhra Pradesh*	*India: Tamil Nadu*
PPA size <0.5% of system capacity	Yes	Yes	Yes	Yes	Yes
Open offer if need capacity	n.a.	No, but very large solicitation	Yes	Yes	Yes
Controlled solicitation if surplus capacity	Yes	n.a.	n.a.	n.a.	n.a.
Milestones on development time afforded SPP	n.a.	Yes	Yes	Yes, if NEDCAP financial guarantees	n.a.
Bid security deposit by SPP	USD 12 per kW	n.a.	USD 20 per Kw	n.a.	n.a.
How renewable technologies are encouraged	Competitive award subsidy	Hierarchy of renewable SPP preference; floor price on renewable power	Floor price on renewable power	Tariff differentiated for base load power and intermittent renewable SPPs	None
Competitive solicitation	Yes	Yes	No	No	No
Standardized PPA	Yes	Yes	Yes	Yes	No, under development

Table 10.4 (cont'd)

Successful Design and Management Practice Features	*Thailand*	*Indonesia*	*Sri Lanka*	*India: Andhra Pradesh*	*India: Tamil Nadu*
Long-term firm PPAs	Yes	Yes	Yes	Yes	Yes
Avoided cost-based tariff	Yes	Yes	Yes	Yes	Yes
Capacity payment for long-term power	Yes	Yes	No	No	No
Allocation of performance risk between seller and buyer	Alteration of capacity payment; utility can refuse delivery	Neutral; originally mutual best efforts	Neutral; mutual best efforts	Non-firm, but utility must accept all power	Non-firm, but utility can refuse delivery
Capacity payment adjustment if seller does not deliver power	Yes	No, capacity payments in peak rate	n.a.	n.a.	n.a.
SPP unit dispatchable	Yes, if firm capacity PPA; 80% minimum annual output purchase obligation	No, as PPA originally conceived, dispatchable without limitations after PPA changed	No	No	No
Wheeling, net metering, or energy banking	Energy banking	Wheeling	n.a.	Energy banking, wheeling	Energy banking, wheeling

2007, the demand for electricity increased by more than 14% each year. It is expected to continue growing at a faster pace than GDP from 2010 to 2030, rising between 15% and 18% per year.[56] Over the last decade, demand for electricity increased by 14.9% per year for 1996–2000, 15.3% for 2001–2005, and 14.1% for 2006–2007. It is predicted to grow about twice the growth rate of the GDP, by 15% in a low-growth scenario and 18% in a high-growth scenario over 2010–2030.[57]

Vietnam's carbon emissions will likely more than double in the period 2000 to 2020, increasing from 102.6 million tons in 2000 to an estimated 233.3 million tons in 2020. The biggest contributor to this increase is the energy sector, anticipated to increase from only 45.9 million tons of carbon emissions in 2000 to 197 million in 2020.[58] During 2010–2030, CO_2 emissions in the power sector are expected to quickly increase to approximately 352 million tons of CO_2 in 2030, which is several times the 45.9 million tons of carbon emissions emitted by the energy sector in year 2000.[59]

The potential for total sustainable renewable generation in Vietnam is significant. Seventy-five per cent of Vietnam is mountainous or hilly. Vietnam has more than 2,200 rivers and streams of longer than 10 km each, which could yield an estimated 8.0–10.0 TWh of power generation potential from a total installed generating capacity of 18,000–20,000 MW.[60] The potential for small hydro is a subset of this hydro potential in Vietnam.

In addition to hydro resources, Vietnam features 3,440 km of windy coast for wind turbine development, over 300 sub-terrain hot-stream sources ranging from 30 °C to 148 °C which can be developed for geothermal power generating, abundant sunshine and agriculture residues for energy use. Geothermal power capacity is estimated at 200 MW and perhaps as high as 400 MW; biomass cogeneration is estimated at 300 MW; wood and agricultural residues for the electric sector are estimated at 50 million tons/annum.[61]

Small hydropower, biomass (in Vietnam utilizing bagasse and rice-husk waste materials) and geothermal resources are cost-effective in Vietnam compared to

---

56. See Nhan T. Nguyen et al., 'Improving the Clean Development Mechanism Post 2012', *CCLR The Carbon & Climate Law Review* 4(1) 76–85(10) (March 2010), <www.ingentaconnect.com/content/lex/cclr/2010/00000004/00000001/art00008;jsessionid=1849v0entekas.alexandra>. (Clean Development Mechanism Post 2012).
57. *Ibid.*
58. Institute of Energy, 'The Pre-Feasibility Study on Nuclear Power Plant in Vietnam', 2006; Institute of Energy, 'The Sixth Master Plan on Electric Power Sector Development', 2007; Institute of Energy, 'Master Plan on New Renewable Energy Development up to 2025', for MOIT, 2008; Climate Focus, 'Renewable Energy Small Power Producers in Vietnam: Finance Consultancy, for MOIT, 2008). (Climate Focus).
59. Clean Development Mechanism Post 2012, fn. 55 *supra* and Nguyen Thanh Nhan, 'Strategies for Promotion of Energy Efficient and Cleaner Technologies in the Power Sector.' Country Report, presented in the National Conference held on 29 Jul. 2005 in Ha Noi, Vietnam.?
60. Economic Consulting Associates, Robert Vernstrom Associates, *EVN Tariffs: Interim Report*, (September 2003), 42.
61. Ministry of Industry, Master *Plan Study on Electric Power Development in Vietnam, Summary Report,* September 2001, at 14, updated October 2002, updated 2003.(Master Plan Study, 2001, 2002, 2003.

conventional fossil fuel resources. Those that are not cost-effective would be so, with modest carbon offset value monetized through the CDM program or otherwise, as discussed below.[62] There also is the potential for solar and wind resource development in Vietnam.[63]

Power supply can be interconnected centrally through much of the nation. As a 1,650 km long country with a width measuring 50–600 km at different points, there is a central transmission grid running the length of the country. This transmission spine supports operation of a national power grid throughout the country, reaching all 64 provinces, 96% of the districts within the provinces, 78% of the communes, and 69% of the households.[64] Demand for power is rising rapidly in Vietnam at one of the fastest rates in the world, and a high growth rate is expected to continue for the next decade. The load forecast reveals that the rate of demand growth ended a decade of approximately 15% annual increases.[65]

With additional supply, additional power demand would materialize. Most significant industrial and commercial establishments in major Vietnam cities have their own private on-site generators. Electricity demand peaks in the evening hours, as well as during the dry season when the existing predominantly hydroelectric capacity of the Electricite de Vietnam (EVN) system has fewer seasonal supply resources upon which to draw.[66] During the rainy season in Vietnam, which runs from mid-June until mid-November, hydroelectric generation constitutes more than 60% of system generation and a lesser percentage in the dry season.[67]

Even with a still developing, and modest amount of centralized power generation, there is the potential for more efficient use of the power consumed. It is estimated that more than 20% of energy in the whole energy sector could be saved by adoption of specific sustainable programs.[68]

3.2 The Legal Framework for Sustainable Energy Technology in Vietnam

The electricity market has been in the gradual process of some deregulation since 2005 when I worked with MOIT. As a socialist republic, Vietnam, and particularly its electric sector, relies on a centrally planned and administered economy. Under

62. Biomass (wood residue and plantation) and some other sources of mini and small hydro suggested as cost-effective generation sources at 2 $/t CO_2 and significant wind energy potential would become economical for producing electricity at 3 $/t CO_2 (Nguyen and Ha-Duong, 2009a, 2009b).
63. PROACT International, Inc., *Renewable Energy Master Plan in the Northern Part of the Socialist Republic of Vietnam,* (July 2002), 69–81.
64. Master Plan Study 2001, 2002, 2003. This does not mean than 69% of households actually subscribe as customers and consume electricity, rather that 69% of households have physical access to electricity.
65. See Clean Development Mechanism Post 2012, fn. 55 *supra*.
66. The dry season runs from December through May (Worley International 2000).
67. Master Plan Study 2001, 2002, 2003.
68. Clean Development Mechanism Post 2012, fn. 55 *supra*.

the government's current energy roadmap, the electricity sector will be opened in phases and fully to retail/household sales after 2022, pursuant to the current plan. Until then, the government sets the retail electricity price at a subsidized level of less than 6 cents/Kwh (USD/kWh equivalent). The power sector is served and administered by the state-affiliated utility, EVN.

Vietnam law is a civil law system which is statutory. Codes and laws are passed by the National Assembly while ordinances are enacted by the Standing Committee of the National Assembly. Implementation of codes, laws and ordinances is supplemented by government decrees and inter-ministerial circulars and decisions which add to the complexity of the Vietnam legal system.[69]

Vietnam does not have an independent law drafting body. Instead, draft laws and governmental decrees are prepared by the ministry in charge, which often does not have a broad overview of other relating laws or regulations already in place. As result, there can be technical confusions and contradictions among laws and regulations. By law, legality of legislative instruments enacted by lower level authorities must yield to those enacted by higher authorities. However, in practice, implementing decrees or circulars can be more important because they provide the detail for interpretation of law provisions and procedures for implementation.

Common law neither applies nor is recognized in Vietnam. Law interpretation must be in a form of a written resolution of the Standing Committee of the National Assembly, which is the only body under the constitution vested with the power to provide interpretation of laws. This Committee has made few interpretations. Administrative government agencies therefore can present different interpretation of legal issues. For many years, these guidelines and instructions, as well as the courts' judgments and decisions, have not yet been easily publicly available. This legal system operates differently from a common law legal system that exists in the US, for example.

The only publicly available source of legal documents is the expensive Official Gazette, which is now printed weekly providing major laws, decrees, circulars, governmental decisions and international treaties. The Gazette does not include all laws. Legal documents of lower authorities, like provincial people's committees and official letters of ministries and government agencies, are not included in the Official Gazette. So the complete level of detailed law is not generally available or organized.

Vietnam during the past decade was distinct from some other Asian countries implementing sustainable technology discussed above:[70]

> There were more approvals for small renewable power project (SPP) development than required in other developing nations.

---

69. See e.g., Constitution 1992, the Law on Promulgation of Legal Documents (No. 17/2008/QH12), and the Law on Promulgation of Legal Documents of People's Council and People's Committee (No. 31/2004/QH11 in force), Loan Nguyen, Hierarchy of Legal Documents of Vietnam, (April 2009). <www.adetef.org.vn/website/index.php?option=com_content&task=blogcategory&id=32&Itemid=60>.
70. Steven Ferrey & Robert Vernstrom, 'RESPP Planning and Preparation', Final Report prepared for Vietnam Ministry of Industry, 2005 (RESPP Planning 2005).

There was no centralized government authority with responsibility to expedite processing of these approvals.

There was no centralized repository of accessible information about the various approvals.

There was a lock-up of certain renewable hydro-power generation sites claimed by certain developers, that could remain undeveloped for years and unavailable to other developers.

There was a more connected role of state-related developers in Vietnam due to history and institutions, compared to other countries.

There was not an established mechanism for neutral international court dispute resolution for the power sector.

In 2004–2006, as part of Vietnam's Renewable Energy Small Power Program (RESPP) planning process, international economic consultant Robert Vernstrom, local legal consultant Nguyen Tuan Minh, and I prepared an RESPP Planning Reports for the Vietnam Ministry of Industry and Trade.[71] In these reports for MOIT,[72] the consultants highlighted the complex approval process to form a company allowed to conduct business in Vietnam and obtain approval for, and successfully complete a small power project in Vietnam compared to other international frameworks. Specific recommendations were made in that report to stream-line and simplify the approval process for both local, and particularly foreign, SPP investment, siting, and approval.

At the beginning of this decade, the legal and permitting complexity confronting sustainable renewable SPP developers, especially those who did not enjoy the advantage of being state-affiliated companies, and even more so those that are foreign private companies, was challenging. In a time of tightly strained capital markets, investors need confidence of a transparent, neutral market with enforceable rules, into which to advance capital. The lending sector is still maturing in Vietnam. There is still a planned transition of the power sector market over time. Long-term Non-Negotiable Standardized Power Producer Agreements (NSPPA) will be frustrated in attracting international investment capital if there is on-going government-initiated regulatory change in the future that could negate or undermine the basis of long-term contract rights for independent power developers of renewable power.

Prior to reforms suggested in 2008, EVN performed the integrated power sector role of both the generation company and the transmission company. In the future, it is planned that it will be disaggregated, in stages, so that different companies perform these functions. However, when the power distribution

71. Nguyen Tuan Minh, 'Legal Assistance to Ministry of Industry on Legal Issues Relating to RESPP Development in Vietnam' (July 2006), (Legal Issues Relating to RESPP 2006); Steven Ferrey & Robert Vernstrom, 'RESPP Planning and Preparation', Final Report prepared for Vietnam Ministry of Industry, 2006 (RESPP Planning 2006).
72. Legal Issues Relating to RESPP 2006, RESPP Planning 2005.

companies are equitized, both the buyer and the transmission function also could be held by or assigned to the local Peoples Committees in Vietnam (PCs).

To anticipate these changes, when I was last in Hanoi, I designed a PPA for the government to be flexible and adaptable to the NSPPA program and contemplated power sector evolution. By splitting the rights and obligations in the original design of this NSPPA, it allows many sectoral variations over time for future evolution and change of the Vietnam power sector under the government Roadmap.

While there is a new Electricity Law in Vietnam, the approval process is still complex. During 2005–2007, I listened to feedback from ERAV and sector stakeholders, as reflected in an Assessment Report, and thereafter an energy sector NSPPA Report. Those reports identify the choices made to allocate and mitigate risks. This NSPPA evolved through four successive redrafted versions, responding to comments and feedback of ERAV and stakeholders, as well as a stakeholder workshop for invitees to discuss the draft NSPPA and its tariff, to facilitate a smooth transition to a competitive renewable power market under the Vietnam Roadmap.

The important role of Vietnam government agencies in this process is to oversee a fair power sector renewable power offer, and implement the final NSPPA. There is a particular challenge in Vietnam, due to the elaborate set of permits and approvals that traditionally are required for successful independent power project development. If Vietnam wants to encourage greater foreign investment in its electric sector, it could make permitted investment in private power projects as straightforward and manageable as in other Asian countries.

3.3 Compared to other Asian Nations

Regarding energy-sector organization, it is interesting to compare Vietnam to some neighbouring countries. Vietnam has many elements in common regarding transferring energy technology for renewable energy and small power potential with some of its Asian neighbours. Like Sri Lanka when it started its SPP program, hydro is a substantial and core element of Vietnam's base power system. Like Indonesia, Sri Lanka, and India, the need for new investment in additional power generation capacity outstrips the available internal capital resources. Therefore, attraction of additional private investment capital can play an important near-term role.

Like Indonesia and Thailand, Vietnam has local access to petroleum and natural gas hydrocarbons to generate fossil fuel–based conventional energy. Vietnam has large off-shore natural gas resources. Crude oil production is estimated potentially to reach 25–30 million tons/annum and natural gas production could rise to 15–30 billion cubic meters/annum. Vietnam produces oil, and is completing its first two refineries, originally scheduled for completion in 2005 and 2010. Vietnam has significant coal reserves in its northern provinces, which produce coal for export, beyond domestic coal requirements. Coal extraction is estimated to be able to reach 15–20 million tons of coal annually by 2020.

The reports in 2005 and 2006[73] identified the institutional impediments and needs for a viable SPP market in Vietnam. There were new laws enacted regarding Electricity in December 2004 and allowing more liberal foreign investment in 2001. Regulatory discretion was given to the Ministry of Industry and Trade to make accommodations to make the SPP process more stream-lined and the sector more transparent and attractive to international developers, lenders and venture capital. ERAV was created as a new energy-sector regulatory agency.

Notwithstanding changes, there is imperfect access to energy and site information relevant to small-scale renewable technologies.[74] Vietnam, by implementing its SPP program a few years later than the other Asian countries discussed above, had the opportunity to analyse what has worked best in other systems, and why, and benefit from the experience of these other Asian countries in designing and implementing its own NSPPA programs.

4 EMBELLISHING THE SUSTAINABLE ENERGY OPPORTUNITIES IN VIETNAM THROUGH INTERNATIONAL CLIMATE CHANGE PROGRAMS

Vietnam is one of the countries most vulnerable to the dangerous impacts of climate change, but also can benefit from prospects for sustainable development.

4.1 The CDN of the Kyoto Protocol

There is no international mechanism in the Kyoto Protocol to ensure compliance of any nation that fails to achieve its reductions. Achievement, at the end of the day, is voluntary and unenforceable.

The Clean Development Mechanism (CDM) allows projects that reduce greenhouses gases in developing nations to earn Certified Emission Reductions (CERs) for each ton of CO_2-equivalent of GHG reduced.[75] Those CERs are then traded or sold to owners of activities in Annex I developed countries which increase that country's emission cap allocated in the Protocol. Credits generate value for a maximum of seven years with two renewals (21 total years), or a maximum of 10 years with no renewal.[76] Some CERs related to forestry projects are deemed temporary for a period upto 60 years,[77] subject to verification on a

73. Legal Issues Relating to RESPP 2006, RESPP Planning 2005.
74. *Ibid.* (Technology recipients/developers lack information and awareness of available domestic and international renewable technologies, which often causes them difficulties in making proper investment decisions.)
75. See Art. 12, Kyoto Protocol.
76. *Ibid.*
77. See UNFCCC Decision 5/CMP.1, UN Doc FCCC/KP/CMP/2005/8/Add.1 (2006), Watchman (ed.), *Climate Change: A Guide to Carbon Law and Practice* (London: Globe Law & Business Publisher, 2008), 96, n. 7.

recurring 5-year basis that burning or logging do not later release carbon from the forest.

Those CERs must be traded or sold to owners of regulated activities in Annex I developed countries to increase the buyer's and country's GHG emission cap allocated in the Protocol.[78] Emission trading can serve as a means to transfer technology and funds to developing countries, as well as fostering least-cost emission mitigation internationally. To date worldwide, renewable energy projects account for 28% of CDM CERs; methane capture and flaring projects producing no electricity, mostly located at large landfills, coal mines, and concentrated animal feeding operations (CAFOs), account for 19% of CERs.[79] Most of the CERs are from industrial emissions mitigation.

While it would seem that the Kyoto CDM mechanism would result in promoting renewable power in developing nations that has not been the case. It requires a structured state regulatory effort to channel private sector investments into renewable power opportunities. This regulatory initiative in some developing nations involves changing laws, developing new regulations, privatizing opportunities for private sector investment in the power sectors.

Since Kyoto CDM CER offsets must be created in developing countries, there is sovereign risk and commercial risk associated with these intangible items.[80] The requirement for CDM CERs also includes the certification by the host developing nation through its Designated National Authority, that the project supports its goals for sustainable development.[81] Sustainable development has been defined as 'development that meets the needs of the present without compromising the ability of future generations to meet their own needs'.[82] Long-term renewable energy developments clearly satisfy this definition, while many of the other CDM projects that have created CERs may be more questionable.

CDM CERs are required to be 'additional' to baseline project emissions.[83] This involves the establishment of an individual emissions baseline, taking account of sector reform initiatives, barriers to expansion, and sector expansion plans.[84]

---

78. UNFCCC.org. Credits earned after 2000 can be used to achieve compliance during first commitment period which begins in 2008. About 2.5% of ERUs and CERs may be carried over to the second phase of implementation after 2012.
79. Michael Wara, 'Measuring the Clean Development Mechanism's Performance and Potential, 55 *UCLA L Rev* 1759, 1779 (2008).
80. For treatment of sovereign risk and commercial in developing countries, see Law of Independent Power, § 3:10. Commercial risk is mitigated by pooling CERs from different technologies and from different countries in financial instruments for trading.
81. Under the Kyoto Protocol this is embodied in a Letter of Approval (LoA) from the host country's Designated National Authority (DNA). See CDM Rulebook. <http://cdmrulebook.org/PageId/92>. (Explaining CDM rules, practice and procedure). After receiving the LoA and being verified by an authorized third party, the CDM project is ready to be certified as creating CERs. This is done by the CDM Executive Board, or for a JI project the JI Supervisory Committee.
82. The World Commission on Environmental Development, 1987.
83. See Kyoto Protocol, Arts 3, 5, 7.
84. CDM Modalities and Procedures, Art. 48.

Early entrants in the CDM protocol established less stringent guidelines of 'additionality' than now demanded.[85]

CDM projects may only be pursued by registration of the credit through Annex 1 countries.[86] The CDM first project was registered on 16 February 2005; by 14 April 2008, three years later, the 1,000th project had been registered. Less than two years after that, the 2,000th registered project milestone was passed, with 2,067 registered projects registered as of 1 March 2010.[87] The CDM remains the only established institutional instrument allowing an active role for the developing world in carbon mitigation activities.

4.2 The Application of CDM in Vietnam

Vietnam ratified the Kyoto Protocol on 25 September 2002. Yet it has not benefited proportionately with other Asian countries. Asia is the focus of most CDM projects: Over 75% of registered CDM projects are currently being carried out in Asia. Vietnam is endowed with an abundance of natural resources and geophysical conditions that can be leveraged to generate significant amounts of renewable energy, including hydropower, wind, geothermal, sun, biogas and biomass,[88] and various forms of waste-derived energy.[89] These indigenous renewable resources can satisfy electric requirements, while leaving Vietnam fossil fuels to earn hard-currency export value.

The Ministry of Natural Resources and Environment of Vietnam (MONRE) was assigned by the Vietnam government as a national focal agency for implementing the Kyoto Protocol in Vietnam. In 2008, Vietnam approved the National Target Program to Respond to Climate Change (NTP), which envisions specific policies and action plans for climate mitigation activities over the short and long term. It includes energy generation, energy efficiency, renewable energy, nuclear energy, reforestation, and landfill methane gas recovery. The government has acknowledged the need for financial and technology support and transfer from developed countries and international funding sources.[90]

As of 2009, about 70% of the CDM projects approved by the DNA in Vietnam deploy small-scale hydropower.[91] Vietnam offers significant potential for small

85. See UNFCCC.org regarding requirements for 'additionality'; Watchman (ed.), fn. 77 *supra*, at 62.
86. Kyoto Protocol, Art. 12, and Marrakech Accords.
87. See Clean Development Mechanism Post 2012. Figures provided by the UNFCCC Secretariat, 'CDM Statistics', <cdm.unfccc.int/Statistics/index.html>.
88. This is principally residue from sugar, rice, agriculture, and wood.
89. This is principally waste from landfills, animal farms, and tapioca starch.
90. Vietnam, National Target Program to Respond to Climate Change, Decision No. 158/QĐ-TTg, (2 Dec. 2008). <www.adetef.org.vn/website/index.php?option=com_content&task=blog category&id=32&Itemid=60>. Institute of Strategy and Policy on Natural Resources and Environment, 'Vietnam Assessment Report on Climate Change' (Ha Noi: ISPNRE, 2009) (20 Nov. 2010). <www.roap.unep.org/pub/VTN_ASS_REP_CC.pdf>.
91. Clean Development Mechanism Post 2012. Many experts argue that the implementation of small-scale power projects under the CDM does not attract much interest to EVN because these projects are often small and generate less profit yet are more time consuming.

industrial projects and scattered mitigation actions, such as demand-side energy efficiency improvements, measures for increased boiler efficiency in small industries, biomass-fired cooking stoves, biogas digesters, solar water heating, geothermal energy, small hydropower, and wind energy.

If forest preservation were available as an eligible CDM project, which it currently is not, Vietnam has significant and increasing forest resources. The country has 48% forest area (16.2 million hectares). It augmented an average of 236,000 hectares of forest per year between 1990 and 2000, equivalent to a 2.5% annual increase, or a growth rate of 2.1% annually from 2000 to 2005.[92] Forest preservation has a good chance of becoming an eligible credit mechanism as climate change evolves in the international context.

This small-scale renewable energy development potential in Vietnam is currently mostly untapped. By the start of 2010, Vietnam only had 20 projects registered with the CDM Executive Board (CDM EB). The first two Vietnam projects registered at the CDM EB in 2007 were the large-scale Rang Dong gas flaring reduction project and the small-scale 2 MW Song Muc hydropower project. As of the end of 2009, 104 CDM projects had been approved by the Vietnam DNA and 122 projects had been submitted for registration to the U.N.FCCC, yet only 14 of these were registered to be officially recognized in the Kyoto process to produce tradable CERs.

These registration statistics do not compare proportionately to other Asian countries, given the resources available in Vietnam. At the end of 2009, Vietnam had only registered with CDM 7 small-scale hydropower projects of less than 30 MW nameplate capacity, compared to 218 registered hydroelectric projects around the Southeast Asia region. CDM activities in renewables have been significantly untapped in Vietnam. Only one non-hydro renewable energy project, the Binh Thuan wind power farm with a nameplate capacity of 30 MW, was registered as of early 2009.

As Figure 10.1 illustrates, Vietnam has only 1% of the total CDM projects registered within the Asia and Pacific region. While not represented in number of projects, Vietnam is even less well represented in the number of tradable credits produced from these projects. In addition, Vietnam also accounts for only 1% of credit volume expected by 2012.[93]

The number of successful projects in Vietnam, and even more so the number of CERs created, is disproportionately small compared to the potential. Vietnam has slightly more than 5% the population that China has, but creates less than 1% of the CERs that China creates. China, alone, dominates CERs in the Asia-Pacific region. Vietnam in CER creation also lags other regional countries with less robust growth in demand for electricity, such as Indonesia, Philippines and Malaysia

---

92. Gesellschaft für Technische Zusammenarbeit (GTZ), 'Designing a REDD compliant benefit distribution system for Vietnam' Executive summary. UN-REDD Programme (30 Nov. 2009). <http://europeandcis.undp.org/poverty/show/5A0EC8BA-F203-1EE9-BABE45003F1E10AC>.
93. See UNEP Risoe Centre on Energy, Climate and Sustainable Development (URC), 'CDM Projects by Host Region', (January 2010). <cdmpipeline.org>.

Figure 10.1 Registered CDM projects, Asia & Pacific, by host party as of February 2010.

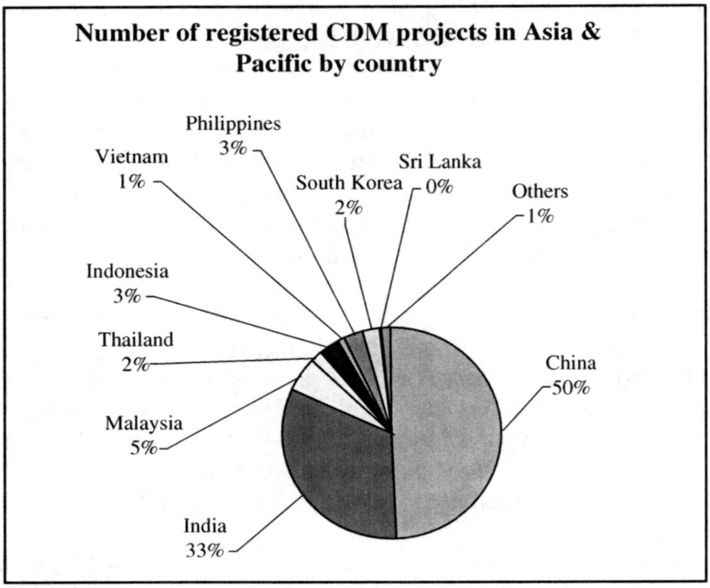

Source: <cdm.unfccc.int>.

(Figure 10.2). CDM projects in Vietnam exceed those in less developed nations of Cambodia, Lao PDR and Myanmar.

Appropriate government policies and actions are necessary for Vietnam to benefit more fully from the CDM process as a way to transfer sustainable power sector technology.[94] In Vietnam, there remain regulatory barriers. There are leverage point suggested by other observers for removing the regulatory, legal, and business barriers in Vietnam to realize a more robust utilization of the CDM program illustrated in Figure 10.3:[95]

Vietnam utilizes a MONRE-led DNA steering committee comprised of members of 14 different Vietnam ministries

Project Designed Document ('PDD') approval which is supposed to take less than two months, requires double or triple this time for Vietnam

A lack of PDD approval criteria and experts for evaluation

Lack of transparent updated information outside of EVN regarding sites for renewable power generation and market access to them, as well as a lack of Vietnam baseline data for CDM documentation

94. Clean Development Mechanism Post 2012.
95. *Ibid.*

Figure 10.2 CERs Created, Asia & Pacific, by Country as of February 2010

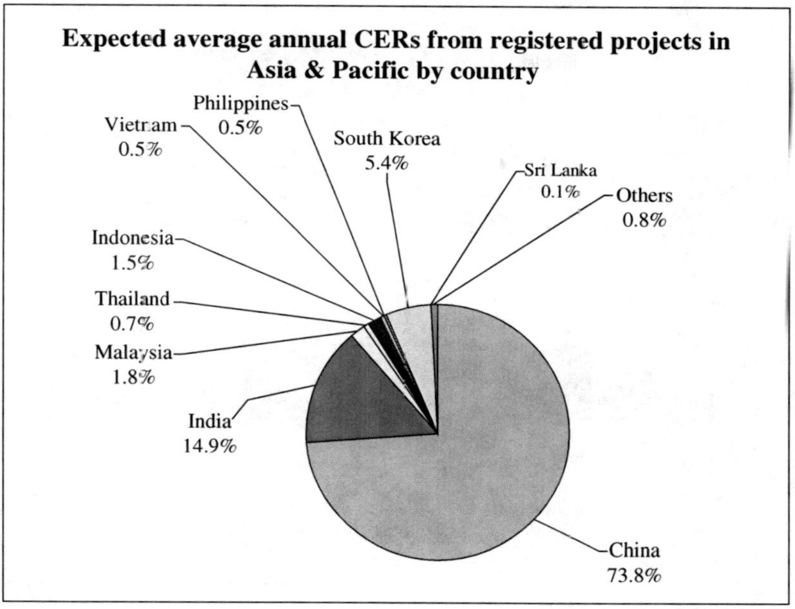

Source: <cdm.unfccc.int>.

A lack of local capacity to pursue projects

Institutional and regulatory barriers to businesses given the government role to keep electricity prices below the marginal cost of producing electricity

Inability to obtain long-term PPAs at the avoided cost of electricity production

The CDM potential in Vietnam is identified as small-scale renewable projects.[96] And this is precisely what is needed for an effective Kyoto CDM program.[97] In Vietnam, small-scale projects may face greater technical and financial barriers than large projects.[98] Long delays in approvals of CDM projects are reported, and a tax on CDM revenues was introduced.[99] The CDM process in Vietnam is set forth in Figure 10.3. Vernstrom and Ferrey recommended in 2005 that the country's CDM process could be revised and stream-lined.

96. *Ibid.*
97. S. Ferrey, 'The Failure of International Global Warming Regulation to Promote Needed Renewable Energy', 37 *B C Envt'l Aff L Rev* 67 (2010).
98. T. Nhan Nguyen & Ha-Duong Minh, 'CO_2 Emissions Mitigation Potential in Vietnam's Power Sector', DEPOCEN Working Paper 2009/22 (25 Nov. 2009). <depocenwp.org/index.php>.
99. Clean Development Mechanism In Vietnam Post 2012.

Figure 10.3 The CDM institutional structure and host country approval procedures for CDM projects in Vietnam.

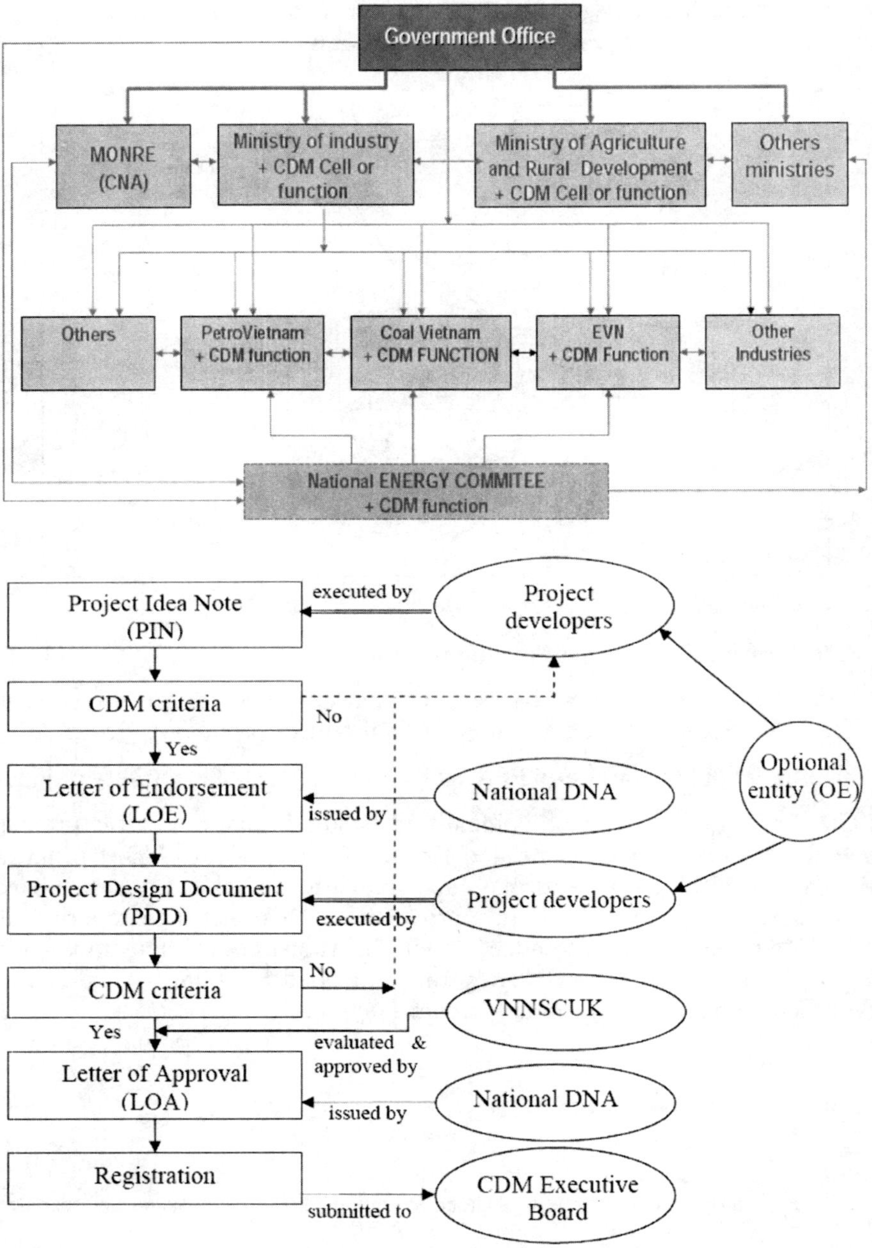

Source: Nguyen K. Hieu, 'CDM Perspective From Vietnam', 2007, at COP 13, Bali; Hoang Manh Hoa, 'Report on the Latest CDM Activities in Vietnam', 2008; N. Nguyen et al., 'The Clean Development Mechanism in Vietnam', *supra*.

5 CONCLUSION: SOLVING THE INTERNATIONAL CLIMATE CHANGE EQUATION

The future of greenhouse gas emissions, linked to global warming, is a function of the choice of electric power production technologies. This choice of technology is critical. The choice of energy technology is one of the few inputs in the three-part global warming algorithm that can be influenced significantly by law and policy. If developing nations do not direct the choice of renewable power options in lieu of conventional reliance on fossil fuel power technologies, the war on global warming cannot be won. Otherwise, the 'business-as-usual' increase in electrification in developing countries will totally swamp the cuts that would be accomplished if the Kyoto Protocol were to achieve its targets in the developed nations that it affects.

Many developing countries do not, alone, have access to sufficient resources and technology to expand their electric infrastructure without technology transfer. Ultimately, the challenge is legal and regulatory: The missing link is the institutional mechanism and model to steer and implement low-carbon power choices through the transfer of sustainable technologies.

When one analyses some of the early efforts to combat global warming emissions in developing nations, there emerges a model of best practices for how to structure a low-carbon high development growth curve for developing nations. Vietnam is now moving forward with a variant of this model with its NSPPA program.[100] To accomplish this transfer of sustainable technology in the electric sector, there needs to be a fair and equitable long-term PPA that memorializes sales of independently produced renewable energy to EVN, which still retains monopoly status to sell retail electric power.[101] In addition, the PPA must contain a tariff for the sale of power that reflects a fair price, with adjustment mechanisms for price over time, for the sale of wholesale power.

The success of these efforts in Vietnam and in other developing nations, especially those which have a vigorous increase in electric-power demand and power-generation facility construction over the next decade, such as Vietnam, is essential for controlling the temperature of the planet. Sustainable technology transfer is a core element of this successful model. Only if all nations succeed in this vision does the carbon-constrained Earth succeed.

---

100. Ferrey, fn. 33 *supra*, at 71–73.
101. For a discussion of elements that go into creation of a PPA, see Law of Independent Power, § 3:77.

Chapter 11

Vietnam: A Case Study for Sustainable Technology Transfer

Tran Viet Dung, Le Thi Nam Giang & Nguyen Nhu Quynh

One of the most important issues in global negotiations in recent years has been the relationship between intellectual property protection and the promotion of technology transfer in developing countries. This debate has largely concerned implications of the Agreement on Trade-Related Aspects of Intellectual Property Rights (TRIPS)[1] under the World Trade Organization (WTO) that affect the interest of developing countries in technological development.

The TRIPS Agreement recognizes the role of technology in social and economic welfare incorporated into the objectives of Article 7 which states:

> The protection and enforcement of intellectual property rights should contribute to the promotion of technological innovation and to the transfer and dissemination of technology, to the mutual advantage of producers and users of technological knowledge and in a manner conducive to social and economic welfare, and to a balance of rights and obligations.

TRIPS provides WTO member countries with flexibilities for approaching different measures within their legal systems and practices, including, but not

1. Agreement on Trade-Related Aspects of Intellectual Property Rights, (Annex 1C to Agreement Establishing the World Trade Organization), 33 I.L.M. 1197, (1994).

Hans Henrik Lidgard, Jeffery Atik & Tu Thanh Nguyen, *Sustainable Technology Transfer*, pp. 251–285.

limited to, granting compulsory licenses and parallel import licenses, in order to promote public interest goals such as health, nutrition and environmental conservation. It also presents the national governments with similar options to tackle the issues regarding abuses of intellectual property rights (IPR) by right holders for the purposes of restraining competition, provided that certain requirements are satisfied.[2]

Overall, the TRIPS Agreement establishes minimum standards of protection and enforcement of intellectual property rights within the WTO system.[3] It also sets up a global legal framework for promoting and encouraging technology transfer, especially technology transfer from developed to developing country members.[4] Nonetheless, how technology transfer occurs in practice depends a great deal upon how developing countries utilize flexibilities defined under the TRIPS Agreement to promote inflows of technology. Developing countries should take advantage of these flexibilities in an appropriate manner to maximize the benefits of domestic competition and minimize the social costs of adopting a strong IP regime required by the TRIPS Agreement.

In this context, Vietnam provides an interesting case study for developing countries for compulsory licensing and parallel import licensing trends in the technology sector. The last two decades have been an extraordinary period for Vietnam. The country has undergone a dramatic transformation from a centrally planned economy to a 'socialist-oriented market economy'[5] under the umbrella of the Doi Moi (Renovation) Policy. Significantly, after a long period of self-isolation, Vietnam openly recognized the importance of international economic cooperation. The development of a market economy together with a proactive economic integration policy has resulted in rapid economic growth in the country over the last 20 years.

Accession to various international economic agreements, especially to the WTO, have also contributed significantly to economic and legal reforms in Vietnam. Combined with the internal push of the Doi Moi policy, this impetus ensures more persistent and powerful liberalization reforms in the country. These efforts used to be almost independent of outside pressure, but now carry a new dimension of international obligation. Indeed, the government is fully aware that the WTO and the international community closely observed Vietnam's performance of its commitments under the international arrangements and the changes they wrought in Vietnam prior to, as well as after its accession to the WTO system. The Vietnamese government is aware that it should take these international obligations into account when it considers reform measures and the development of national legislation and this makes it more globally oriented. In the course of

---

2. Article 8 of the TRIPS Agreement.
3. Article 1(1) of the TRIPS Agreement.
4. Tu Thanh Nguyen, *Competition Law, Technology Transfer and the TRIPS Agreement: Implications for Developing Countries* (Cheltenham, England: Edward Elgar Publishing, 2010), 32.
5. This term is used in Art. 15 of the Constitution of the Socialist Republic of Vietnam, adopted by the National Assembly in 1992, and amended in 2001 (Constitution 1992), to characterize the new model of economic structure.

integration, not only is the economic structure of Vietnam adjusting to international standards, but its national legal system and foreign trade practices are also undergoing transformation in response to the changes and requirements of the international community.

In the course of the above-mentioned developments Vietnam's policy makers have openly acknowledged the importance of establishing a solid legal framework for the protection of competition as well as intellectual property and regard it as fundamental to the process of industrializing and modernizing the country. According to the Resolution on the Strategy for the development and improvement of Vietnam's legal system to the year 2010 and direction for the period up to 2020,[6] developing and improving the laws regarding ownership and freedom of business activities, the creation of comprehensive markets, and intellectual property protection in compliance with the TRIPS Agreement are considered to be the top priorities.

In order to comply with the substantive minimum standards of IP protection under the TRIPS Agreement, further promote creative activities, facilitate technology transfer, and enhance the competitiveness of the economy, Vietnam has made tremendous efforts to improve and develop the laws in this area. In 2005, the National Assembly adopted a new Civil Code (Civil Code 2005)[7] and Intellectual Property Law,[8] to replace previous regulations under the Civil Code of 1995.[9] The newly established IP framework is considered to be complete and consistent with the TRIPS Agreement's requirements.[10] The basic IPR-related competition framework was developed under the Competition Law.[11] Relevant regulations on

---

6. Resolution No. 48-NQ-TW of the Politburo of the Communist Party of Vietnam on the Strategy for the development and improvement of Vietnam's legal system to the year 2010 and direction for the period up to 2020, dated 24 May 2005.
7. Law No. 33/2005/QH11 of the National Assembly, dated 14 Jun. 2005.
8. Law No. 50/2005/QH11 of the National Assembly on Intellectual Property, dated 29 Nov. 2005 and Law No 36/2009/QH12 of the National Assembly amending and supplementing a number of articles of the Law No 50/2005/QH11, dated 19 Jun. 2009 (Intellectual Property Law).
9. The Civil Code of the Socialist Republic of Vietnam of the National Assembly, dated 28 Oct. 1995 (Civil Code 1995). Prior to 2005, the IP framework was governed under the umbrella of the Civil Code 1995. Hence, the regulations of the Civil Code 1995 were only limited to protection of copyright and related rights, patents, utility solutions, industrial designs, trademarks and appellations of origin/geographical indications. In 2005, the National Assembly promulgated amendments to the Civil Code, which reaffirmed the basic civil principles of intellectual property rights (Part VI of the Code), as well as an Intellectual Property Law governing all aspects of intellectual property rights. The Civil Code 2005 and Intellectual Property Law have formed the current IP framework in Vietnam and replaced all previous regulations. English translations of these two legislations may be found on the web sites of the National Office of Intellectual Property of Vietnam and WIPO's 'Wipolex' National laws web page. <www.noip.gov.vn/web/noip/home/en>; and <www.wipo.int/wipolex/en/>.
10. See WTO General Council Meeting 7 Nov. 2006, 'Report of the Working Party on the Accession of Vietnam', WT/ACC/VNM/48, 27 Oct. 2006, Trade-Related intellectual Property Rights (TRIPS), s. 2. Substantive standards of protection, including procedures for the acquisition and maintenance of intellectual property rights, paras 390–471.
11. Law No. 27/2004/QH11 of the National Assembly on Competition, dated 14 Dec. 2004 (Competition Law).

compulsory licensing and IPR exhaustion/parallel importation were also adopted as part of the legal regime on intellectual property. It is expected that efficient enforcement of these legal tools will serve to promote competition and create better access to imported technology in Vietnam.

Like most other emerging economies, Vietnam is still a net importer of technology. Therefore, central issues for the government of Vietnam are (i) how to assess or acquire technology (especially IPR-intensive technology) at low cost, and (ii) how to intervene in markets with legitimate measures to help Vietnamese companies develop indigenous innovation based on imported technology.

From the intellectual property law perspective, the State's competent authorities may grant compulsory licenses on grounds of the protection of public interests, and in particular, for public health needs. Compulsory licenses may be granted to enable production of generic versions of patented pharmaceutical products. They may also be utilized (where needed), as a threat or bargaining tool by the State's competent authorities in negotiations with IPR holders to achieve voluntary licensing.

Vietnam's legal provisions on parallel imports, meanwhile, can help ensure adequate access to IP protected imports. The regulation of parallel trade involves balancing the interests of local consumers and distributors, particularly with regard to pharmaceutical and agrichemical products. Parallel imports of products protected by IPRs is particularly important in the public health sector of Vietnam, where market prices for medicines are often much higher than prices in developed countries. International exhaustion of IPRs can provide the public with greater access to patented pharmaceutical products at affordable prices. Hospitals, pharmacies, and patients may acquire pharmaceutical products at lower price from other markets through parallel trade. In addition, parallel imports can also be used to access basic inputs to agricultural production (such as pesticides and fertilizers) at lower prices than those charged locally by IPR holders. These reduced costs could contribute to improving poor farmers' incomes and livelihoods. This factor is critical for Vietnam, a country with an almost 80% rural population.[12]

Another aspect of concern to the government of Vietnam is establishing and developing mechanisms to effectively control and correct abuses of IPRs by rights holders. Compulsory licensing may be used in this context as a remedy against abuse by IPR holders of market power and the application of anti-competitive practices to restrain technology transfer and competition in the Vietnamese market. This can be effectively settled under competition law.

Hence, the development of IPR and IPR-related competition regulations as well as their enforcement has been more or less a 'learning by doing' process as these are relatively new areas of law in Vietnam. Accordingly, there are still many issues, both theoretical and practical to be elaborated and improved to work efficiently and serve the purposes of economic development in a developing economy.

---

12. General Statistics Office of Vietnam, 'The 2009 Population and Housing Census: Dissemination of completed census results' <www.gso.gov.vn/default_en.aspx?tabid=599&ItemID=9788>.

This chapter aims to provide an overview of the development of IP policy in Vietnam in relation to compulsory licensing and parallel importation with a focus on technology transfer. It aims to demonstrate how Vietnamese policy makers deal with the larger issue of links between intellectual property protection, attracting investment, protecting competition and improving social welfare. The position of the chapter is that intellectual property protection affects the inventive behaviour of firms and technology transfer in varied ways. The effects of such behaviour depends upon factors such as national IP strategies, socio-economic conditions, cooperation of IP holders with the competent State authorities and other stakeholders, coordination between intellectual property law and competition law, and cooperation between competent State authorities.

1 LAW AND PRACTICE ON COMPULSORY LICENSING

The regulation on compulsory or 'non-voluntary' licensing was referred to in Vietnam's first IPR legislation, the Ordinance on Protection of Industrial Property (1989)[13] and further elaborated under the Civil Code 1995[14] in Article 802. However, IPR is still a novel practice in Vietnam and to date; State authorities have never made full use of its enforcement possibilities or granted compulsory licenses. There has been only one occasion, the *Tamiflu* case, where the State authorities have approached the compulsory license for the purpose of the protection of public health. The case, however, provides interesting insights into the country's general policy toward the compulsory licensing as well as possible avenues for development and application of this legal instrument under the current economic and legal situation in Vietnam.

1.1 The *Tamiflu* Case

Tamiflu (oseltamivir phosphate) is an antiviral medicine for the treatment of influenza invented and patented by Gilead Sciences, a US based company. The medicine prevents the influenza virus from spreading inside a human body and is designed to be active against all clinically relevant influenza virus strains. Tamiflu is the primary antiviral medicine used for the prevention and treatment of avian influenza, commonly known as 'bird-flu'. In 1996, Hoffmann–La Roche Ltd purchased a license for the exclusive rights to develop and market Tamiflu from Gilead Sciences and became the only pharmaceutical company authorized to manufacture and distribute Tamiflu worldwide.[15]

---

13. Ordinance No 13-LCT/HDNN8 of the National Council on the protection of industrial property rights, dated 28 Jan. 1989 (Ordinance on Protection of Industrial Property).
14. Footnote 9 *supra* (the Civil Code 1995 was replaced by the Civil Code 2005).
15. Hoffman-la Roche Ltd 'Factsheet Tamiflu', (Factsheet), <www.roche.com/med_mbfstamiflu.pdf>.

The first recorded instance of human infection with avian influenza virus (H5N1) in Vietnam was recorded in December 2003. The pandemic then developed rapidly not only in Vietnam but also in other countries in the Asia, such as the Republic of Korea, Japan, Thailand, Cambodia, Laos, Taiwan, Indonesia, and China.[16] Vietnam was among the most severely affected by the H5N1 virus, where outbreaks rapidly extended to all parts of the country. By November 2005 Vietnam's Ministry of Health noted 3 phases of the pandemic with 91 patients, 41 of which died.[17]

According to the World Health Organization (WHO) epidemiology report during period 2004–2006, the overall case-fatality rate was as high as 56%, of which the mortality is highest in the elderly.[18] The Ministry of Health estimated that if a pandemic of avian influenza occurred, 10% of Vietnamese population could be infected and 1% of the population could die.[19] Therefore, the country was required to stockpile a large amount of Tamiflu to deal with the spread of avian influenza. However, Roche had difficulty in meeting the strong demand for the patented medicine.

On the 26 October 2005 Drug Administration of Vietnam (DAV) requested that the Ministry of Health (MOH) approve an urgent action plan for the local production of Tamiflu. The proposal received the in-principle approval of the MOH.

Initial meetings between DAV and Roche to negotiate easing of the latter's patent monopoly in Vietnam to permit local pharmaceutical companies to produce generic versions of Tamiflu failed. Roche repeatedly refused to license Tamiflu production to other Vietnamese producers. As a consequence, the MOH threatened to grant compulsory licensing for Tamiflu production on the basis of Article 802 of the Civil Code 1995, which empowers the government to take the intellectual property of a private entity, subject to compensation being paid to the rights holder. Article 802 provides:

> Upon application by persons who have the need to use the inventions, utility solutions and industrial designs, the competent State authority may order the [patent] holders thereof to transfer the right to use these inventions, utility solutions and industrial designs to such persons on the basis of reasonable remuneration, if any of the following circumstances hold true:
>
> 1. The holders have failed to use the industrial property, or have used it in a manner not in accordance with the needs of the economic or social development of the country and without reasonable motivation.

16. WHO, 'Avian Influenza: Assessing the Pandemic Threat' January 2005-WHO/CDS/2005.29., <www.who.int/csr/disease/influenza/H5N1-9reduit.pdf>.
17. Ministry of Health, 'Report on the Situation of Avian influenza' (Official Letter 9823/BYT-VP1), (Hanoi, 30 Nov. 2005).
18. WHO, Epidemiology of WHO-confirmed human cases of avian influenza A(H5N1) infection in Weekly Epidemiological Record (WER), vol. 81, 26, 249–260, <www.who.int/wer/wer8126.pdf>. Pursuant to the report, the highest case-fatality rate reported was 73% in 2004.
19. See Official Letter 9823/BYT-VP1, fn. 17 *supra*.

2. The persons who need to use such inventions, utility solutions or industrial designs have negotiated with the holders in different ways and offered a reasonable price, but the latter still refuses to conclude a contract for the transfer of the right to use such objects.
3. The use of such industrial property is necessary to meet the needs of national defence, national security, health or other urgent needs of society.[20]

Pursuant to point 1 and 2 of Article 802, the MOH could allow a third party to use a patented invention without authorization of the patent holder. Mr Cao Minh Quang, Director of the DAV, publicly announced that 'If Roche does not allow [Vietnam to produce the drug], and in case Vietnam declares the state of emergency, we can still produce [it] without Roche's agreement'.[21]

During the same period other countries in the region also considered granting compulsory licenses for the production of generic versions of Tamiflu. On 25 November 2005 after talks with Roche broke down, the Taiwan Intellectual Property Office (TIPO) issued a decision to grant a compulsory license to Taiwanese pharmaceutical companies to produce generic versions of Tamiflu in consideration of the possible outbreak of avian influenza in Taiwan.[22]

Roche softened its stance under this pressure and agreed to discuss sub-licensing arrangements with Vietnamese companies interested in producing generic versions of Tamiflu. However, Roche noted that sub-licenses would only be issued to third parties that 'can realistically produce substantial amounts of the medicine for emergency pandemic use, in accordance with appropriate quality specifications, safety and regulatory guidelines'.[23]

At the third negotiation meeting held on 8 November 2005, Roche officially agreed to grant a sub-license of production antiviral drug Tamiflu to the MOH. Full details of the voluntary license agreement between Roche and MOH were not publicly disclosed, however, it has been announced that Roche would provide relevant technology and raw ingredients to selected pharmaceutical producers in Vietnam so that they could start producing Tamiflu 2 months from the date of execution of the agreement.[24]

The Tamiflu case provides several important observations on the policy approach of the Vietnamese government in relation to the protection of holders

---

20. Article 802 of the Civil Code 1995. (The provisions of Art. 802 were superseded by provisions of the Intellectual Property Law, fn. 9 *supra*.)
21. Forbes, 'Vietnam wants to produce Roche bird flu drug', <www.forbes.com/feeds/afx/2005/10/27/afx2302760.html>.
22. Kathrin Hille, 'Taiwan employs compulsory licensing for Tamiflu', *Financial Times*, 25 Nov. 2005. <www.ft.com/cms/s/cebeb882-5dcb-11da-be9c-0000779e2340, Authorised=false.html?_i_location=http%3A%2F%2Fwww.ft.com%2Fcms%2Fs%2F0%2Fcebeb882-5dcb-11da-be9c-0000779e2340.html&_i_referer=>.
23. Factsheet, fn. 15 *supra*.
24. Drug Administration of Vietnam, Minute of the Meeting 7 Nov. 2005 between MOH and Hoffman-Le Roche (Official Letter 8186/QLD-DK), dated 9 Nov. 2005.

of IPRs and the protection of public health, as well as the development of a legal framework for technology transfer.

Avian influenza was a dangerous and fast spreading pandemic. Under both international and national laws, Vietnam was entitled to apply compulsory licensing. The 2001 Declaration on the TRIPS Agreement and Public Health (Doha Declaration) stipulates that the TRIPS Agreement 'can and should be interpreted and implemented in a manner supportive of WTO Members' right to protect public health and, in particular, to promote access to medicines for all'.[25] In addition, the Doha Declaration explains that a national government 'has the right to determine what constitutes a national emergency or other circumstances of extreme urgency, it being understood that public health crises, including those relating to HIV/AIDS, tuberculosis, malaria and other epidemics, can represent a national emergency or other circumstances of extreme urgency'.[26] Nevertheless, Vietnamese authorities decided to negotiate with Roche for a voluntary license. This policy shows that the economic interests of inventors and foreign investors, including the rights and interests of patent holders, are highly regarded by Vietnam's authorities.

It should be noted that Vietnamese IP authorities have asserted on many occasions that the primary purpose of Vietnam's intellectual property laws and regulations is to provide individuals and organizations with economic incentives to take part in research and development (R&D) activities leading to new products or processes. By granting inventors an exclusive right over the use of their invention, patent holders will be able to receive a return on their investment in R&D. This raises a country's technology base by attracting local and international funds, through support of local research efforts, and by encouraging the growth of production using new technology. It is expected that strengthening the national IP system will lead to an increase in foreign direct investment (FDI), direct manufacturing and joint-ventures, which are considered to be some of the most important development tools for modernization and industrialization.[27] As such, imposing compulsory licenses on avian flu medicines pursuant to Article 802 could 'take away' incentives for other companies to undertake the difficult and costly R&D for new antivirals and vaccines against other diseases affecting Vietnam.

Given the above-mentioned policy approach, voluntary licensing, which comprises a situation where patent holders license their patents to a third party for purpose of local production in exchange for the latter paying licensing fees and agreeing to certain conditions, was regarded by Vietnamese authorities as the most balanced solution for the issue as it would allow foreign investors like Roche to control the situation and protect its economic interests in the Vietnamese market. Had a compulsory license been issued in the case of Tamiflu, Roche might have lost not only the patent holder's exclusive right to produce and sell the medicine to

25. WTO 'TRIPS and Health, Frequently Asked Questions: Compulsory Licensing of Pharmaceuticals and TRIPS'. <www.wto.org/english/tratop_e/trips_e/public_health_faq_e.htm>.
26. *Ibid.*
27. Decision No. 153/2004/QD-TTg of Prime Minister issuing the 'Strategic Orientation for Sustainable Development in Vietnam', 17 Aug. 2004.

the government for its stockpile but also the opportunity to sell the raw ingredients to local producers for manufacture.

From the industrial development perspective, a voluntary license facilitates technology transfer and the upgrade of local manufacturing capacity, while compulsory licensing does not facilitate the assignment of technology to recipients of the license. This is of paramount importance for a developing country like Vietnam as the technological level of local industries is still low. The grant of a non-voluntary or compulsory licensing by State authorities may not produce desirable results where local producers do not have sufficient industrial capacity to produce 'high tech' products like Tamiflu.

It should also be noted that the *Tamiflu* case occurred just prior to Vietnam's accession to the WTO. Therefore, the government of Vietnam sought to maintain a good international image as an investment friendly country in order to receive support from WTO member countries. Balancing the interests of foreign investors, on the one hand, with the public interest, on the other, was priority for the Vietnamese government. Therefore, Vietnamese authorities were very cautious in approaching an extreme measure such as compulsory licensing.

From the legal perspective, it is observed that the grant of compulsory license in 2005 by Vietnamese authorities might face a number of legal issues due to the then weak legal framework on the subject matter. One of the most outstanding issues was that Article 802 and its implementation regulations did not define the method to determine 'adequate remuneration' for patent holders in a case of compulsory licensing. The absence of special criteria for calculating such remuneration or compensation for patent holders might create serious problems of uncertainty and inconsistency with regard to international treaties of which Vietnam is a member.[28] This could have resulted in a challenge by Roche if the authorities decided to grant a compulsory license for the Tamiflu patent. Other than stating a basis for granting compulsory licenses, Article 802 did not mention certain other important features of compulsory licensing, such as 'public non-commercial use', 'national emergency' or 'extreme urgency'. The term 'public non-commercial use' shows the nature of the transaction, or the relation to the purpose of the use of the license (i.e. for public benefit), while the term 'emergency' or 'urgency' suggests that the government may take extraordinary action immediately in relation to the patent to settle the issue. It is internationally accepted that under such circumstances prior commercial negotiations with patent holders may be waived. The absence of these terms in Article 802 considerably diminished the government's flexibility for issuing compulsory licenses.

The law also did not provide a clear procedure for implementing the rule. The coordination mechanism between the relevant authorities, e.g. the MOH and the Ministry of Science and Technology (the government agency authorized to

28. Paris Convention for the Protection of Industrial Property of 20 Mar. 1883, as revised, 828 U.N.T.S. 305 (Viet Nam Accession dated 14 Jul. 1967; Agreement between the United States of America and The Socialist Republic of Vietnam signed by the US, 13 Jul. 2000, signed by Viet Nam 14 Jul. 2000, in effect 10 Dec. 2001 (U.S. Vietnam BIT).

grant compulsory licenses in Vietnam), was not defined by law of regulation. As such, it was quite uncertain how the agencies could coordinate in order to grant a compulsory license for Tamiflu.

Nonetheless, the threat of granting a compulsory license by the authorities led to positive results for their negotiations with Hoffman-La Roche. The latter had to compromise and accept the voluntary license scheme. This case shows that regulations on compulsory licensing are important and necessary in the national legal framework as a reasonable application can bring positive results for society, especially in the field of public health protection.

1.2 Recent Legislative Developments

Since the *Tamiflu case* the legal framework for application of compulsory licensing in Vietnam has been improved significantly. The matter is now regulated under the Intellectual Property Law[29] and the regulatory guidelines for detailed implementation of the law.[30]

The Intellectual Property Law provides principles for the State policy for the protection of IPRs in general, and the application of compulsory licensing. Pursuant to these principles, the exercise of IPRs must not infringe the interests of the State, the public interest or the legitimate rights and interests of other organizations and individuals, and must not breach other relevant provisions of law.[31] The State recognizes and protects IPRs held by organizations and individuals on the basis of ensuring the equal benefits to both rights holders and the public interest.[32] The State also has the right to prohibit or limit IPR holders from the use of or exercise of their rights, or compel them to grant licenses to other organizations, individuals for one or more of their rights under appropriate circumstances.[33]

The Intellectual Property Law expressly confirms that a compulsory license of patents shall be non-exclusive and is limited to a scope and period sufficient to attain the aim of the compulsory license.[34] The compulsory licensee under such situation shall neither assign licensed rights (except with the assignment of his/her business premise) nor grant a sub-license to others.

---

29. Footnote 8 *supra*.
30. Decree 88/2010/ND-CP of the Government detailing and guiding a number of articles of the Intellectual Property Law concerning rights to plant varieties, 16 Aug. 2010 (Decree 88/2010/ND-CP); Decree 103/2006/ND-CP of the Government making detailed provisions and providing guidelines for implementing some articles of the Intellectual Property Law concerning industrial property, 22 Sep. 2006 (Decree 103/2006/ND-CP), Circular No. 01/2007/TT-BKHCN of Ministry of Science and Technology guiding the implementation of the government's Decree 103/2006/ND-CP, 14 Feb. 2007 (Circular 01/2007/TT-BKHCN).
31. Article 7.2 of the Intellectual Property Law.
32. Article 8.1, *ibid.*
33. Article 8.2, *ibid.*
34. Article 146.2, *ibid.*

The basis for granting compulsory licenses of patents[35] under the Intellectual Property Law is more comprehensive than that under Article 802 of the Civil Code 1995, including (i) the use of patent for public non-commercial purposes (e.g., national defence, security, people's healthcare, nutrition and other urgent public demands); (ii) failure to exploit the patent (after 4 years from the filing date of the patent application or 3 years from the date of patent grant); (iii) refusal to license (where the patent holder refuses, after a reasonable period of negotiations, to enter into a voluntary licensing agreement on reasonable commercial terms); and (v) the patent holder is determined by competent State authority to have committed anti-competitive practices prohibited under competition law.[36] In addition, compulsory licenses may be granted on the basis of certain conditions, such as where a new invention requires the use of a pre-existing patented invention to be functional and the holder of principal invention fails to satisfy the request of the holder of dependent invention without reasonable justification.[37]

The recognition by the Intellectual Property Law that certain licensing practices or conditions pertaining to IPRs may restrain competition and impede the transfer of technology is a significant legal development as it establishes a legal ground for implementing competition law in the area of intellectual property. The anti-competitive practices of patent holders prohibited by the Competition Law are now subject to the remedy of compulsory licensing in Vietnam.

In addition to the above determinations concerning 'adequate remuneration' in case of compulsory licensing is clearly regulated.[38] Accordingly, a patent holder upon the issue of a compulsory license is entitled to receive adequate remuneration or compensation determined on the basis of the economic value of licensed rights taking into account a number of factors, namely (i) the price of transfer of the rights to use such an invention on a contractual basis, the funding from State budget to create the invention (if any) (ii) possible profits from the use of the patent, (iii) the remaining term of the patent, (iv) the necessity of the transfer of the rights to use such an invention and other factors directly determining the economic value of the licensed rights.[39] The law also stipulates that remuneration shall not exceed 5% of the net sales price of the products manufactured using the relevant invention.[40] The proper amount of compensation may be determined by a relevant evaluation organization.

The current legal framework has filled up the loopholes under Article 802 of the Civil Code to include the necessary components for the application of compulsory licensing as required by TRIPS Agreement.

---

35. Pursuant to the Intellectual Property Law, compulsory licensing is applicable not only to patented inventions but also to protected plant varieties. However, for purposes of this chapter the authors only analyse the regulations on compulsory licensing in relation to patented inventions.
36. Article 145.1 of the Intellectual Property Law.
37. Article 137.2, *ibid.*
38. Article 24 of Decree 103/2006/ND-CP.
39. *Ibid.*
40. *Ibid.*

1.3 Preliminary Evaluation

Vietnam in connection with the fulfilment of its established national development strategy and in order to meet the requirements of economic integration has developed a relatively comprehensive legal framework for compulsory licensing. However, Vietnam has yet utilized fully and efficiently the flexibility under the TRIPS. This is due to a variety of reasons, such as lack of awareness or understanding about the available flexibilities, lack of legal expertise on related issues in government agencies, and to certain degree, inappropriate or inadequate laws to address the issues.

Some legal provisions still need to be further amended and improved, in particular the normative requirements regarding justification for issuance of compulsory licenses for public non-commercial purposes. Circular No. 01/2007/TT-BKHCN guiding the implementation of the Intellectual Property Law requires that an applicant for a compulsory license must provide documents proving that (i) at the time of application submission there is a practical demand for the use of the patent for the public, non-commercial, defence, national security, disease prevention and treatment or people's nutrition purposes, or for other urgent societal needs society; (ii) the patent holder has failed to use the invention; and (iii) such non-use affects achievement of the above-mentioned goals.[41] Pursuant to this regulation, the State authority may not issue a compulsory license for public non-commercial purpose (i.e. defence, national security, disease prevention and treatment or people's nutrition purposes) if the patent holder has utilized the invention in practice, even with limited degree of application for the society. It is assumed that such regulatory requirements create obstacles in a practical sense to enforcement of the rule of compulsory licensing for the purpose of public non-commercial use of Article 145.1(a) of the Intellectual Property Law because it is too easy for patent holder to prove his/her 'use of the invention'.

This might be viewed as unnecessary caution on the part of lawmakers and/or a misunderstanding of the purpose of compulsory licenses for public interest or use inventions on behalf of the State. The purpose of compulsory licensing applications for public non-commercial purposes is for the benefit of society and the State, but not as a sanction against non-use by the holder. By adopting this provision, Vietnamese lawmakers failed to make use of the relevant flexibility set forth under TRIPS as well as the provisions of the Doha Declaration. It should be noted that Article 31(b) of TRIPS recognizes the rights of WTO Members to grant compulsory licenses in cases of national emergency or other circumstances of extreme urgency, or in cases of public non-commercial use, including their use on behalf of the State, but not subject to any conditions. It should also be noted that the Doha Declaration expressly states that WTO Members are entitled to determine a national emergency or other circumstances of extreme urgency when granting compulsory licenses. It may also be observed that the relevant laws of many countries, including developing economies, such as Korea, Malaysia, India, and Brazil, permit the grant of compulsory license

41. Article 50.2 (b) of Circular 01/2007/TT-BKHCN.

for public non-commercial purpose without requiring the any ground similar to that of Circular No. 01/2007/TT-BKHCN.[42]

The Intellectual Property Law recognizes the use of compulsory licenses as a remedy in the application of competition law rules to anti-competitive practices by rights holders.[43] Nonetheless, there are no further guidelines regarding which anti-competitive practices regulated under the Competition Law are subject to compulsory licensing. The Competition Law itself does not contain specific provisions addressing IPR-related competition issues. This may create some uncertainty in the application of the Competition Law to IPRs and in particular, to technology transfer-related anti-competitive practices. One might advocate that Vietnam should not only incorporate provisions on compulsory licensing into its Intellectual Property law but should also specify the grounds for issuance of compulsory licenses in order to avoid ambiguity or uncertainty and ensure its widest possible use. In particular, the further development of IPR-related competition law is required in order to supplement the provisions of the Intellectual Property Law.

Another significant barrier to the use of compulsory licensing is the absence of clear, straightforward administrative procedures to establish and delineate clear decision-making processes and responsibilities. It may be necessary to establish a mechanism to enable State agencies to discuss and take joint decisions and/or delegate power to issue compulsory licenses to the State agency in charge of specific areas of State administration.

The setting of adequate remuneration as required by Article 31(h) of TRIPS should also be further developed to make the process more transparent, predictable and administrable. These may consist of such things as the adoption of royalty guidelines, to reduce uncertainty and to facilitate speedier decision-making.

Finally, Vietnam needs to improve the professional capacity of its State authorities and its technological infrastructure so that the compulsory licensing can be applied efficiently. Competent authorities should be equipped with the appropriate knowledge and skills so that they can be applied efficiently and expeditiously when needed. Local industries must also develop their technological capacity to be able to implement licensed technology in the production lines and be further able to develop the technology (where possible).

2 LAW AND PRACTICE ON PARALLEL IMPORTS

Exhaustion of IPRs was initially recognized in Vietnam's Civil Code (1995) and then in Governmental Decree No. 63/1996/ND-CP which provided detailed

42. See WIPO Committee on Development and Intellectual Property (CDIP), Fifth Session, Geneva, 26–30 Apr. 1010, 'Patent related flexibilities in the multilateral legal framework and their legislative implementation at the national and regional levels', CDIP/5/4 REV. 18 Aug. 2010, Annex I 'Provisions of Law on Compulsory Licensing, <www.wipo.int/meetings/en/doc_details.jsp?doc_id=142068>.
43. Article 145.1(d) of the Intellectual Property Law.

regulations for industrial property.[44] Nonetheless, neither instrument clarified whether parallel imports were lawful. In 2000 when the *Tribeco* case was brought to the Vietnamese administrative authorities and a series of parallel import-related questions were raised by IPR enforcement agencies, the question was clarified by Circular No. 825/2000/TT-BKHCNMT[45] where the legal status of parallel importation was accepted in Vietnam. Four years later, Decision No. 1906/2004/QD-BYT[46] regarding the parallel import of medicines for the prevention and cure of human diseases was promulgated in the face of heated debate. An international exhaustion regime providing legality for parallel imports still faced a strong opposition in the process of drafting the Intellectual Property Law. The concept of international exhaustion was eventually codified into law. Nonetheless, from both the academic and practical perspectives, parallel import-related matters are still new to Vietnam. A handful of parallel trade cases, however, have been brought to administrative agencies but have not made their appearance in the courts. Analyses of these cases may aid in understanding the legal and practical aspects of parallel importation in Vietnam.

2.1 From the *Tribeco* Case to the Circular No. 825/2000/TT-BKHCNMT

The *Tribeco* case was not a parallel trade case but rather, dealt with the conditions triggering IPR exhaustion. The case, nevertheless, gave an impetus to the issuance of the first legal document dealing with parallel importation, the Circular No. 825/2000/TT-BKHCNMT.

The *Tribeco case*, concerned claims by the Saigon Joint Stock Soft Drink Company (TRIBECO Company) against Tan My Private Enterprise (hereunder called Tan My) with respect to an alleged infringement of industrial design rights owned by TRIBECO concerning its soft drink bottles.[47] The facts of the case are as follows: some of TRIBECO Company's distributors did not return empty bottles to TRIBECO Company (as concluded in distribution agreements between TRIBECO Company and its distributors) paying instead, to the company, an amount equivalent to the amount owed for their return.[48] Tan My purchased the empty bottles from

44. Decree No. 63/1996/ND-CP of the Government on detailed regulations concerning industrial property, 24 Oct. 1996 (Decree No. 63/1996/ND-CP).
45. Circular No. 825/2000/TT-BKHCNMT of Ministry of Science, Technology and Environment guiding implementation of Decree No. 12/1999/ND-CP of 6 Mar. 1999 of the Government of the handling of administrative violations in the field of industrial property, 3 May 2000 (Circular No. 825/2000/TT-BKHCNMT).
46. Decision No. 1906/2004/QD-BYT of Minister of Health promulgating the Regulation on parallel import of medicines for the prevention and cure of human diseases, 28 May 2004 (Decision No. 1906/2004/QD-BYT).
47. The Protection Title of Industrial Design Number 46 (granted 1990).
48. Official Notice No. 184/TTra of Ministry of Science, Technology and Environment sending to the Standing Vice Minister Chu Hao the Submission dealing with TRIBECO Company's complaint against the conclusion of the Director of Science, Technology and Environment Department of Long An Province, 16 Jul. 2001 (Official Notice 184/Ttra).

TRIBECO Company's distributors and small bottle-gathering groups, filled them with cacao milk and affixed their 'An Binh' trademark on the bottles, placing them on the market.[49] In 1999, TRIBECO Company filed a claim with the Inspectorate of the Department of Science, Technology and Environment of Long An province against Tan My for the infringement of their industrial design rights, requesting a settlement of administrative violations under Decree No. 12/1999/ND-CP.[50]

The Inspectorate eventually concluded that Tan My did not infringe TRIBECO Company's industrial design rights because it reused bottles put onto the market by TRIBECO Company.[51] TRIBECO Company did not agree with the decision and brought a claim to the Director of the Department of Science, Technology and Environment of Long An province. The Director subsequently upheld the decision of the Inspectorate.[52] TRIBECO Company then appealed to the Minister of Science, Technology and Environment. The Inspectorate of the Ministry of Science, Technology and Environment, having competence to deal with the case,[53] held that the bottles used by Tan My were not the ones sold or put onto the market by TRIBECO Company or with its consent. Tan My, had, therefore, infringed TRIBECO Company's industrial design rights. Accordingly, the Ministerial Inspectorate proposed that the Minister overturn the provincial decisions and handle the administrative violation of Tan My on the basis of Decree No. 12/1999/ND-CP.[54] Finally, the Minister of the Ministry of Science, Technology and Environment concluded in Decision No. 1206/QD-BKHCNMT that Tan My's reuse of the bottles constituted 'an infringement of industrial property rights'.[55]

---

49. Ibid; see also s. 2 of Official Notice No. 2051/BKHCNMT-TTra of the Ministry of Science, Technology and Environment dealing with TRIBECO Company's complaint, 23 Jul. 2001 (Official Notice No. 2051/BKHCNMT-Ttra); s. 3 of the Letter of Complaint No. 19/ CV.2000 of Tan My Private Enterprise 3 Nov. 2000 requesting the Department of Science, Technology and Environment of Long An Province to annul the Department's Memorandum suspending Tan My Private Enterprise from reusing TRIBECO Company's bottles.
50. Decree No. 12/1999/ND-CP of the Government of the handling of administrative violations in the field of industrial property, 6 Mar. 1999.
51. Official Notice No. 07/CV.Ttra of the Inspectorate of the Department of Science, Technology and Environment of Long An Province responding to TRIBECO Company's complaint against Tan My Private Enterprise and the Department of Science, Technology and Environment of Long An Province.
52. Decision No. 24/QD-KCM of the Director of Science, Technology and Environment Department of Long An Province, 1 Dec. 2000 (Decision No. 24/QD-KCM).
53. As of 5 Aug. 2002 the Ministry of Science, Technology and Environment was converted into the Ministry of Science and Technology pursuant to Resolution 02/2002/QH11 of the National Assembly.
54. Official Notice No. 99/Ttra of the Inspectorate of the Ministry of Science, Technology and Environment sending to the Inspectorate of Science, Technology and Environment Department of Long An Province, 12 May 2000; Official Notice No. 211/Ttra of the of the Ministry of Science, Technology and Environment sent to TRIBECO Company, 20 Sep. 2000.
55. Decision No. 1206/QD-BKHCNMT of the Minister of the Ministry of Science, Technology and Environment dealing with the complaint relating the Decision 24/QDKH of the Director of Science, Technology and Environment Department of Long An Province, 19 Jul. 2001 (Decision 1206/QD-BKHCNMT). It is worth noting that in the course of settlement of cases, Vietnamese administrative authorities usually issue official notices and decisions.

Here, the respective Inspectorates of the Department of Science, Technology and Environment of Long An Province and of the Ministry of Science, Technology and Environment made determinations regarding whether Tan My's alleged infringements were excepted on the basis of the exhaustion of industrial property rights provided under Article 803.2 of the Civil Code 1995.[56] Specifically, whether or not had the bottles used by Tan My been those sold or put on the market by TRIBECO Company, or with TRIBECO Company's authorization, would exhaustion of rights have been a defence to Tan My's infringement.[57] The inspectorates relied on the provisions of distribution contracts between TRIBECO Company and its distributors in order to determine whether TRIBECO Company had put the industrial design right-protected bottles on the market.[58] Those agencies, however, based their conclusions upon different contractual provisions and came to differing conclusions. The provincial Inspectorate based its decision on Article 2 of the distribution agreement (which stated that 'party B [the distributor] must purchase number of bottles [. . .] in order to put the soft drink bottles into circulation'); but the ministerial Inspectorate relied on Article 5 (which stated that 'after receiving the goods, party B [the distributor] must return the empty bottles, and party B must compensate party A [TRIBECO Company] for damage caused by party B or customers of party B').[59] The final decision of the Ministry of Science, Technology and Environment in *Tribeco* held that Tan My infringed TRIBECO Company's IPR rights because 'it used TRIBECO Company's industrial design-protected bottles in its business *without TRIBECO Company's consent, and TRIBECO Company did not sell or authorize its distributors to sell the bottles on the market*'.[60]

Official notices are often made at the beginning of the settlement and convey issuer's primary opinions on the matter at issue to related parties whereas decisions are often made at the end of the process and legally valid.

56. In Vietnam, exhaustion was initially included in the Civil Code 1995 for patents, utility solutions and industrial designs at Art. 803.2.
57. Decision No. 24/QD-KCM, fn. 52 *supra*; Art. 3 of Decision No. 1206/QD-BKHCNMT, fn. 55 *supra*.
58. Regarding the provincial inspectorate, see Decision No. 24/QD-KCM, fn. 52 *supra*, 2; regarding the ministerial inspectorate, see Official Notice No. 184/TTra 16 Jul. 2001 of the Ministry of Science, Technology and Environment sending to the Submission dealing with TRIBECO Company's complaint against the conclusion of the Director of Science, Technology and Environment Department of Long An Province Standing Vice Minister Chu Hao, s. III.1; Official Notice No. 2051/BKHCNMT-Ttra 23 Jul. 2001 of the Ministry of Science, Technology and Environment dealing with TRIBECO Company's complaint, s. 2.
59. Contracts between TRIBECO Company and its distributors were standardized. See e.g., distribution contract No. 45/HDDL and No. 295/HDDL between TRIBECO Company and Mr Huynh Tan Long; the distribution contract No. 312/HDDL between TRIBECO Company and Mrs Tran Thanh Minh (those are the distributors that Tan My Private Enterprise bought the bottles from). (Translation by the author and emphasis added.)
60. Article 3 of Decision No. 1206/QD-BKHCNMT, fn. 55 *supra*. (Translation by author, emphasis added.) An investigation into the facts of TRIBECO Company indicates an incomplete investigation by the Ministry of Science, Technology and Environment in this case. Art. 2 of the distribution contracts show that exhaustion of rights occurred when TRIBECO Company sold

Tribeco is the seminal case on the application of the principle of IPR exhaustion in which two conditions triggering exhaustion were expressly emphasized. The final decision of the Ministry of Science, Technology and Environment went beyond the existing legal normative documents since the word 'consent' had not been used in those documents. Questions arising from this case together with difficulties faced by IPR enforcement agencies, particularly market management offices and customs offices,[61] made it necessary that governmental authorities clarify whether exhaustion occurred if imported goods had already been put on the overseas market by an IPR holder or with his consent. In response, to that question, the Ministry of Science, Technology and Environment introduced provisions on parallel imports in the Circular No. 825/2000/TT-BKHCNMT. Section 8 (d), Part II of the Circular states:

The use of, or carrying out of, commercial activities (importing, selling, stocking for sale, offering for sale, and advertising for sale) with respect to products that *have been put on the market by owners of industrial property objects (including an overseas market)* [does not infringe industrial property rights].[62]

The words 'importing' and 'including an overseas market' in the section constituted recognition of an international exhaustion regime and permitted parallel imports of industrial property right-protected products. This recognition was clearly explained in two illustrative cases referred to in the Circular:

Commercial activities relating to products containing industrial property right-protected elements that are supplied (distributed, sold) by others and the main supplier being *the owner of the industrial property objects (the person possessing protection title or the licensee) shall not be considered as an infringement of industrial property rights, regardless of whether the products have been supplied in Vietnam or abroad).*

---

the bottles to its distributors. Art. 5 of the distribution contracts must be regarded as a separate agreement and cannot prevent IPR exhaustion. The Distributors' contractual breaches by failure to return the empty bottles constitute contract infringements whose remedy is the compensation of TRIBECO Company as set forth in Art. 5. At this point however, the Company no longer had the right to prevent third parties from using the industrial design-protected bottles. Therefore Tan My was entitled to use the bottles. Nevertheless, the act of filling the empty bottles 3 with cacao milk, affixing the 'An Binh' trademark on the bottles, and putting them on the market constitutes an infringement upon TRIBECO Company 's industrial design rights. The Ministry of Science, Technology and Environment was correct in understanding that these two conditions triggered industrial design exhaustion and the termination of Tan My's industrial design infringement. Regretfully, the Ministry only referred to Art. 5 of the distribution contracts and did not put Art. 5 into context with the other provisions of the contracts, particularly Art. 2. This caused its arguments for the conclusion to be unconvincing.

61. Official Notice No. 30 of Hanoi's Market Management Office provided to the NOIP and Market Management Office of Tan Binh District, Ho Chi Minh City, 1 Jul. 1999; Office Notice of No. 137-QLTT of Vietnam Market Management Department provided to the NOIP and Ho Chi Minh City's Market Management Office, 5 Jul. 1999; and Official Notice No. 5799/TCHQ-GSQL of the General Customs Office provided to the NOIP and local customs, 12 Oct. 1999.

62. Section 8(d) of Circular No. 825/2000/TT-BKHCNMT. (The translation by author, emphasis added.)

Parallel import: Parallel import is the importation of products containing industrial property right-protected components from a source supplied by licensees or persons or companies/firms subordinate to the mother company in the distribution chain . . ., rather than being supplied by the right holder, shall not be considered as an infringement of the industrial property rights.[63]

Under the Circular No. 825/2000/TT-BKHCNMT, the term 'parallel import' was used and parallel import officially adopted, in a Vietnamese legal normative document. Nonetheless, there remain some questions. First, the word 'supplied' was inaccurate when used to refer to products which have been put on the market because 'supplied' is only one of the acts of putting those products on the market. Moreover, the definition of parallel import did not explain the nature and characteristics thereof. Parallel import essentially is an importation of IPR-embodied goods without the intellectual property right holder's authorization. Furthermore, the non-exhaustive list of suppliers in the Circular ('licensees or persons or companies/firms subordinate to the mother company in the distribution chain . . .') is also not sufficiently descriptive of, or an accurate list of entities in the supply chain. In addition, the term 'including an overseas market' was put unobtrusively in brackets indicating the lawmakers' hesitation and experimentation.[64] Those defects and the low legal validity of the Circular limited its practical application.[65]

One year after the promulgation of the Circular No. 825/2000/TT-BKHCNMT, issues regarding parallel importation were regulated in a legal document promulgated by the Government,[66] in Decree No. 06/2001/ND-CP.[67] However, it should be noted that the Circular No. 825/2000/TT-BKHCNMT and the Decree No. 06/2001/ND-CP have been repealed.[68] Currently determinations regarding IPR exhaustion and parallel importation are based on Intellectual Property Law 2005 and legal documents providing detailed regulations for implementation of the Law (as analysed below).

2.2 *Honda* Case

In 2001, Honda Giken Kogyo Kabushiki Kaisha (Honda Japan) filed a complaint with Vietnam's National Office of Intellectual Property (NOIP) alleging infringement of its Patent No. 1980 by the General Corporation of Transportation

63. *Ibid.* (The translation by the author, emphasis added.)
64. Usually, content that is placed in brackets is auxiliary and not binding.
65. In the hierarchy of Vietnamese legal normative documents, at the national level, the Constitution contains the highest legal validity and circulars have the lowest legal validity. See Art. 2 of the Law No. 17/2008/QH12 on Enactment of Legal Normative Documents, 3 Jun. 2008 (Law on Enactment of Legal Normative Documents).
66. Articles 115, 116 of the Constitution 1992 (amended in 2001); Arts 2, 59, 69 of the Law on Enactment of Legal Normative Documents.
67. Section 23 of Decree No. 06/2001/ND-CP on amendments and supplements to some provisions of Decree No. 63/1996/ND-CP/CP.
68. Article 36.2 of the Decree No. 106/2006/ND-CP.

Mechanics of Vietnam (hereunder called Honda Vietnam).[69] The respondent argued that it imported patented motorbike saddles from a company in China for assembly on motorbikes for sale in the Vietnamese market.[70] Within its function of providing evaluation and instruction to related parties and/or IPR enforcement authorities[71] the NOIP issued Official Letter No. 722/KN dated 12 July 2001 responding to the complainant as follows:

> According to Article 805 of the Civil Code, if the structures of motorbike saddle described in the materials provided by the complainant are used by persons other than Honda Giken Kogyo Kabushiki Kaisha, and if those persons are not licensed for the use of Patent No. 1980, and if they do not belong to those mentioned in Articles 801 and 803 of the Civil Code, and if such use is carried out after the granting date of Patent No. 1980 [. . .], such use will be considered as an act of infringement of Patent No. 1980.[72]

The complicated wording and the word 'if' made this Official Letter ambiguous. It did not particularly describe the particulars of an act of importing the patented products, nor make a clear explanation of legal matters concerning this act.[73] Interestingly, the NOIP referred to Article 803 of the Civil Code 1995 (which did not clarify the legitimacy of parallel import) instead of the Circular No. 825/2000/TT-BKHCNMT or the Decree No. 06/2001/ND-CP (recognition of the parallel import doctrine) notwithstanding the fact that the latter were in force. If the NOIP had based its instruction on the Circular No. 825/2000/TT-BKHCNMT and the Decree No. 06/2001/ND-CP, it would have simply been required to explain that the importation of motorbike saddles into Vietnam for assembly on and sale of the motorbikes in Vietnam would not constitute an infringement if the products had been put on the Chinese market by Honda Japan or with its consent.

This lack of clarity appeared to be unreasonable but was derived from the policy at the time of the Vietnamese government for preventing imports of motorbikes and motorbike accessories from China.[74] In fact, low quality motorbikes and motorbike accessories imported from China had been causing serious social

69. NOIP, Complaints Record of 2001 (Hanoi, 2001).
70. NOIP, *Complaint of Patent Infringement by Honda Giken Kogyo Kabushiki Kaisha*, in Complaints Record of 2001.
71. NOIP is not competent to handle acts of IPR infringement. See 'The Functions and Tasks of NOIP', <www.noip.gov.vn>; Arts 1 and 2.6 of Decree No.54/2003/ND-CP of the Government on the function, duty, and rights of Ministry of Science and Technology, 19 May 2003.
72. NOIP, *Complaint of Patent Infringement by Honda Giken Kogyo Kabushiki Kaisha (Japan)*, Complaints Record of 2001.
73. Pham Hong Quat, *How to Comply With The TRIPS and WTO Law: The New Challenges To Vietnam's Patent Legislation from WTO Dispute Settlement Practice*, Doctoral Dissertation, Nagoya University, Japan, December 2007, 189–190. <http://iprenforcement.most.gov.vn/ckfinder/userfiles/files/How%20to%20Comply%20with%20the%20TRIPS%20and%20WTO%20Law(2).pdf>, 25 Jan. 2011.
74. Instruction No. 3470 TCT/NV6 of the Taxation Department 16 Sep. 2002; Instruction of the Government Office No. 5110/VPCP-KHTH requested the Ministries of Finance, Trade, Industry, Planning and Investment, Transportation, Department of Customs and the other related agencies

concerns such as traffic problems in urban areas as well as a series of severe traffic accidents. In addition, domestic motorbike producers suffered from the low quality and pressures of cheap prices motorbikes and motorbike accessories from China.[75] The industry, particularly Honda Vietnam, requested that the government prevent entry of those products into Vietnam.[76] The protection of consumers' interests and the development of the domestic industry called for a protectionist policy for those products. In these circumstances, the instruction from the NOIP on parallel importation was appropriate. This argument is based on the overall goals of the country in the Strategy for Socio-Economic Development for the period of 2001–2010:[77]

> To bring [the] country out of underdevelopment; improve noticeably the people's material, cultural and spiritual life; and lay the foundation for making ours basically a modern-oriented industrialized country by 2020. To ensure that the human resources, scientific and technological capacities, infrastructure, and economic, defense and security potentials are enhanced.[78]

And the implementation of the above-mentioned goals based on development approaches are:

> [...] to *improve the quality of the people's life* [...]; to consider *economic development the central task*...; to closely link building of an independent and autonomous economy with proactive international economic integration.[79]

The case required a harmonization of law and policy on parallel importation and legalization of the issue in a higher legal normative document rather than that in the existing Circular and Decree.

2.3 Recent Developments

2.3.1 The Decision No. 1906/2004/QD-BYT

Four years after the first provisions on parallel importation were introduced, in order to stabilize the domestic pharmaceutical market and to block monopolization

---

to strengthen the measures to restrict the importation of motorbikes and motorbike accessories, 16 Sep. 2002; Instruction No. 38/2002/QD-TTg of the Prime Minister requested the Ministry of Industry to impose the severe sanctions on the acts violating the importation policy on the domestically production of motorbikes, allowing the prohibition of the importation of motorbike accessories in the appropriate circumstances, 14 Mar. 2002; Instruction No. 15/TB-VPCP 25 Jan. 2002 of the Government Office; Instruction No. 136/TB-VPCP of the Government Office 8 Aug. 2002; Circular No. 69/2002/TT-BTC of the Ministry of Finance, 1 Sep. 2002.

75. Report of Honda Vietnam, WIPO National Roving Workshop on the Enforcement of Intellectual Property Rights, Hanoi, 5–6 Jul. 2004.
76. *Ibid.*
77. See the Politburo of the Communist Party of Vietnam, *Văn kiện Đại hội Đảng toàn quốc lần thứ IX* [Documents of Communist Party Congress IX].
78. *Ibid.*, s. 1, Part II (emphasis added).
79. *Ibid.*, s. 2, Part II (some contents omitted and emphasis added).

by large multinational pharmaceutical manufacturers[80] the Ministry of Health issued the Decision No. 1906/2004/QD-BYT[81] dealing with the parallel import of medicines for the prevention and cure of human diseases.[82] While the repealed Circular No. 825/2000 TT-BKHCNMT constituted legal recognition of parallel imports with respect to all industrial property right-embodied goods, the Decision No. 1906/2004/QD-BYT regulates parallel import of only one type of goods, namely medicines. However, this Decision creates a number of conflicts and raises some questions. First, it mainly focuses on the aspect of State management with respect to parallel import of medicines for prevention and cure of human diseases.[83] Moreover, the definition of parallel import of medicines in Article 3 of the Regulation promulgated together with the Decision No. 1906/2004/QD-BYT does not clearly set forth the nature and characteristics of parallel imports, i.e. importation without a right holders' authorization.[84] Instead, Article 3 refers to: (i) a condition of parallel importation of medicines ('have the same specific names'); and (ii) economic rationale of parallel trade in general ('set the prices of such medicines in a country lower than those in another country'). These two conditions are more or less already mentioned in Article 1 of the Regulation.

Under the Regulation on parallel imports of medicines, in Vietnam, the parallel import of medicines is permitted in two circumstances.[85] The first, concerns the import of medicines which have the same specific names as medicines which have already been granted registration numbers in Vietnam but are manufactured by different manufacturers within the same pharmaceutical-manufacturing company or group. For example, two manufacturers A and B in the same pharmaceutical-manufacturing company or group manufacture the same product S. Product S of manufacturer A has been granted a registration number in Vietnam and is currently on sale in the Vietnamese market at price G1. Product S of manufacturer B has not been

---

80. For many years, the Vietnamese pharmaceutical market has been dominated by three 100% owned foreign capital companies, Diethelm Vietnam, Mega Lifesciences VN Ltd, and Zuellig Pharma Vietnam Ltd. See Vietnam's Competition Administration Department, *Báo cáo Pháp luật cạnh tranh điều chinh hành vi phản cạnh tranh trong hệ thống phân phối dược phẩm tại thị trường Việt Nam*, [Legal Report on Anti-Competitive Practices in the Distribution Chain of Pharmaceuticals in the Vietnamese Market], (Hanoi, 2009), 127.
81. Decision No. 1906/2004/QD-BYT, fn. 46 *supra*.
82. Decision No. 110/2005/QD-TTg of the Prime Minister on Approval to the Plan on 'National stock of medicine for the prevention and cure of people's diseases', 16 May 2005. The use of the doctrine of parallel import of medicines was confirmed by s. 3 of this Decision.
83. Regulation on parallel imports of medicines for prevention and cure human diseases promulgated together with the Decision No. 1906/2004/QD-BYT (Regulation on parallel import of medicines 2004). There are six articles in the total of 10 articles of the Regulation addressing parallel importation of medicines 2004 related to state management. See Arts 4–10 of the Regulation on parallel import of medicines 2004.
84. Ibid., Art. 3 of the Regulation on parallel import of medicines 2004 stipulates that '[p]arallel import of medicines means the imports of medicines which have the same specific names as medicines already granted registration numbers for circulation in Vietnam when the foreign pharmaceutical companies set the prices of such medicines in a country lower than those in another country'.
85. Article 3 of the Regulation on parallel import of medicines 2004.

granted a registration number in Vietnam and is currently on sale in another country at price G2. If price G2 is lower than the price G1, a Vietnamese importer may purchase product S in the latter country for sale in Vietnam at price G3, provided that price G3 is lower than price G1 (G3<G1).[86] The second circumstance concerns the import of medicines which have the same specific names as medicines which have already been granted registration numbers in Vietnam and are manufactured by the same manufacturer in the same country but are supplied from different countries. For example, manufacturer X manufactures product S, which has been granted a registration number and is currently on sale in the Vietnamese market at price G1. Product S is, however, sold by manufacturer X in country A at the price G2. If the price G2 is lower than price G1, a Vietnamese importer may import product S from country A for sale in Vietnam at price G3, provided that G3 is always lower than G1 (G3<G1).[87] These examples of parallel imports of medicines in Decision No. 1906/2004/QD-BYT are simple cases. In fact, parallel imports of medicines are more complicated and relate to different manufacturers and distributors in different countries. Given that, the current provision of only two specific cases of parallel imports is insufficient.

As in most other developing and least developed countries, Vietnam remains very much in need of imported medicines since the capacity of its domestic pharmaceutical industry has not met its health care requirements.[88] Moreover, '[m]edicines were unaffordable for the lowest paid unskilled government workers, thus being unaffordable the large percentage population who earn less than this benchmark'.[89] Nonetheless, because of concerns regarding quality control and under pressure from market-dominated pharmaceutical companies, the parallel import of medicines for prevention and cure of human diseases is strictly controlled by the Ministry of Health regardless of the lawfulness of this activity. As a result, the value and volume of parallel imported medicines into Vietnam are very limited.

2.3.2 The Intellectual Property Law

Although parallel import had been regulated, it was still a matter of debate in the process of drafting Intellectual Property Law 2005.[90] There were a number of

86. Article 3.1 of the Regulation on parallel import of medicines 2004.
87. Article 3.2 of the Regulation on parallel imports of medicines 2004.
88. Legal Report on Anti-Competitive Practices, fn. 77 *supra*, 98–99. With respect to pharmaceuticals, the WHO ranked Vietnam in between levels 2 and 3 in terms of development level of pharmaceutical industry. The country produces some generic medicines, exports a few pharmaceutical products, and imports most of medicines for special treatment. The value of domestically manufactured pharmaceuticals is gradually increasing; nonetheless, the value of imported pharmaceuticals is still relatively higher than the value of domestically manufactured pharmaceuticals.
89. Anh Tuan Nguyen et al., 'Medicine prices, availability, and affordability in Vietnam', *Southern Med Rev* (September 2009), <www.southernmedreview.org/index.php/smr/article/viewArticle/11>.
90. The Legal Department of Ministry of Science and Technology, The Second Submission of the Draft of Intellectual Property Law' (Hanoi, 7 Feb. 2005), 5–6. Regarding the debate surrounding the issue of parallel importation in the process of making Intellectual Property Law, see Pham Hong Quat, fn. 73 *supra*, 184–185, 193.

views on the issue. Some opposed the inclusion of provisions on parallel importation in the Law because the issue is unclear in term of its legal basis and two-sided economic effects.[91] And some wished to include specific provisions regarding parallel importation in the Law in order to comply with international conventions that Vietnam has already entered into or to which it planned to accede to, particularly the TRIPS Agreement.[92] The regime of international exhaustion and parallel import, were ultimately included in Article 125.2(b) of the Law in order to harmonize four objectives, namely accelerating international integration, enhancing IPR protection and enforcement, and developing domestic industries, as well as protection of consumers' interests.[93] It provides:

> Owners of industrial property objects as well as organizations and individuals granted the right to use or the right to manage geographical indications *shall not have the right to prevent others from performing the acts of circulating, importing, exploiting utilities of products that having been lawfully put on the market, including overseas markets*, except for products put on the overseas markets not by the mark owners or *the person who is authorized by the trademark owner to put the products on the market*.[94]

By use of the wording '[o]wners of industrial property objects [. . .] including overseas markets', Vietnam asserts the adoption of the doctrine of international exhaustion of industrial property rights. Similar to the wording of Circular No. 825/2000/TT-BKHCNMT, 'importing' and 'including overseas markets' are used in Article 125.2(b). Parallel importation, thus, is lawful under Vietnam's current Intellectual Property Law. In our view, however, the phrase 'except for products put on the overseas markets not by the mark owners or the person who is authorized by the trademark owner to put the products on the market' makes Article 125.2(b) ambiguous. Because of this caveat, Article 125.2(b) may be understood as having two meanings. Under the first, parallel importation applies to all industrial property right-embodied goods (including patents, industrial designs, layout-designs of semiconductor integrated circuits, geographical indications, and trademarks). Under the second, parallel importation does not apply to trademarked goods.

In order to determine which interpretation is more accurate, the last part of the provision ('except for products put on the overseas markets not by the mark owners or the person who is authorized by the trademark owner to put the products on the market') should be read in combination with the first part ('[o]wners of industrial

91. The Legal Department of Ministry of Science and Technology, 'The First Submission of the Draft of Intellectual Property Law' (Hanoi, 10 Oct. 2005), 2–3; see also 'The Second Submission of the Draft of Intellectual Property Law', *ibid.*, 5–6.
92. The 'The Second Submission of the Draft of Intellectual Property Law', *ibid.*, 14. Even the proponents of the provision on parallel imports in the Intellectual Property Law did not have the same view on the issue. Specifically, some of them asserted that the regime of international exhaustion should be applicable to all categories of intellectual property rights while the others maintained that this regime was suitable for some categories only.
93. 'The Second Submission of the Draft of Intellectual Property Law', *ibid.*, 5–6.
94. Article 125.2(b) of the Intellectual Property Law (emphasis added).

property objects [. . .] including overseas markets'), in which the former must be seen as an affirmative clause. In other words, the former should be interpreted such that permitted parallel importation applies to all industrial property right-embodied goods. The last part accordingly should be understood as meaning that the legitimacy of parallel import cannot apply to trademarked products when they have been put on an overseas market either by the trademark owner or his licensees. Therefore, as with other industrial property right-embodied goods, parallel importation of trademarked goods is lawful. This argument, moreover, is supported by a historical consideration of the provisions on IPR exhaustion in the pre-Intellectual Property Law regulations. As indicated above, international exhaustion was previously adopted in the Circular No. 825/2000/TT-BKHCNMT and the Decree No. 06/2001/ND-CP.[95]

Under the current Intellectual Property Law, parallel importation applies to not only industrial property right-embodied goods but also plant variety right-embodied products. Article 190.2 of the Intellectual Property Law states that:

> Rights to plant varieties shall not be applicable to acts related to materials of protected plant varieties which have been sold or otherwise brought to the domestic or overseas markets by protection certificate holders or their licensees.[96]

The legitimacy of the parallel importation of copyrighted goods however, is uncertain. As provided under Article 20.1(d) of the Intellectual Property Law, copyright holders have the right 'to deliver, import the originals or the copies of the work'. The problem is that it is unclear whether parallel import of originals or copies of the work is legal or not. Parallel importation of copyrighted goods is not included in the list of copyright infringements in Article 28 (Copyright infringement) of the Intellectual Property Law. Therefore, there is no provision on the issue in either the Intellectual Property Law or the legal documents providing detailed regulations

95. On 29 Apr. 2009, the author discussed Art. 125.2(b) with Mr Pham Dinh Chuong a member of the Drafting Committee for the Intellectual Property Law, the former Director of the NOIP and is presently the Director of Scientific Institute of Intellectual Property. Mr Chuong stated that the intention of the Drafting Committee in Art. 125.2(b) was the adoption of the doctrines of international exhaustion and parallel imports for all industrial property objects. Furthermore, Mr Pham Dinh Chuong agreed that Art. 125.2(b) is ambiguous. In addition, the WHO lists Vietnam as one of the countries that apply the regime of international exhaustion. See Sisule Musungu & Cecilia Oh, 'The Use of Flexibilities in TRIPS by Developing Countries: Can They Promote Access to Medicines?', in WHO Study on IPRs, Innovation and Public Health (Geneva: WHO and South Centre, 2006), <www.who.int/intellectualproperty/studies/TRIPS_flexibilities/en/index.html>. Furthermore, the most recent WIPO survey on the patent exhaustion regimes of WIPO members refers to Art. 125.2(b) Intellectual Property Law to assert that Vietnam applies international exhaustion to patents (it must be noted that Art. 125.2(b) applies to all categories of industrial property rights, not only to patent). See WIPO Committee on Development and Intellectual Property, 'Patent Related Flexibilities in the Multilateral Legal Framework and Their Legislative Implementation at the National and Regional Levels' CDIP/5/4', fn. 42 *supra*, Annex II, 42.
96. Article 190.2 Intellectual Property Law (emphasis added).

for its implementation. In this case, the rights of copyright holder may prevail.[97] They have the exclusive right to commercialize the copyrighted goods including importation and exportation of these goods.

It should be stressed that since the invalidation of Circular No. 825/2000/TT-BKHCNMT no specific provision on parallel import-related matters is in effect in Vietnam. According to the government's regulation making plan, those matters are to be included in a Circular guiding Decree No. 97/2010/ND-CP.[98] Moreover, there is an absence of regulations on relation between parallel importation with other closely related areas such as technology transfer and competition in the Vietnamese law on parallel importation. Put differently, any there is no interaction amongst those legal fields in the area.

2.3.3 Kingmax Case

In this case, a batch of new RAM memory bearing the trademark KINGMAX was imported into Vietnam from Hong Kong by Chi Duc Service and Trade Company Ltd., (Chi Duc). According to Official Notice No. 1420/SHTT-TTKN dated 06 September 2007 of the NOIP, those products, manufactured by Kingmax Semiconductor Inc., Taiwan, were widely sold on the world market, and 'have long existed on the Vietnamese market through importation of different companies'.[99] The NOIP granted trademark protection to the trademark 'KINGMAX' affixed to computer memory and computers to Vien Son Applicable Technique Development Company Ltd, (Vien Son).[100] Based on this registration Vien Son, exclusive distributor of the Kingmax Company, requested that the Ho Chi Minh City Customs Office initiate a suspension of customs procedures for Chi Duc's imported batch of computer memory because the goods bore a counterfeit trademark.

Upon the request of Vien Son, the Customs Office suspended import proceedings for the goods and requested the views of the Ho Chi Minh City Department of Science and Technology. The Department of Science and Technology held that Chi Duc had to prove the computer memory were put on the market by the trademark owner or with his authorization, otherwise the dispute would be handled under administrative procedures provided in Article 15 of the Decree No. 106/2006/ND-CP.[101] Ho Chi Minh City Customs Office then made a written record and

97. Carlos Correa, *Trade Related Aspects of IPRs – A Commentary on the TRIPS Agreement* (Oxford: Oxford University Press, 2007), 87; Carolyn Deere, *The Implementation Game: The TRIPS Agreement and the Global Politics of Intellectual Property Reform in Developing Countries* (Oxford: Oxford University Press, 2009), 76.
98. This means that in the near future, there will be no separate regulation on parallel importation. This proposed Circular is to be issued by the Ministry of Science and Technology guiding implementation of Decree No. 97/2010/ND-CP of the Government of the handling of administrative violations in the field of industrial property, 21 Sep. 2010.
99. Official Notice No. 1420/SHTT-TTKN of the NOIP responding to Official Notice No. 905/QLTT-NV of 27 Aug. 2007 of Ho Chi Minh City Market Management Department, 6 Sep. 2007.
100. Protection Title No. 77256 24 Nov. 2006.
101. Article 15 of Decree No. 106/2006/ND-CP of the Government providing for sanctioning of administrative violations in the domain of industrial property, 22 Sep. 2006 (Decree No. 106/2006/ND-CP).

requested that the competent agencies sanction Chi Duc's importation of counterfeit mark goods.[102] In response to Vien Son's complaint, Chi Duc requested that the NOIP annul the trademark KINGMAX granted to Vien Son. Chi Duc argued that: (i) it had imported KINGMAX branded computer memory since 2000; and (ii) Vien Son did not have right to file an application for registration of the KINGMAX trademark. Chi Duc argued that the purpose of Vien Son's registration was 'to seek profit and prevent other companies from legally doing business'.[103] In other words that Vien Son's registration was solely to prevent other importers from importing RAM memory and computers bearing the trademark 'KINGMAX' into Vietnam. The NOIP accepted Chi Duc's arguments and issued Decision No. 1352/QD-SHTT (dated 29 November 2007) annulling the protection rights granted to Vien Son. Vien Son brought an action against Decision No. 1352/QD-SHTT in the Ho Chi Minh City People's Court. The Court accepted the complainant's request.[104] The Court of Appeal of the Supreme People's Court at Ho Chi Minh City, however, overruled the court of first instance's judgment and upheld Decision No. 1352/QD-SHTT.[105] It should be noted that the courts did not examine legal issues concerning parallel importation but rather, only the validity of Decision No. 1352/QD-SHTT.

On the filing date of application, Vien Son were entitled to trade 'office equipments, computer parts, electronic parts, antitheft devices, cameras, mobile phones, telecommunication equipments, and warranty services for these products' (as listed in the Certification of business registration No. 4102014000 issued by the Department of Plan and Investment of Ho Chi Minh City dated 14 February 2003) and did not function as a manufacturer of computer memory and computers.[106] The Court of Appeal, thus, based its decision on Article 87.2 the Law on Intellectual Property 2005 to overrule the judgment of the Court of first instance.[107] The Article 87.2 states that:

> Organization or individuals lawfully trading in products produced by others shall have the right to register marks for such products, provided that the

102. Ho Chi Minh City Customs Office transferred the violation dossier to the president of the Ho Chi Minh City People's Committee for administrative sanctioning according to his competence with the sanctioning level of 821.991.240 VND as provided in Art. 15 of Decree No.106/2006/ND-CP. The reason was that the provincial-level Customs Offices may only impose a maximum sanctioning level of 20.000.0000 VND. However, the president of the Ho Chi Minh City People's Committee may impose a maximum sanctioning level of 100.000 VND. The Committee, thus, asked the Customs Office to submit the case to the General Department of Customs.
103. Official Notice No. 1420/SHTT-TTKN, fn. 96 *supra*.
104. Judgment No. 1737/2008/HC-ST of the Ho Chi Minh City People's Court on instituting legal proceedings against a decision annulling validity of trademark protection title, 22 Oct. 2008.
105. Judgment No. 05/2009/HC-PT of the Court of Appeal of the Supreme People's Court at Ho Chi Minh City on instituting legal proceedings against a decision annulling validity of trademark protection title, 16 Jan. 2009.
106. See Decision No. 1352/QD-SHTT of the NOIP dated 29 Oct. 2007 annulling the validity of the Certificate of trademark registration, 2.
107. See Judgment No. 05/2009/HC-PT, fn. 105 *supra*, 6.

producers neither use such marks for their products nor object to such registration.

The Court found that 'until 30 August 2007, the Kingmax Company still uses the trademark 'KINGMAX' for its products and Vien Son are not transferred the right to use that trademark'.[108] Consequently, it concluded:

> The Decision No. 12325/QD-SHTT of the NOIP on issuing the Certificate of trademark registration No. 77256 for a company [Vien Son] having no right to trademark registration does not comply with the Law on Intellectual Property [2005]. . . . The NOIP lawfully acted as it issued the Decision No. 1352/QD-SHTT to annul the validity of the Certificate of trademark registration No. 77256. Therefore, [we] accept the NOIP and Chi Duc's requests, amend the Court of first instance's judgment, and dismiss Vien Son's claim. [We] uphold the Decision No. 1352/QD-SHTT of the NOIP dated 29 October 2007.[109]

The *Kingmax* case indicates that when dealing with imported goods bearing trademarks identical with or similar to protected trademarks in Vietnam, competent agencies determine whether the goods are counterfeit or parallel imported products. In cases relating to parallel import, the core issue is the parallel importers' burden of proof. As mentioned above, after the nullification of Circular No. 825/2000/TT-BKHCNMT, there is no provision on the requirements for proof in parallel import cases. The burden, in practice, is still on parallel importers. They must supply evidence to show where the goods come from and who put the goods on the market. They must demonstrate that the imported goods have been put on the market by the IPR owner or with his authorization in order to show that IPR exhaustion has occurred and the IPR owner no longer has the right to control the IPR-embodied goods-related transactions. On the other hand, when the IPR owner successfully responds to the parallel importer's evidence and succeeds in sustaining its burden of going forward, the goods will be treated as counterfeit.

The NOIP's annulment of the trademark protection granted to Vien Son and the affirmation of its position by the Court of Appeal of the Supreme People's Court reinforces and clarifies the policy on parallel import in the Intellectual Property Law. Accordingly, parallel importation into Vietnam of the memory bearing the trademark KINGMAX having been proved to have already been put on the overseas market into Vietnam was sustained.

2.4 Preliminary Evaluation

The Vietnamese regulation on parallel importation was initially established more than ten years ago and codified in the Intellectual Property Law as the requirement

108. *Ibid.*, 7.
109. *Ibid.*

of practice. Nevertheless, not many parallel import-related cases have been brought to or adjudicated and competent authorities, especially the courts have not had a chance to deal with this kind of case.

In order to utilize the advantages of parallel import to the fullest Vietnam might bring attention the issue in order to raise public awareness of the positive and negative effects of parallel import. In addition, legislation relating to the issue should be amended and supplemented to remedy some of the shortcomings of recent provisions on parallel imports. The regulations should reflect the nature and characteristics of parallel import; avoid reliance on administrative solutions, and harmonize the application of the doctrine of parallel imports in relation to related areas such as technology transfer and competition. Parallel importation may be regarded as a channel to foster technology transfer and a tool to address IPR-related anti-competitive practices. Furthermore, Vietnam should review its legislation to in order to reduce the negative effects of parallel importation by improving law and enforcement mechanisms against counterfeit parallel imported goods; law and technical systems for the efficient management of parallel importation; and build Vietnam's capacities to aid officials and agencies in combating counterfeit and low quality goods.

3 COMPETITION LAW AND POLICY ON TECHNOLOGY TRANSFER

Competition law and policy is a new area of legal development in Vietnam. Under the traditional central planning economic system, competition had no role to play either in theory or practice. Therefore there was no occasion to formulate competition policy. Accordingly, Vietnam did not have a competition policy until the late 1980s when the government started to implement the Doi Moi Policy. With the recognition of a market oriented economy, the government began efforts to develop its private sector and introduced competition into economic life. Since then, some kinds of competition regulations together with various policy measures have been gradually introduced to promote the competitive environment in Vietnam.

In 2004, the Competition Law was adopted by the National Assembly and became the fundamental legal framework governing competition in Vietnam. The central goal of the Competition Law is the promotion of a competitive business environment and the socio-economic development of Vietnam as well as the protection of public interests, the legitimate rights and interests of the state, and enterprises and consumers in Vietnam.[110] The Competition Law regulates both (i) unfair competitive practices and (ii) anti-competitive practices, i.e. practices in restraint of competition, carried out by enterprises (including foreign companies doing business in the country).

110. Article 4 of the Competition Law.

Both anti-competitive agreements and abuse of dominant/monopoly market position are subject to regulation of the Competition Law.[111] Accordingly, enterprises are prohibited from executing business agreements for the purposes of (i) collusive tender, (ii) market division (in order to prevent, hinder or restrain other enterprises to enter into a market or to develop their business), and (iii) elimination of the competition of other enterprises.[112] Other anti-competitive agreements, such as agreements on price fixing or market sharing, agreements to restrict output, commodities and service supplies, restraint of production and sales, restraints of technical developments, technology or investment, and agreement to impose trading conditions on other parties (which have no direct connection with the subject of the agreement), are also prohibited if the parties to such agreements have combined market share of 30% or more on the relevant market.[113] Hence, they may be exempted from the possible sanction if such agreement reduces production costs in a manner beneficial for consumers.[114]

Meanwhile, enterprises that have a dominant[115] or monopoly position in a relevant market are prohibited from conducting anti-competitive practices such as (i) selling below cost, (ii) price fixing, restricting production or distribution, (iii) restricting the market for technical or technological developments, (iv) applying discriminatory commercial conditions, (v) imposing conditions for signing contracts, (vi) bundling unrelated obligations into a contract or preventing other enterprises from entering the market.[116] Enterprises that hold monopoly market position are subject to additional restrictions such as prohibitions on imposing disadvantageous conditions on customers or abuse of a monopoly position to unilaterally change or rescind a signed contract without a legitimate reason.[117]

Generally, the Competition Law does not contain specific provisions regulating IPR-related competition issues. However, given the scope of application of the Competition Law which includes all business activities related to competition, the law (at least theoretically) can be directly applied to regulate IPR and technology

111. Articles 8, 13 and 14 of the Competition Law. See also Competition Law, Technology Transfer and the TRIPS Agreement, fn. 4 *supra*, 212–230.
112. Article 9.1 of the Competition Law.
113. Article 9.2 of the Competition Law.
114. Article 10 of the Competition Law. Such anti-competitive agreements may be subject to exemption if they are satisfied one of the following conditions: (i) optimizing the business structure, (ii) promoting technical or technological progress, improving the quality of goods and services, (iii) promoting uniform applicability of quality standards and technical norms of certain types of products, (iv) unifies commercial practices except price-related-conditions, (v) increases the competitiveness of medium and small-sized enterprises, or (vi) increases the competitiveness of Vietnamese enterprises in the international market.
115. An enterprise considered to have a dominant position if it holds 30% or more of the relevant market or is capable of significantly restraining competition includes; a group of enterprises acting together if they possess a combined market share of 50% or more (for two enterprises), 65% or more (for three enterprises) or 75% or more (for four enterprises) in the relevant market. See Art. 11 of the Competition Law.
116. Article 13 of the Competition Law.
117. Article 14 of the Competition Law.

transfer activities that restrain competition. There is no in-practice assessment of IPR-related competition rules since no technology transfer-related competition law case has been adjudicated in Vietnam since adoption of the Competition Law in 2004. To that end, the *Zuellig Pharma Vietnam* case, which concerns the abuse of exclusive distribution rights for pharmaceutical products, may provide some important points on dealing with the intersection of IPR and competition law/policy in Vietnam.

4 *ZUELLIG PHARMA VIETNAM* CASE

Zuellig Pharma Vietnam Limited (Zuellig Pharma Vietnam) was a subsidiary of Zuellig Pharma Specialty Solutions Group, a Singapore based multinational company specializing in pharmaceutical distribution. In May 2001, Zuellig Pharma Vietnam received permission from the Vietnamese licensing authorities to import, export, and distribute pharmaceuticals in the Vietnamese market.[118] The Vietnamese pharmaceutical market, during that period was not elastic due to the strict control over the numbers of pharmaceutical distributors exercised by the Government. Zuellig Pharma Vietnam quickly gained a large market share after executing exclusive distribution agreements with over 27 foreign major pharmaceutical companies to distribute their medicines in Vietnam.[119] It is worth noting that most pharmaceutical products available in Vietnam up to the present day are still predominantly imported from foreign sources.

There was an unprecedented rise in pharmaceutical prices in Vietnam during the period from 2003 to 2004. According to the Ministry of Health, the increase of pharmaceutical price, especially the rise of imported patented pharmaceutical prices, was mainly attributable to the de facto monopoly of Zuellig Pharma Vietnam in the pharmaceutical distribution market.[120] According to Mr Nguyen Van Thanh, then Deputy Head of Drug Administration of Vietnam, '[t]he result of inspection by Vietnam Drug Administration on the distribution chains of Zuellig Pharma Vietnam showed that each year Zuellig Pharma Vietnam increased the price of pharmaceuticals in average between 2% and 12%, some specific pharmaceuticals by 30%–60%, leading to the rise of pharmaceutical prices'. Statistics shows that during this period, out of 500 pharmaceutical products that were increased in price, 157 were distributed by Zuellig Pharma Vietnam.[121]

118. Official Letter No. 110/CV-BQL-HN of Hanoi Industrial and Export Processing Zone, 7 Dec. 2001; Official Letter No. 5375/QLD-CL of Drug Administration of Vietnam.
119. Zullig Pharma, Company Profile, <www.zuelligpharma.com/offices_vietcam_pro&ser.html>.
120. Decision No. 1353/QD-BYT of the Ministry of Health on the plan on implementation of some urgent solutions to stabilize pharmaceutical prices, 16 Apr. 2004.
121. Hanoi Moi Online, <www.hanoimoi.com.vn/forumdetail/Doi-song/14585/quot2727841i-giaquot-zuellig-pharma-vi7879t-nam-b7883-273inh-ch7881-ho7841t-2737897ng.htm>. (Translation was by the authors.)

The fact that Zuellig Pharma Vietnam enjoyed exclusive rights of distribution for foreign patented pharmaceutical products can now be subject to scrutiny under the current competition law. From a competition law perspective, the key issues in the case were whether pharmaceutical products which were exclusively distributed by Zuellig Pharma Vietnam could constitute a 'separate relevant market' and whether Zuellig Pharma Vietnam has a dominant market position under the Competition Law. If this were established, the acts of raising prices of pharmaceutical products by Zuellig Pharma Vietnam would be considered to be anti-competitive practices in breach of the current Competition Law and could be remedied by state competent authorities.

The Ministry of Health while conducting the investigation on the increase of price of pharmaceutical products also concluded that one of the main reasons of the increase was due to the abuse of its market position by Zuellig Pharma Vietnam as the exclusive distributor of imported pharmaceutical products. However, the State competent authorities were unable to neither conduct any proceeding nor apply corrective measures against any excessive pricing practices by Zuellig Pharma Vietnam for a lack of any legal basis to do so. The Competition Law which governs the abuse of market power by enterprises in Vietnam was, at the time, only in the drafting process and was not adopted by the National Assembly until the end of 2004. It should be noted that at that time the parallel import of pharmaceuticals in Vietnam was not permitted.[122]

The *Zuellig Pharma Vietnam* case illustrates the importance of competition law enforcement in respect of IPRs to control excessive pricing over imported pharmaceuticals by exclusive distributors.

4.1 Preliminary Evaluation

Admittedly, establishing a competition law framework is an important development in Vietnam. The Competition Law together with the Intellectual Property Law can create a mechanism to promote socio-economic development and facilitate the process of technology transfer in Vietnam. The mechanism is, to a large extent, a work in progress with a number of shortcomings and ambiguities that are obstacles to effective enforcement.

One of these areas concerns the framework regulating collusive conduct under the Competition Law. Current regulations do not make a clear distinction between agreements between competitors (horizontal agreements) and non-competitors (vertical agreements). Under Article 9 of the Competition Law, listed anti-competitive agreements are prohibited if the 'combined market share' of the parties to the agreement is at least 30% or more in the relevant market. These categories cover a broad range of activities: price fixing and market sharing

122. Decision No. 1906/2004/QD-BYT of providing the regulation on parallel import of pharmaceuticals was adopted by the Ministry of Health only on 28 May 2004. See Part 4 of this chapter.

agreements, agreements to restrict output, commodities and service supplies, restraints of production and sales, restraints of technical developments, technology or investment, and agreement to impose trading conditions on other parties. This means that the parties to these agreements must operate at the same level of trade level in order to calculate their 'aggregate' or 'combined' market share.

Thus, it can be deduced that these prohibited agreements are regulated only if they are horizontal agreements. Vertical anti-competitive agreements therefore, may only be investigated under provisions relating to abuses of a dominant/monopoly position, because the term 'market share of a participating party' is a substantive precondition to competence to consider vertical agreements and parties at different levels do not share a relevant market. However, agreements to impose trading conditions on other parties having no direct connection with the subject of the agreement are usually found in vertical agreements.

This jurisdictional gap can create problems for the authorities which may have the effect of hindering the enforcement of the law. In addition, the absence of clear distinctions between horizontal agreements and vertical agreements may lead to situations whereby anti-competitive licensing agreements between both competitors and non-competitors will be subject to the same level of scrutiny, despite the fact that horizontal agreements between competitors are more likely to cause competition law problems.[123]

From the technology transfer perspective, anti-competitive licensing agreements other than market entry foreclosure agreements[124] may not be restricted by the competition law. Such agreements may fall under the scope of other types of anti-competitive agreements (as defined under Articles 8.1 to 8.5 of the Competition Law), which are prohibited only if the combined market share of participating parties is under 30% in the relevant market.[125] However, if the above cap is exceeded, the parties to relevant licensing agreements may still obtain an exemption if they can prove that the agreements encourage the dissemination of technology or technical and technological improvement, and enhance quality of products under Article 10.1 of the Competition Law. They are not allowed to obtain the above-mentioned exemption only when their actions constitute abuse of dominant/monopoly market position.[126] As consequence, licensors and and/or licensees may execute licensing agreements providing restrictions to technology transfer such as price fixing, limitation of output, allocation of market, etc. provided that such practices do not constitute abuse of a dominant/monopoly market position under the Competition Law.[127] The above shortcomings of the Competition Law can hinder State authorities from controlling anti-competitive

123. See Tu Nguyen Thanh, fn. 4 *supra*, 219–221.
124. Anti-competitive agreements which have market entry foreclosure effects are per se prohibited under Art. 8.6 and 8.7 of the Competition Law.
125. Article 9 of the Competition Law.
126. Articles 13 and 14 of the Competition Law.
127. Articles 11 and 14 of the Competition Law.

practices of enterprises, which harm customer welfare and hinder technology transfer.

Competition Law and Intellectual Property Law are not harmonized in cases of 'refusal to license' by patent holders. Under the Intellectual Property Law, compulsory licenses for patents can be issued if parties cannot reach a voluntary licensing agreement on reasonable commercial terms and conditions within a reasonable period of negotiation time.[128] Among six categories of acts prohibited as abuses of a dominant position under the Competition Law,[129] restrictions on the distribution of goods and the prevention of new competitors entering the market are closely associated with refusals to license. However, the Competition Law does not directly set forth that these acts constitute a prohibition on refusals to license. As such, a refusal to license may not be regarded as abuse of a dominant position under the Competition Law. In international practice, the granting compulsory licenses on the basis of refusal to license by patent holder it is widely regarded to be a competition law offence and justified if such refusal is determined to be an abuse of a dominant position. Currently, the absence of a clear statement in the Competition Law implies that the compulsory licensing based on refusal to license only can be determined under the Intellectual Property Law. However, such a determination regarding a failure to agree on reasonable terms and conditions within a reasonable period of time under the Intellectual Property Law may be subject to challenge under the WTO dispute settlement mechanism.[130] Therefore the grant of a compulsory license based on a refusal to license will be made more difficult to challenge if State authorities can make the decision referencing both competition and intellectual property laws. Therefore, it is imperative that competition law should also regulate this issue.

5 CONCLUSION

The TRIPS Agreement mandates minimum standards for WTO country members for protecting intellectual property. As a member of the multilateral trading system, Vietnam is required to respect these standards. However, Vietnam can use the flexibilities provided by the TRIPS Agreement to strengthen its national laws and regulations and enforce them efficiently in order to promote technology transfer and protect sustainable development. This has been understood and supported by

128. Articles 145.1 (c) and 195.1 (b) of the Intellectual Property Law.
129. Article 13 of the Competition Law.
130. European Commission, TBR Investigation Report, Executive Summary, 'Examination Procedure Concerning an Obstacle to Trade, within the Meaning of Council Regulation (EC) No. 3286/94, Consisting of Measures Adopted by the Separate Customs Territory of Taiwan, Penghu, Kinmen and Matsu Affecting Patent Protection in Respect of Recordable Compact Discs, Report to the Trade Barriers Regulation Committee' (30 Jan. 2008).<http://trade.ec.europa.eu/doclib/docs/2008/january/tradoc_137632.pdf>. See Tu Thanh Nguyen, fn. 4 *supra*, 204–211.

Vietnamese policy makers.[131] Adoption of a comprehensive and TRIPS compatible Intellectual Property Law in 2005, which includes among others the norms on compulsory licensing and parallel import, is regarded as an important development toward utilization of the flexibilities of TRIPS Agreement.

There have been cases where the doctrines in respect of compulsory licensing and parallel import have been utilized in Vietnam by private parties and State authorities. The *Tamiflu* case, in particular proved the effectiveness of the compulsory licensing doctrine as a 'threat' and/or 'bargaining' tool to access technology for the public interest. Nonetheless, to date, a complete system governing the use and enforcement of these tools continues to present difficult issues for Vietnam for to many reasons

Intellectual property laws and regulations intellectual property alone cannot resolve all aspects of access to technology, nor can they effectively prevent IPR abuses in the area of restrictions of competition or technology transfer. The Intellectual Property Law needs to be supported by effective intellectual property related competition laws and policy.

Competition law can play an active role in curbing abusive exploitation of market power in connection with the exercise of IPRs in the areas of the anti-competitive effects of licensing agreements, especially those containing exclusivity or restrictive clauses. Licensing agreements can provide positive effects upon societal development because they facilitate dissemination and exploitation of technology and know-how. Nevertheless, the transfer of patented technology may involve excessive and unnecessary restrictions on competition (e.g., territorial exclusivity, exclusive dealing, tying requirements, and grant-back requirements). These sorts of contractual arrangements and limitations on market conditions may unduly restrict competition and need they require control and regulation.

Competition law may also be a means to deal with issues concerning excessive pricing by IPR holders. The Intellectual Property Law permits intellectual property right holders and their licensees to market their technology and/or technology-embodied products in particular territory at prices they consider appropriate. The Intellectual Property Law does permit State authorities to control excessive pricing on the market through compulsory licensing on narrowly specified grounds, including but not limited to disease prevention/treatment or peoples nutrition. Hence, special justification for such public emergency situation is required. However, under the Competition Law, State authorities can control and correct anti-competitive pricing abuses by IPR holders with a dominant market position.

If a reasonable and balance between intellectual property protection and fair competition can be set up, it will encourage not only innovation and competition but also economic growth.[132] Vietnamese legislators must design and adopt

---

131. See Resolution No. 48-NQ-TW, fn. 6 *supra*.
132. US Federal Trade Commission, 'To Promote Innovation: The Proper Balance of Competition and Patent Law and Policy' (Washington, October 2003). <www.ftc.gov/os/2003/10/innovationrpt.pdf>; see also Competition Law, Technology Transfer and the TRIPS Agreement, fn. 4 *supra*, 292; Mark Lemley, 'A New Balance between IP and Antitrust', Stanford Law and Economics Online Working Paper No. 340 (April 2007), <http://ssrn.com/abstract=980045>.

guidelines for applying Competition Law in relation to intellectual property and technology transfer. In particular, these guidelines should include clear provisions for the application of compulsory licensing and protection of parallel importation to remedy intellectual property rights related anti-competitive practices.

Vietnam also needs to develop its infrastructure for enforcement of its legislation, in particular the by building professional capacity and cooperative mechanisms among the various State authorities with related competence. Application of intellectual property law and competition law in relation to intellectual property consists of complex interactions. There is still a serious lack of knowledge, experience and legal expertise on these subjects in Vietnam among lawyers, policy makers, and the commercial world. Constraints in professional capacity prevent State authorities and private parties from effectively exercising their rights and powers or observing their obligations. There should be also a simple and effective mechanism of coordination between State authorities to deal with enforcement so that the entire procedure from fact finding and case analyses to decision making and enforcement can be carried out quickly and effectively. Coordination and cooperation among State authorities is essential to effective enforcement of the law.

Lastly, Vietnam should also establish an efficient review mechanism to evaluate the actual effects of compulsory licensing and parallel importation to ensure that its enforcement is truly efficient and beneficial for society.

Chapter 12

From Reality to Law: Sustainable Technology Transfer – An Outlook

Christoffer Wong, Hans Henrik Lidgard, Jeffery Atik & Tu Thanh Nguyen

The preceding chapters of this book have, all in their own way, sought to relate the law as laid down in international instruments – whether hard or soft – to the realties faced by the States, people, enterprises, climate, environment, biological heritage etc., that these instruments are meant to affect. While much of these discussions point to shortcomings in the law, other criticisms are directed not towards the law as such, but rather to its implementation and enforcement, or the lack of means for settling disputes regarding the law.

This concluding chapter will take another look at the law, having gained some insight into the realities of its functioning and especially at what needs to be done in order to improve the process of Sustainable Technology Transfer.

1 TRENDS

Activities during the last decade indicate an increased understanding of the common goal to eradicate poverty and its importance. Efforts are required from all. The reason is not just that activities are compelled from a human rights perspective or that equity and fairness so dictates, but also that sensible development benefits all – a form of new Marshall Plan.

Hans Henrik Lidgard, Jeffery Atik & Tu Thanh Nguyen, *Sustainable Technology Transfer*, pp. 287–299.

Have all the efforts made during the last decade resulted in improvements? The views diverge:

UNCTAD Secretary-General Supachai Panitchpakdi expressed the results of all efforts to improve the situation of LDCs when he noted that only two countries have graduated from the LDC group in the last 40 years, which is 'a very dismal record' and 'a total failure from all sides'.[1] If countries only go on the traditional development path the world will end up where it has always been. According to the Secretary-General, new development paths are required to boost the 'enabling role' of the State in domestic economic growth; to foster greater reliance on domestic sources of growth; to expand domestic markets; to reduce commodity dependence; and to deal effectively with external policies that affect LDCs, as their economies are very open to the forces of global markets. He added that steps are needed to attract foreign investment to LDCs, focused not on extracting natural resources, but rather on expanding manufacturing or towards more advanced processing of goods.[2]

On the other hand, the UN 2010 Human Development Report[3] paints a rosier picture of global development. The Report measures a country's success not by money alone, but gauges whether people have long and healthy lives, whether they have the opportunity to be educated, and whether they are free to use their knowledge and talents to shape their own destinies. On such a measure it seems that many countries have indeed made gains in health, life expectancy, schooling and overall living standards despite only modest growth in income. The 2010 Global Wealth Report[4] also suggests that substantial achievements have been made in the last decades.

WTO Director-General Pascal Lamy told the Fourth United Nations Conference on the Least Developed Countries in Istanbul on 9 May 2011[5] that the progress of the LDCs is unquestionable. Their average growth rate of 7% over the past decade exceeds the world average and their trade grew twice as fast as world trade. The LDCs account for one fifth of all WTO Members and receive close to half of the technical assistance budget. Official Development Assistance in the LDCs has progressed, from 5 billion dollars in the early 2000s to 12 billion in 2009. 'So on the whole, we are on the right path: the LDCs' share in world trade has doubled in the space of 10 years. And yet, the LDCs still account for only 1% of world trade. In other words, the contribution of international trade to the development of the LDCs remains inadequate.'

1. Statements by Supachai Panitchpakdi, Secretary-General of UNCTAD (2008–2010),UNCTAD Trade and Development Board, Forty-ninth executive session, 8 Jun. 2010. <www.unctad.org/Templates/webflyer.asp?docid=13354&intItemID=3549&lang=1>.
2. *Ibid.*
3. Human Development Report 2010, 'The Real Wealth of Nations: Pathways to Human Development', Published for the United Nations Development Programme (UNDP) (November 2010). <http://hdr.undp.org/en/reports/global/hdr2010/>.
4. Credit Suisse, 'Global Wealth Report 2010' (2010 Global Wealth Report), 3. <www.credit-suisse.com/news/doc/credit_suisse_global_wealth_report.pdf>.
5. <www.wto.org/english/news_e/sppl_e/sppl191_e.htm>.

Progress is difficult to measure and is also difficult to determine that progress is evenly distributed. Some developing countries have emerged with unprecedented speed, whereas several of the LDCs have only seen minor changes. Much work remains.

The experience of emerging East Asian economies points to the crucial role that import of sophisticated technology and management practices has played in their successful industrialization. Access to and adoption of new technologies to local condition are central to the long-run growth of these economies. This experience now needs to be reused and applied to the poorest nations. The legal infrastructure and the mechanisms to supervise and understand their development needs to be improved.

2 LAW OF TECHNOLOGY TRANSFER

2.1 Definitional Dilemmas

2.1.1 The Developing Country/Developed Country Dyad

Across many platforms of law and development there is a continuum from least developed to developing to developed countries, with some intermediate statuses as well. Clearly, a country's status on this continuum is determinative of its rights and/or obligations under many treaties, including the central instruments considered in this book. Is it that developing countries – or least developed countries as the case may be – do have rights as a direct corollary of the fact that the developed countries have an obligation? It may be argued that, rather than being possessors of rights, the developing countries' position can be seen as a kind of liberty (i.e. non-obligation) or even a kind of immunity (i.e. others cannot impose obligations on them).[6] The question boils down to whether it is only developed countries that have an obligation to assist with technology transfer, or that this obligation rests on all countries but the LDCs and consequently that countries like China, India, Brazil and other emerging economies should take a more active role in technology dissemination and reporting under TRIPS Article 66.2.

These are not, however, questions specific to TRIPS. As the discussions on climate change within the UNFCCC regime and biodiversity and benefit sharing under CBD show ambiguities with regard to the most fundamental legal concepts abound. Furthermore, one main thread that runs through the contributions in this book has concerned the 'real' effects of provisions in treaties and other international instruments that formally do not create a legally-binding obligation under international law: i.e. the hard law v. soft law debate. We may thus say – One law, a multitude of realities.

---

6. The terminology has partly been borrowed from Hohfeld's classical analysis of the concept of rights, expounded in, e.g., Wesley N. Hohfeld, 'Some Fundamental Legal Conceptions as Applied in Judicial Reasoning', 23 *Yale Law Journal* 16 (1913).

A central problem that has been raised in connection with the definition debate concerns the meaning of 'developing countries'. The criteria for 'graduation' from 'developing' to 'developed' or from 'least developed' to 'developing' is also a part of this definitional issue. However, the comparmentalization of countries into different categories reveals a more fundamental problem. While categories may make (some) sense in designing schemes for the distribution of rights and duties in the pursuit of common goals, the categories are irrelevant for the identification of the goals themselves – e.g. biological diversity, good climate/environment or poverty reduction.

Thus, once it is agreed, for instance, that it is our goal to achieve 'the stabilization of greenhouse gas concentrations in the atmosphere at a level that would prevent dangerous anthropogenic interference with the climate system',[7] then it makes sense to speak of different countries having different obligations in the pursuit of this goal. But all countries share in this respect the same goal regardless of which categories they belong to. This is the kind of new outlook on the international community that more and more has become the context in which discourse in international law is being conducted – that of common, but differentiated, responsibilities.

TRIPS is the chief example of where definitional ambiguities provide for considerable elasticity in the distribution of undertakings – and where these ambiguities potentially contribute to the disappointing performance of fulfilling TRIPS' call for technology transfer. The intended beneficiaries of TRIPS Article 66.2 are fairly clear: we know the identity of the LDC WTO Members from UN inventories. TRIPS Article 67 is less clear – as the intended recipients of the technical and financial cooperation called for in Article 67 include both LDCs and 'developing country' WTO members. Since the category of 'developing country' WTO members is not defined, be it within the WTO or by the UN, many, if not most, WTO members can 'opt in' to the benefits of Article 67 by self-identifying as a 'developing country' member.

What is less clear is which WTO Members are 'developed country' members for purposes of determining which states bear the obligations created by Articles 66.2 and 67. While it would be impossible for certain states to avoid categorization as 'developed countries' – such as Europe, Japan or the United States – many WTO members are effectively conceded the discretion to avoid self-identifying as a 'developed country', and so evade any responsibility (under TRIPS) to provide the incentives required by Article 66.2. Here, the ambiguity of term permits many WTO members to 'opt out' from the demands of Article 66.2 by rejecting the self-identification of 'developed country' member.

The general technology transfer obligation found in the Convention on Biological Diversity is binding on all contracting parties – and certain enhanced privileges are granted to 'developing countries.' Like TRIPS, the CBD structure does not provide a definition of which CBD contracting parties are 'developing

7. Article 2, United Nations Framework Convention (1992) on Climate Change.

countries' and are so entitled to exercise these privileges. Again, the liberty to self-identify as a 'developing country' may permit many CBD contracting parties to claim benefits of CBD non-developing country contracting parties. Note here that there is no consistency requirement of classification across these regimes: a country (perhaps Brazil) might not be a 'developing country' WTO member (given its capacities and level of development) for purposes of TRIPS Article 67, but might be 'developing country' CBD contracting party.

The UNFCCC has similar – if not more serious – ambiguities as to the identity of the parties subject to a technology transfer obligation, where it uses the terms 'developed' and 'developing' parties in its technology transfer provisions, a classification scheme which co-exists with the more precisely defined Annex I and Annex II lists of countries. Conceptually, there may be a middle ground of countries which, for UNFCCC purposes, might be neither developed nor developing, although experience under the UNFCC has taught that countries (such as China and India) may cling to a 'developing country' self-identification in order to avoid more stringent undertakings. And the terms 'developed country' and 'developing country' operate throughout the UNFCCC regime (together with the more precise lists from Annex I and Annex II) to generally delimit the 'common but differentiated responsibilities' provided in the climate change regime.

The terminological difficulties present in these three families of instruments may not be accidental. First, by providing many states the option of self-identification as 'developing countries', states retain significant control over the international obligations these instruments create. Second, this permits shifts in status that may accompany eventual (and still hoped for) graduation. Finally, while the dyadic set 'developing/developed' seem to cleanly divide obligations between two mutually-exclusive sets of states, a better understanding – and one consistent with 'common but differentiated responsibilities' across all of these instruments – is that the definitional ambiguities permit a distribution of the intensity of obligation. A state may be more or less developed country – and so have a greater or lesser responsibility to transfer technology.

2.1.2 The Scope of the Obligations to Transfer Technology to Developing Countries

The interdependence of States in not least the economic and ecological and environmental fields means that it is no longer possible for States to exist in splendid Westphalian isolation, choosing when and in which matter they wish to participate in the international arena – there is simply no possibility of an 'opt out' for climate change or a global financial crisis. From its origin as a law of nations whose primary purpose is to maintain peaceful relations, international law has grown to encompass norms on international cooperation in an interdependent world.[8] Given this outlook, international law should be more concerned with defining

8. For this kind of outlook see, e.g., the most recent edition of the standard work, Pierre-Marie Dupuy & Yann Kerbrat, *Droit International Public*, 10th edn (Paris: Dalloz, 2010), 2–3.

goals than stipulating categories. Problems arise however when goals stand in conflict with other interests.

Various international instruments impose legal obligations upon states to encourage or facilitate the transfer of technology. This book has extended coverage of the obligations imposed by the Convention on Biological Diversity regime, the UN Framework Convention on Climate Change regime, and TRIPS. As noted in prior discussions, there is considerable ambiguity with respect to the very terms that are used to express these obligations. Further, this ambiguity is more than latent; it is frequently an intended result, either masking continuing disagreement about the scope of the obligations or creating flexibilities demanded by various parties.

The very term technology transfer is filled with shifting meanings. When 'technology transfer' is further qualified – as we have done in this book – with the phrase 'sustainable', it is a significant challenge to define. While it is easier to categorize a particular project or initiative as an exemplar of 'technology transfer' once observed, it is much harder to state the essence of the category in a way that can be easily used to prescribe policy responses.

The phrase 'sustainable technology transfer' is hardly the only mysterious term arising in the term of this coverage. Technology transfer lends itself to differentiated obligations. Some states owe the obligation to transfer technology; other states are the intended beneficiaries of such obligations. Different instruments use different techniques to identify a particular State's status for purposes of stipulating its obligation (or entitlement) with respect to technology transfer.

The UNFCCC contains, of course, a specialized technology transfer mandate: in Article 4(5), it calls for the transfer of 'environmentally sound technologies and know-how'. This term, which is thickly ambiguous, generally addresses the content of the obligation, but it also depends on the identity of the developing country recipient; for a given developing country, a particular technology may or may not be environmentally sound.

The vagueness of terms such as 'technology' and 'transfer' has been criticized as being obstacles to an effective fulfilment of the obligations imposed by international agreements. The argument goes like this: If almost everything is capable of being counted as, for example, technology, then the obligation is devoid of real meaning. In order, for instance, to enable the least developed countries to create a 'sound and viable technological base', it is argued that a precise definition is necessary. In this view, only when obligations are clearly defined can state parties be held accountable for failure to fulfil or falling short of such obligations.

This may admittedly be an appropriate analysis if we are dealing with ordinary contracts between private individuals – more or less strangers to each other – who use the contract as a means to establish rights and duties with respect to each other. Under such circumstances it is only rational to require that contract terms be as precise as possible. However, agreements such as TRIPS are not such ordinary contracts. TRIPS does not deal with single isolated transactions; the Agreement creates rather a framework of rules governing the relationships between and among states parties that is meant to survive on a long-term basis.

In particular, provisions such as Article 66.2 are not based on single transfers in which each of the 'contracting parties' seeks to maximize its own utility to be derived from 'the deal'.[9]

The aims and special character of such agreements provide thus further data required to give terms in the agreement a more precise meaning. After all, it is just a general principle of international law that 'a treaty shall be interpreted in good faith in accordance with the ordinary meaning to be given to the terms of the treaty in the context and in the light of its object and purposes'.[10]

However, to be even more radical, one may argue that the point here is not to strive towards more precise definitions. To begin with, in some cases the very intention of a provision is that the concept used should remain imprecise, allowing a margin of appreciation either way. More interestingly, however, vague concepts can be used to facilitate a process of dynamic norm creating.

The naïvely optimistic view here is thus one in which the lack of precise definitions – far from being obstacles to the fulfilment of obligations – can be used instead to enhance the process of constitutive norm development. In particular, when dealing with multi-layered concepts such as 'sustainable technology transfer' precise definitions of each constituent element of the term are likely to be unhelpful when it is the meaning of the complex concept that really matters. Furthermore, one should not underestimate the 'visionary' aspect of even technical concepts.

2.2 Nature of Legal Obligations

The definitional ambiguities are just part of a general muting of the obligations created to transfer technology to the developing world. In no instance is the obligation created by any of the central instruments reviewed here hard – notwithstanding the consistent use of the term 'shall' in TRIPS Article 66.2, CBD Article 16, and UNFCCC Article 4(5). In each instance, ambiguity exists as to precisely which countries owe any obligation to transfer technology to developing countries, as discussed above. Considerable uncertainty also attaches as to which actions, programmes and initiatives constitute technology transfer, further weakening the legal force of the obligations.

Moreover, the texts of the various obligations contain phrases that convert the otherwise imperative 'shall' into something quite a bit more hortatory. TRIPS Article 66.2 calls on developed country WTO members to 'provide incentives' for the 'promoting and encouraging' technology transfer – which is quite distant from a technology transfer mandate. CBD Article 16 calls on Parties to 'provide and/or

9. It is, of course, a totally different question as to why States Parties are motivated to include as a general rule a provision such as Art. 66.2 at all. It is a common view that the obligation to transfer technology is a quid pro quod for other undertakings assumed by developed countries.
10. Article 31.1 Vienna Convention (1969) on the Law of Treaties.

facilitate access for and transfer' technologies, a clumsy formulation. UNFCCC Article 4(5) calls on certain parties to 'take all practicable steps to promote, facilitate and finance, as appropriate, the transfer' of technology. The qualifiers 'practicable' and 'as appropriate' effectively eliminate any binding effect of Article 4(5).

Again, the softness is certainly deliberate, and as such may simply represent the outer bounds of the developing countries' willingness to commit resources to technology transfer. An optimistic view encompasses progressive development – anticipating harder, more challenging and more effective technology transfer mandates to follow in future years.

2.3 Fragmentation

As this book amply demonstrates, there is no unitary law of technology transfer in the context of 'development'. Despite common normative sources – such as the right to development, the right to health, environmental justice and the right to access the common heritage of humankind – that motivate the urge to transfer technology from developed to developing countries, precise obligations are distributed among various instruments with varying degree of hardness, shifting classifications of transferring and recipient states, and different substantive coverage. The law of technology transfer suffers from extreme fragmentation.

Fragmentation creates the possibility of both undercounting and overcounting initiatives that arguably discharge technology transfer obligations. By obscuring which policies and programme constitute technology transfer, various activities, including certain forms of laudable development aid, may be inappropriately characterized as technology transfer.

Many developed countries – including all of Europe – are active parties of TRIPS, CBD and the UNFCCC climate change regime. Many of these countries commit themselves to transfer technology as part of larger programmes of development assistance. Fragmentation may introduce odd and accidental, if not perverse distortions, as an ideal technology transfer initiative should 'qualify' under all three instruments.

Fragmentation refers to a horizontal fracturing of norms among diverse instruments. There is a relevant vertical dimension as well: the retention of national discretion to effect technology transfer without regard to international mandates. Countries can and do effect development aid generally beyond their legal obligations. And technology is effectively transferred without the need for compulsion. Indeed self-interest does, to some small degree, promote meaningful technology transfer. Still the presence of technology transfer obligations reflects the broad judgment that states, if left to national discretion, will not adequately address the needs of development.

Developing countries can also take important unilateral action. The flexibilities contained in TRIPS can be understood as re-establishing national discretion, to be enjoyed by least developed and perhaps developing countries, to actively draw technology against the resistance of IP owners.

3 A WAY FORWARD

As is clear from the preceding remarks there are numerous items to address how to on the approach to further development discussion. More could be said. In a larger context, much of the discussion in this book indicates that in order to impact on Sustainable Technology Transfer there is a need to address matters such as a balanced dispute resolution systems, corruption and criminal law etc.

The emerging standpoint in the Vietnam Conference was that further fragmentation and new institutions were not the solution to the problems we encountered. Such novel measures have a tendency to distract from the overall goals and are often cumbersome and create a less focused approach. A more constructive approach is probably to look for solutions within existing frameworks and improve on them. As our focus throughout has been on measures which will help developing and especially least developed states to catch up with state-of-the-art technology in order to achieve economic advancement, we have had a special eye on how their development has been reported.

One major conclusion was that the TRIPS Agreement contains far reaching obligations to LDC's, but that they are vague and enigmatic. Only few developed countries report and these reporting countries overload their reports with matters unrelated to the provisions of Article 66.2. The reports are not focused on the LDC's needs, nor do they primarily deal with the incentives they provide to their own companies to foster technology transfer. There is a general lack of imagination and creative thinking in how to approach the obligations of Article 66.2. Therefore, the reports do not provide a basis for extracting best practices recommendations which ignores a crucial reason for the reporting.

The fact that there is no clear definition of the notions, 'technology', 'technology transfer' and 'technology assistance', also provoked a discussion. Is a workable technology required or the latest state-of-the-art technology that will allow poorer countries to leap-frog into a developed country status?

Nor was it clear who carries the relevant obligations. The notion 'LDC' is precise, but it is unclear which countries are 'developed'. There is no statistical reporting mechanism on actual transfers. Nor was it clear that Article 66.2 activities must be reported separately from technical assistance or development aid actions directed to developing countries in general. Equally clear is that reporting under TRIPS does not require that measures undertaken have any effect, or that such effects be monitored. It was not possible to establish what the results of these international agreements and the ensuing activities have led to and whether there is indeed a flow of technology benefiting the developing world, especially the LDCs.

It is true that these reports are becoming more sophisticated when it comes to addressing LDC's needs. Several countries have, after the 2008 WTO workshop discussion, adapted to new requirements and have now submitted more specific reports. The developed world is amenable to a robust dialogue between developed and LDC Members in order to target concrete activities in a way that is responsive to the self-identified technology transfer interests and needs of LDC Members.

Based on this promising attitude our discussions as presented in earlier chapters leads us to the following propositions.

3.1 Article 66.2 Reporting

The problem with the reports as presented during the first decade is that they are long and un-focused, dealing with a great variety of problems that may, or may not, have any connection to the situation of least developed countries.

It is easy to agree with remarks in the WTO Council by some developing countries that Article 66.2 reporting should have a clear focus on incentives to their own companies to foster technology transfer. It has repeatedly been emphasized that the concrete technology is primarily controlled by private interests. Private interests normally act in their own self-interest and are driven by profit maximization. It cannot be expected that they philanthropically contribute to sustainable technology transfer. There must be incentives in line with their requirements or private interests will not be involved in active and sustainable contributions. Several of the country reports indicate that they support their own industries by matchmaking, and providing support for initial feasibility studies. Some countries go further and share in the initial risk-taking in the form of repayable loans.

This is precisely the sort of activities that should be reported under Article 66.2. Nothing else. Such reports will be much shorter and focused. They will only deal with long-term, sustainable measures directed to LDCs, partly because these activities are normally not provided in the self-interest of the reporting country. Similar activities in relation to developing countries and emerging economies are often dictated by a hope that they will lead to favourable trade relations, a more farfetched concept in their relationships with LDCs.

Reports should not only be detailed as to exactly what incentives are provided, but also explain how the project has been carried out and exactly what has been achieved. This will lead to precisely the Bank of Science and Technology for LDCs which will facilitate LDCs' access to technologies and technological know-how that the 2011 UNCTAD conference requested. Perhaps even more important is that this bank of knowledge should serve as inspiration for developed countries when they consider how to better live up to their obligations under Article 66.2.

Reports in this sense should be forthcoming from all countries that could be regarded as developed. Whether EU Member States provide individual reports or if they are coordinated under the EU umbrella is of no importance, but it should be made clear that they all must contribute and their creative measures (or lack thereof) will be made public. But the obligation is not only incumbent on countries which are now submitting reports. The only countries excluded from reporting obligations are LDCs. BRIC countries and emerging economies should have an equal responsibility to actively support the developing world without necessarily acting in their self-interest. For example, China is a major trading partner in Africa and it should provide reports on what it is doing to incentivize its industry to transfer technology to the LDCs. It is reasonable to expect that China, rapidly becoming a leading economic world power, has a great deal to contribute in this respect.

Another matter of significance is how to handle incoming reports. At present they are received by The TRIPS Council and discussed in meetings. The deliberations are made public through minutes from the meetings and the results are subject to, for example, academic scrutiny. There are no real remedies for failures to report. Moon suggests that a Monitoring Mechanism Group should be appointed with representatives from the WTO membership and independent experts.[11] The group should have informational and evaluation functions. She also provides specific suggestions for how future reports could be streamlined much in line with propositions advanced at the Vietnam conference. Perhaps rather than creating yet another institution, initially it might be helpful to improve the reporting format and request the secretariat to summarize the outcome in preparation of the TRIPS Council meetings.

It has also been suggested that failure to meet reporting obligations under Article 66.2 could be referred to the WTO dispute system. It might well be desired that such failure be pursued by the TRIPS Council in WTO dispute settlement as a way to secure adherence to the obligations in the TRIPS system, much like the EU Commission can address failure to meet obligations under EU rules. This is, however, not possible. Only WTO members have such standing in WTO dispute settlement and it is hardly conceivable that countries like US would agree to transfer such rights to a super national power. That leaves the states to take action. Rumour has it that a group of LDCs are preparing such an action. If successful, a powerful remedy would be in the hands of the LDCs, but it is a risky gamble. If a DSU panel would refuse to hear such a matter or, even worse, not accept it on substantive grounds, the requirements of Article 66.2 would be significantly weakened.

3.2 Article 67 Reporting

Article 67 stipulates that developed-country Members shall provide technical and financial cooperation in favour of developing and least developed country Members. This includes assistance in the preparation of laws and regulations on IPR and establishment or reinforcement of domestic offices and agencies relevant to these matters, including the training of personnel. That is something completely different from the reporting requirement under Article 66.2, but it is still important to create an environment ready to receive and deal with technology. Here it is that the administrative assets of the developed world can be put to good use.

Article 67 may be read as only covering the intellectual property regime, the prime focus of the TRIPS Agreement. There is little reason for such a narrow

11. Suerie Moon, 'Meaningful Technology Transfer to LDcs: A Proposal for a Monitoring Mechanism for TRIPS Article 66.2', Policy Brief No 9, May 2011, <http://ictsd.org/i/publications/106434/>.

reading. The same restriction should then be applied to Article 66. It should only relate to technology covered by IPR protection. No developed country has in its reporting reduced its remarks to technology solely applicable to IPR protection. On the contrary, in the 2008–2010 reporting, reference to IPRs in relation to substantive technology transfer was only made on a few occasions. Most projects appear to be unrelated to IPR protection.

If it is important to report on concrete technology transfer activities, it appears equally important to report on technical assistance efforts. Today they are undertaken on an ad hoc basis. In some developing countries it appears as if the developed world is almost competing in order to be allowed to provide assistance, whereas others obtain much less support.

If technical assistance reports under Article 67 could be organized in a similar way as Article 66.2 reporting, they could presumably be condensed and to the point. They would serve as a bank of inspiration both for recipient countries and for the providers and they would promptly show if measures are spread in an optimal way or focused on too few in an unbalanced manner.

The reports should provide information on what projects have been carried out. They should relate to training activities and assistance in building up viable and sustainable infrastructure. The SIDA project on empowering legal education in Vietnam is a good example of a technical assistance project which should be carefully reported as a showcase in this field. However, it should not ignore the developing world's needs for assistance in specific matters such as in bringing cases to the WTO Dispute Settlement System, or in the many arbitration panels resulting from the rules on Foreign Direct Investment. Such activities, if performed in a systematic way, should also qualify.

3.3 General Development Aid

Transfer of Production Technology and Technical Assistance are not the only matters covered by present Article 66.2 reporting. Much of the information rather qualifies as general Development Aid or specific reporting relating to a special sector. This is perhaps so that reporting officers may cut and paste from different reports submitted to different institutions if they believe that they have any significance. This is understandable. The less the country has to say about Production Technology and Technical Assistance the more it adds to its reporting from general development aid activities. Everyone acts as they are measured and so do officers in governmental institutions, politicians and countries. A long report indicates substantial contribution to the overall aim of poverty reduction and countries are keen to show that they make an active contribution.

First of all, such reporting has no place in the focused reporting required under the TRIPS Agreement. Secondly, all countries that provide general development aid should be allowed to account for it and receive praise for efforts done. Countries are reporting their different activities in different fora. It is probably a good thing if the international organizations could combine all such efforts into a comprehensive yearly publication, where efforts are published in a comparable way to show best practices, avoid fragmentation and to stimulate countries to put in more efforts to reach the overall Millennium Goals. But that is another story, one that did not make it to our discussion on Sustainable Technology Transfer in Vietnam in October 2010.

In conclusion, the concept of technology transfer is not so much an empirical or legal concept as it is a political (normative) concept capturing visions of what can be done with technology transfer. The kind of problems technology transfer is invoked to solve (or contribute to solving) is connected to broad issues such as sustainable development and global environmental justice. Putting the adjective 'sustainable' in front of technology transfer may mean more than we currently can imagine.

Bibliography

Abate, R.S. 'Kyoto or Not, Here We Come: The Promise and Perils of the Piecemeal Approach to Climate Change Regulation in the United States', 15 *Cornell J.L. & Pub. Pol'y* 369 (2006).

Abate, R.S. & A.B. Greenlee. 'Sowing Seeds Uncertain: Ocean Iron Fertilization, Climate Change, and the International Environmental Law Framework', 27 *Pace Envtl L Rev* 555 (2010).

Abbott, F. & J.H Reichman. 'Access to Essential Medicines: Lessons Learned since the Doha Declaration on the TRIPS Agreement and Public Health, and Policy Options for the European Union' June 2007 PE 381.392 (prepared for the International Trade Committee, European Parliament). <www.law.duke.edu/news/pdf/a-r_directorate_report.pdf>.

Abbott, F.M. 'Innovation and Technology Transfer to Address Climate Change: Lessons from the Global Debate on Intellectual Property and Public Health', ICTSD, Issue Paper No. 24 (June 2009) (Abbott). <http://ictsd.org/i/publications/50454/>.

Abbott, K.W. & D. Snidal. 'Hard and Soft Law in International Governance' 54 *Int'l Organ.*, 421 (2000).

Adam, A. 'Technology Transfer to Combat Climate Change: Opportunities and Obligations', 9 *J. High Tech. L.* 1 (2009).

Ahdieh, R.B. 'Dialectical Regulation' 38 *Conn. L. Rev.* 863 (2006).

Amann, D.M. 'Calling Children to Account: The Proposal for a Juvenile Chamber in the Special Court for Sierra Leone', 29 *Pepp. L. Rev.* 167 (2001).

Amann, D.M. 'Abu Graib', 153 *U. Pa. L. Rev.* 2085 (2005).

Andersen, S.O., K. Madhava Sarma & K.N. Taddonio, *Technology Transfer for the Ozone Layer: Lessons for Climate Change* (London: Earthscan, 2007).

Anderson, A.M. & B. Razavi. 'The Globalization of Intellectual Property Rights: TRIPS, BITS, and the Search for Uniform Protection', 38 *Ga J Int'l & Comp L* 265 (Winter 2010).

Anderson, R.D. 'Competition Policy and Intellectual Property in the WTO: More Guidance Needed?', in *Research Handbook on Intellectual Property and Competition Law*, ed. J. Drexl (Cheltenham: Edward Elgar, 2008).

Andreen, W., et al. 'Cooperative Federalism and Climate Change: Why Federal, State, and Local Governments Must Continue to Partner', Center for Progressive Reform, 29 May 2008. <http://progressiveregulation.org/articles/Cooperative_Federalism_and_Climate_Change.pdf>.

Aoki, R. & J. Small. 'Compulsory Licensing of Technology and the Essential Facilities Doctrine', 16 *Info Econ & Pol'y* 13 (2004).

Averchenkova, A. 'The Outcomes of Copenhagen: The Negotiations & the Accord', UNDP Environment & Energy Group Climate Policy Series, 2010.

Ayoob, E. Note, 'Recent Development: The Anti-Counterfeiting Trade Agreement', 28 *Cardozo Arts & Ent L J* 175 (2010).

Banerjee, B. 'The Limitations of Geoengineering Governance in a World of Uncertainty', 4 *Stan J L Sci & Pol'y* 15 (published online 2011).

Barbosa, D.B., M. Chon & A.M. von Hase. 'Slouching towards Development in International Intellectual Property', 2007 *Mich St L. Rev* 71 (2007).

Barton, J.H. 'New Trends in Technology Transfer Implications for National and International Policy', International Centre for Trade and Sustainable Development, Issue Paper No. 18 (2007) ('New Trends in Technology Transfer').

Barton, J.H. 'Statement to the United States Senate Finance Committee Hearing on International Enforcement of Intellectual Property Rights and American Competitiveness', (15 Jul. 2008) <http://finance.senate.gov/imo/media/doc/071508jbtest.pdf>.

Battisti, D. Climate Engineering with Aerosols – Predictable Consequences?, MIT Earth System Initiative Symposium-Climate Engineering Symposium, (2 Dec. 2009). <http://techtv.mit.edu/genres/32-science/videos/4636-climate-engineering-with-aerosolspredictable-consequences>.

Baylis, E.A. 'Parallel Courts in Post-Conflict Kosovo', 32 *Yale J. Int'l L.* 1 (2007).

Bélanger, C. 'Concepts in Social Science and History', <www2.marianopolis.edu/quebeckhistory/events/concepts.htm>.

Berman, P.S. 'Global Legal Pluralism', 80 *S. Cal. L. Rev.* 1155 (2007).

Blakeney, M. & L. Blakeney. 'Stealth Legislation? Negotiating the Anti-Counterfeiting Trade Agreement (ACTA)', 16(4), *Int'l T L R*, 85 (2010).

Boyle, Alan. 'Soft Law in International Law-Making', in *International Law*, ed. Malcolm D. Evans (Oxford: Oxford University Press, 2010).

Brahmbhatt, M. & A. Hu. 'Ideas and Innovation in East Asia', Policy Research Working Paper No. 4403, World Bank (November 2007).

Brand, U. & G. Christoph. 'Post-Fordist Governance of Nature: The Internationalization of the State and the Case of Genetic Resources – A Neo-Poulantzian Perspective', 15 *Rev Int'l Political Econ* 567 (2008).

Branstetter, L.G., R. Fisman & C.F. Foley. 'Do Stronger Intellectual Property Rights Increase Technology Transfer? Empirical Evidence from US Firm-Level Panel Data', 121 *Q J. Econ* 321 (2006).

Branstetter, L.G., et al. *Intellectual Property Rights, Imitation and Foreign Direct Investment: Theory and Evidence* (Cambridge, MA: National Bureau of Economic Research Working Paper no. 13033, 2007).

Brewer, P.G. 'Evaluating a Technological Fix for Climate', 104 *PNAS* 9915 (2007).

Bronin, Sarah C. 'Solar Rights', 89 *B.U. L. Rev.* 1217 (2009).

Brovkin, V., et al. 'Geoengineering Climate by Stratospheric Sulfur Injections: Earth System Vulnerability to Technological Failure', 92 *Climatic Change* 243 (2009).

Brown, D.A. 'Thinking Globally and Acting Locally: The Emergence of Global Environmental Problems and the Critical Need to Develop Sustainable Development Programs at State and Local Levels in the United States', 5 *Dick. J. Envtl. L. & Pol'y* 175 (1996).

Brownlie, I. *Principles of Public International Law*, 6th edn (Oxford University Press, 2003).

Brunnée, J. & S. Toope. 'Environmental Security and Freshwater Resources: Ecosystem Regime Building', 91 *Am J Int'l L* 26 (1997).

Brunnée, J. & S. Toope, *Legitimacy and Legality in International Law. An Interactional Account* (Cambridge: Cambridge University Press, 2010).

Bruyninckx, H. 'Sustainable Development: The Institutionalization of a Contested Policy Concept', in *Palgrave Advances in International Environmental Politics*, ed. Michele Betsill, Kathryn Hochstetler & Dimitris Stevis (Basingstoke: Palgrave Macmillan, 2006).

Burk, D.L., K. Barovsky & G.H. Monroy. 'Biodiversity and Biotechnology', 260 *Science* 1900 (25 Jun. 1993).

Burke-White, William W. 'International Legal Pluralism', 25 *Mich. J. Int'l L.* 963 (2004).

Burleson, E. '*Energy Policy*, Intellectual Property, and Technology Transfer to Address Climate Change', 18 *Transnat'l L. & Contemp. Probs.* 69 (2009).

Burleson, E. 'Climate Change Displacement to Refugee', 5 *J Envtl L & Litig* 19 (2010).

Burleson, E. & W. Burleson. 'Innovation Cooperation: Energy Biosciences and Law', 2011 *Ill. L. Rev.* 101 (forthcoming 2011).

Burns, W.C.G. 'Climate Geoengineering: Solar Radiation Management and its Implications for Intergenerational Equity' 4 *Stan J L Sci & Pol'y*, 37.

Cabral, N.J. 'The Role of Renewable Portfolio Standards in the Context of a National Carbon Cap-and-Trade Program', 8 *Sustainable Dev. L. & Pol'y* 13 (2007–2008).

Carlson, A.E. 'Implementing Greenhouse Gas Emissions Caps: A Case Study of the Los Angeles Department of Water and Power', 55 *UCLA. L. Rev.* 1479 (2008).

Carstensen, P. 'Post-Sale Restraints via Patent Licensing: A "Seedcentric" Perspective', 16 *Fordham Intel Prop, Media & Ent L J* 1079 (2006).

Cendrowska, T. 'Enabling US-Chinese Cooperation in Standards and Conformity Assessment', *ASTM Standardization News*, <www.astm.org/cgi-bin/SoftCart.exe/SNEWS/APRIL_2005/cendrowska_apr05.html?E+mystore>.

Chen, Y. & T. Puttitanun. 'Intellectual Property Rights and Innovation in Developing Countries', 78 *J Deve Econ* 474 (2005).
Chinkin, C.M. 'The Challenge of Soft Law: Development and Change in International Law', 38 *Int'l & Comp L Q.* 850 (1989).
Chow, D. 'Anti-Counterfeiting Strategies of Multi-National Companies in China: How a Flawed Approach Is Making Counterfeiting Worse', 41 *Geo J Int'l L* 749 (2010).
Climate Focus. 'Renewable Energy Small Power Producers in Vietnam: Finance Consultancy, for MOIT, 2008).
Coe, D.T., E. Helpman & A.W. Hoffmaister. 'North-South R&D Spillovers', 107 *Econ J* 134 (1997).
Comin, D. & M. Mestieri. 'The Intensive Margin of Technology Adoption', Harvard University and NBER Working Paper 11-026 (May 2010).
Conca, K. 'Environmental Governance after Johannesburg: From Stalled Legalization to Environmental Human Rights?', 1 *J Int'l L & Int'l Rel* 121 (2005).
Cooper, D.E. Note, 'The Kyoto Protocol and China: Global Warming's Sleeping Giant', 11 *Geo. Int'l Envt'l L Rev* 401 (1999).
Correa, C.M. *Intellectual Property Rights and the Use of Compulsory Licenses: Options for Developing Countries* (Geneva: South Centre T.R.A.D.E working paper no. 5, 1999).
Correa, C.M. *Trade Related Aspects of IPRs – A Commentary on the TRIPS Agreement* (Oxford: Oxford University Press, 2007).
Cotropia, C.A. 'Compulsory Licensing under TRIPS and the Supreme Court of the United States' Decision in *eBay v. MercExchange*', in *Patent Law and Theory: A Handbook of Contemporary Research*, ed. Toshiko Takenaka & Rainer Moufang (London: Edward Elgar Publishing Ltd, 2009).
Cottier, T. & M. Oesch (eds). *International Trade Regulation – Law and Policy in the WTO, the European Union and Switzerland – Cases, Material and Comments* (Bern, Switzerland: Staempfli Publishers Inc., 2005).
Coughlin, M.D. 'Using the Merck-INBio Agreement to Clarify the Convention on Biological Diversity', 31 *Colum J Transnat'l L* 31 358 (1993–1994).
Cox, K.R. 'Spaces of Dependence, Spaces of Engagement and the Politics of Scale, or Looking for Local Politics', 17 *Pol. Geography* 1 (1998).
Cox, K.R. 'Representation and Power in the Politics of Scale', 17 *Pol. Geography* 41 (1998).
Davidson, O.G. 'Solar Farms That Sip Desert Water', *Phoenix Sun*, 17 Jul. 2009, <http://thephoenixsun.com/archives/4030>.
de Búrca, G. & J. Scott (eds). *Law and New Governance in The EU and US* (Oxford and Portland, OR: Hart Publishing, 2006).
de Sepibus, J. 'The Liberalisation of the Power Industry in the European Union and Its Impact on Climate Change – A Legal Analysis of the Internal Market in Electricity', NCCR, Switzerland, Trade Regulation Working Paper No. 2008/10 (1 May 2008), 2–4. <http://papers.ssrn.com/sol3/papers.cfm?abstract_id=1137823>.

Deere, C. *The Implementation Game: The TRIPS Agreement and the Global Politics of Intellectual Property Reform in Developing Countries* (Oxford: Oxford University Press, 2009).

Delacey, B.J., et al. 'Government Intervention in Standardization: The Case of WAPI', 25 (2006). <http://ssrn.com/abstract=930930>.

Docherty, B. & T. Giannini. 'Confronting a Rising Tide: A Proposal for a Convention on Climate Change Refugees', 33 *Harv Envtl L Rev* 349 (2009).

Drahos, P. 'Trade-Offs and Trade Linkages: TRIPs in a Negotiating Context'. Notes from a talk given at Quaker House Geneva (12 Sep. 2000), <www.quno.org/economicissues/intellectual-property/intellectualLinks.htm#OCCASIONAL>.

Dryzek, J.S. *Deliberative Global Politics: Discourse and Democracy in a Divided World* (Oxford: John Wiley & Sons, 2006).

Dupuy, P.-M. & Y. Kerbrat. *Droit International Public*, 10th edn (Paris: Dalloz, 2010).

Eaton, J. & S.J. Kortum. 'Trade in Ideas: Patenting and Productivity in the OECD', 40 *Journal of International Economics* 251 (1996).

Eeckhout, P. *External Relations of the European Union – Legal and Constitutional Foundations* (Oxford: Oxford University Press 2004).

Egede-Nissen, B. & H.D. Venema. IISD, 'Desperate Times, Desperate Measures: Advancing the Geoengineering Debate at the Arctic Council' (2009). <www.arcticgovernance.org/desperate-times-desperate-measures-advancing-the-geoengineering-debate-at-the-arctic-council.4782242-142902.html>.

El-Ashry, M.T. 'Framework for a Post-Kyoto Climate Change Agreement', 8 *Sustainable Dev. L. & Pol'y* 2, 2–3 (2007–2008).

Elliott, L.M. *The Global Politics of the Environment*, 2nd edn (Oxford: Palgrave Macmillan, 2004).

Engel, K. 'AALS Symposium: State and Local Climate Change Initiatives: What Is Motivating State and Local Governments to Address a Global Problem and What Does This Say About Federalism and Environmental Law?', 38 *Urb. L.* 1015 (2006).

European Commission. 'Pharmaceutical Sector Inquiry: Final Report' (Sector Inquiry) (8 Jul. 2009), <http://ec.europa.eu/competition/sectors/pharmaceuticals/inquiry/staff_working_paper_part1.pdf>.

European Commission. 'General Overview of Active WTO Dispute Settlement Cases Involving the EU as Complainant or Defendant and of Active Cases under the Trade Barriers Regulation (9 Nov. 2010). <http://trade.ec.europa.eu/doclib/docs/2007/may/tradoc_134652.pdf>.

Express Pharma, *Compulsory Licensing under the Indian Patent Law* (2006), <www.expresspharmaonline.com/20060915/management06.shtml>.

Ezrachi, A. & D. Gilo. 'Are Excessive Prices Really Self-Correcting?', 5 *J Comp L & Econ* 249 (2009).

Falk, R.A. 'Casting the Spell: The New Haven School of International Law', 104 *Yale L.J.* 1991 (1995).

Ferrey, S. 'Nothing But Net: Renewable Energy and the Environment, MidAmerican Legal Fictions, and Supremacy Doctrine', 14 *Duke Env L. & Pol F* 1 (2003).

Ferrey, S. 'Small Power Purchase Agreement Application for Renewable Energy Development: Lessons from Five Asian Countries', World Bank with support from United Nations Development Program, 2004. <http://siteresources.worldbank.org/INTEAPREGTOPENERGY/Resources/Power-Purchase.pdf>.

Ferrey, S. 'Renewable Orphans: Adopting Legal Renewable Standards at the State Level', 19 *Electricity J* (March 2006). <www.sciencedirect.com/science?_ob=ArticleURL&_udi=B6VSS-4J624J6-2&_user=10&_coverDate=03%2F31%2F2006&_rdoc=1&_fmt=high&_orig=search&_origin=search&_sort=d&_docanchor=&view=c&_searchStrId=1548878851&_rerunOrigin=google&_acct=C000050221&_version=1&_urlVersion=0&_userid=10&md5=2ff156d42721d836c65887eea142d292&searchtype=a>.

Ferrey, S. 'The Failure of International Global Warming Regulation to Promote Needed Renewable Energy', 37 *B C Envt'l Aff L Rev* 67 (2010).

Ferrey, S. *The Law of Independent Power: Environmental Law Series*, § 2.1, 28th edn (Clark Boardman Callaghan, 2010).

Ferrey, S. & R. Vernstrom. 'RESPP Planning and Preparation', Final Report prepared for Vietnam Ministry of Industry, 2006 (RESPP Planning 2006).

Fikentscher, W. *The Draft International Code of Conduct on the Transfer of Technology – A Study in Third World Development*, IIC Studies in Industrial Property and Copyright Law, vol. 4 (Basel: Verlag Chemie, 1980).

Fink, C. & K.E. Maskus (eds). *Intellectual Property and Development: Lessons from Recent Economic Research* (Washington, DC and New York: co-publication of the World Bank and Oxford University Press, 2005).

Flynn, S.M. 'Using Competition Law to Promote Access to Knowledge', in *Access to Knowledge in the Age of Intellectual Property*, ed. G. Krikorian & A. Kapczynski (New York: Zone Books, 2010).

Footer, M.E. 'Agricultural Biotechnology, Food Security and Human Rights', in *Biotechnology and International Law*, ed. Francesco Francioni & Tullio Scovazzi (Oxford: Hart Publishing, 2006).

Footer, M.E. 'The Role of Soft Law Norms in Reconciling the Antinomies of WTO Law', *Society of International Economic Law (SIEL) Inaugural Conference 2008 paper*. <http://ssrn.com/abstract=1159929>.

Foray, D. 'Technology Transfer in the TRIPS Age: The Need for New Types of Partnerships, between the Least Developed and Most Advanced Economies', ICTSD Issue Paper No.23 at 10 (May 2009), <http://ictsd.org/downloads/2009/07/foray_may2009.pdf>.

Ganslandt, M. 'Intellectual Property Rights and Competition Policy', in *Intellectual Property, Growth and Trade*, ed. Keith E. Maskus (Amsterdam: Elsevier, 2008), 233.

Gantz, D. 'Doi Moi, the VBTA and WTO Accession: The Role of Lawyers in Vietnam's No Longer Cautious Embrace of Globalization', 41(3) *Int'l Lawyer*, 873–890 (Fall 2007).

Gellings, C. & R. Lordan. 'The Power Delivery System of the Future', *Electricity Daily* (January–February 2004) at Conclusion.

General Statistics Office of Vietnam. 'The 2009 Population and Housing Census: Dissemination of completed census results' <www.gso.gov.vn/default_en.aspx?tabid=599&ItemID=9788>.

Gerstetter, C., D. Marcellino & E. von Sperber. 'Technology Transfer in the International Climate Negotiations – The State of Play and Suggestions for the Way Forward', 3 *Climate and Carbon L Rev* 3 (2010).

Gervais, D.J. *The TRIPS Agreement: Drafting History and Analysis*, 2nd edn (London: Sweet & Maxwell, 2003).

Gervais, D.J. 'Intellectual Property, Trade & Development: The State of Play', *74 Fordham L R 505* (2005).

Gervais, D.J. (ed.), *Intellectual Property, Trade and Development: Strategies to Optimize Economic Development in a TRIPS Plus Era* (New York: Oxford University Press, 2007).

Gesellschaft für Technische Zusammenarbeit (GTZ). 'Designing a REDD compliant benefit distribution system for Vietnam' Executive summary. UN-REDD Programme (30 Nov. 2009). <http://europeandcis.undp.org/poverty/show/5A0EC8BA-F203-1EE9-BABE45003F1E10AC>.

Gibson, C.S. 'Globalization and the Technology Standards Game: Balancing Concerns of Protectionism and Intellectual Property in International Standards', 22 *Berkeley Tech L. J.* 1407 (2007).

Gibson, C.S. 'A Look at the Compulsory License in Investment Arbitration: The Case of Indirect Expropriation', 25 *Am U Int'l L. Rev* 367 (Winter 2010).

Gibson, C.S. 'Latent Grounds in Investor-State Arbitration: Do International Investment Agreements Provide New Means to Enforce Intellectual Property Rights?', in *Yearbook on International Investment Law & Policy 2009-2010*, ed. Karl. P. Sauvant (New York: Oxford University Press, 2010).

Giuseppina D'Agostino, *Healing Fair Dealing? A Comparative Copyright Analysis of Canadian Fair Dealing to UK Fair Dealing to US Fair Use* (Toronto: York University, CLPE Research Paper 28), 2007.

Goepel, M. 'Formulating Future Just Policies: Applying the Delhi Sustainable Development Law Principles', 2 *Sustainability* 1695 (2010).

Goff, T.J. 'Regulation of Digital Copyrights and Trademarks at the U.S. Border: How the Proposed Anti-Counterfeiting Trade Agreement and the Enacted U.S. Pro-IP Act Will Destabilize the Current System', 16 *Sw J Int'l L* 207 (2010).

Goldsmith, J.L. & E.A. Posner, *The Limits of International Law* (New York: Oxford University Press, 2005).

Gordon, R. 'Climate Change and the Poorest Nations: Further Reflections on Global Inequality', 78 *U Colo L Rev* 1559 (2007).

Govindasamy, B., et al. 'Impact of Geoengineering Schemes on the Terrestrial Biosphere', 29 *Geophysical Res Letters* 18-1 (2002).

Gray, R.D. & J. Schuster. 'The East Asian Financial Crises-Fallout for Private Power Projects', *Viewpoints* Note No. 146 (World Bank, August 1998), 1. <http://rru.worldbank.org/documents/publicpolicyjournal/146gray.pdf>.

Guzman, A.T. & T. Meyer. 'International Soft Law', 2 *J. Legal Analysis* 171 (2010).

Haas, E.B. *When Knowledge is Power: Three Models of Change in International Organizations* (Berkeley: University of California Press, 1990).

Hagerty, C.L. & J.L. Ramseur. Deepwater Horizon Oil Spill: Selected Issues for Congress, CSR Report for Congress, 30 Jul. 2010. <www.fas.org/sgp/crs/misc/R41262.pdf>.

Hall, L. 'Technology Transfers under the United Nations Framework Convention on Climate Change', 2005 *Colo. J. Int'l. Envtl. L. & Pol'y* 59 (2005).

Halperin, D. *The Bayh-Dole Act and March-In Rights* (Washington DC: paper prepared for Consumer Project on Technology, 2001).

Hansen, J., et al. 'Target Atmospheric CO_2: Where Should Humanity Aim?', 2 *Open Atmos. Sci. J.* 217 (2008).

Hargreaves, I. 'Digital Opportunity: A Review of Intellectual Property and Growth' (May 2011), <www.ipo.gov.uk/ipreview-finalreport.pdf>, 10.

Hegerl, G.C. & S. Solomon. 'Risks of Climate Engineering', 325 *Science* 955 (2009).

Helfer, L.R. 'Regime Shifting: The TRIPS Agreement and the New Dynamics of International Intellectual Property Lawmaking', 29 *Yale J Int'l L* 1 (Winter 2004).

Helfer, L.R. 'Constitutional Analogies in the International Legal System', 37 *Loy LA Int'l & Comp L Rev* 193 (2004).

Hendrik, G. & C. Bodenhausen. *Guide to the Application of the Paris Convention for the Protection of Industrial Property* (Geneva: United International Bureaux for the Protection of Intellectual Property (BIRPI) 1968 – WIPO publication 611).

Hille, K. 'Taiwan employs compulsory licensing for Tamiflu', *Financial Times*, 25 Nov. 2005. <www.ft.com/cms/s/cebeb882-5dcb-11da-be9c-0000779e2340, Authorised=false.html?_i_location=http%3A%2F%2Fwww.ft.com%2Fcms %2Fs%2F0%2Fcebeb882-5dcb-11da-be9c-0000779e2340.html&_i_referer=>.

Hoekman, B.M., K.E. Maskus & K. Saggi. 'Transfer of Technology to Developing Countries: Unilateral and Multilateral Policy Options', *World Bank Policy Research Working Paper 3332*, (World Bank, 2004).

Hohfeld, W.N. 'Some Fundamental Legal Conceptions as Applied in Judicial Reasoning', 23 *Yale Law Journal* 16 (1913).

Horton, J.B. 'Geoengineering and the Myth of Unilateralism: Pressures and Prospects for International Cooperation', 4 *Stan J L Sci & Pol'y* 56 (published online 2011).

Hovenkamp, H. *The Antitrust Enterprise: Principle and Execution* (Cambridge, MA: Harvard University Press, 2005).

Howse, R. 'The Canadian Generic Medicines Panel: A Dangerous Precedent in Dangerous Times', (2000) 3 *J World Int Prop* 493 (2000).

Human Development Report 2010. 'The Real Wealth of Nations: Pathways to Human Development', Published for the United Nations Development Programme (UNDP) (November 2010). <http://hdr.undp.org/en/reports/global/hdr2010/>.

Hutchison, C. 'Does TRIPS Facilitate or Impeded Climate Change Technology Transfer into Developing Countries', 3 *U. Ottawa L. & Tech. J.* 517 (2006).

IEA. 'Energy-Technology Perspectives – Scenarios and Strategies to 2050, 2006' <www.iea.org/publications/free_new_Desc.asp?PUBS_ID=2078>.

Iles, A. 'Rethinking Differential Obligations: Equity under the Biodiversity Convention', 16 *Leiden Journal of International Law* 249 (2003).

Institute of Energy. 'The Pre-Feasibility Study on Nuclear Power Plant in Vietnam', 2006.

Institute of Energy. 'The Sixth Master Plan on Electric Power Sector Development', 2007.

Institute of Energy. 'Master Plan on New Renewable Energy Development up to 2025', for MOIT, 2008.

Inter-American Development Bank, *Advisory Facility on Investor – State Dispute Settlement: Plan of Operations* (RG-T1515). <www.iadb.org/document.cfm?id=1741656>.

Intergovernmental Panel on Climate Change, Fourth Assessment Report: Climate Change 2007 (AR4), ed. Rajendra K. Pachuari & Andy Reisinger (Geneva, Switzerland: 2007), Foreword and Preface. <www.ipcc.ch/publications_and_data/ar4/syr/en/contents.html>.

International Energy Agency. 'World Energy Outlook 2004', (IEA World Energy Outlook 2004). <www.worldenergyoutlook.org/2010.asp>.

Ishii, T. *Compulsory License System: History and Present Problems, Osaka Institute of Technology* <www.oit.ac.jp/ip/property_lab/ishii/data/pdf/slide_e_03.pdf>.

ISO. *World Standards Cooperation for the Information Society*, ISO BULL., Comment (October 2003).

Ivus, O. 'Do Stronger Patent Rights Raise High-Tech Exports to the Developing World?', 81 *Journal of International Economics* 38 (2010).

Jefferies, C.P. 'A Preliminary Evaluation of the Proposed Text', in *International Technology Transfer: The Origins and Aftermath of the United Nations Negotiations on a Draft Code of Conduct*, ed. S. Patel et al. (The Hague: Kluwer Law International, 2001).

Jens, B. 'Making Sense of Global Civil Society', 12 *Eur J Int'l Rel* 371–395 (2006).

Jones, K.T. 'Scale as Epistemology', 17 *Pol. Geography* 25 (1998).

Judd, D.R. 'The Case of the Missing Scales: A Commentary on Cox', 17 *Pol. Geography* 29 (1998).

Jutta, B. 'COPing with Consent: Law-Making under International Environmental Agreements' 15 *Leiden J Int'l L* 1 (2002).

Kaminski, M. 'The Origins and Potential Impact of the Anti-Counterfeiting Trade Agreement (ACTA)', 34 *Yale L J Int'l L* 247 (Winter 2009).

Karkkainen, B.C. '"New Governance" in Legal Thought and in the World: Some Splitting as Antidote to Overzealous Lumping', 89 *Minn. L. Rev*. 471 (2004).
Kaswan, Alice. 'Climate Change, Consumption, and Cities', 36 *Fordham Urb. L.J.* 253 (2009).
Kelly, M.J. 'Pulling at the Threads of Westphalia: '"Involuntary Sovereignty Waiver" – Revolutionary International Legal Theory or Return to Rule by the Great Powers', 10 *UCLA J. Int'l L. & Foreign Aff.* 361 (2005).
Kennedy, David. 'When Renewal Repeats: Thinking Against the Box', 32 *NYU J. Int'l L. & Pol.* 335 (2000).
Kibreab, G. 'Climate Change and Human Migration: A Tenuous Relationship?', 20 *Fordham Envtl L Rev* 357 (2009).
Kim, L. *Technology Transfer and Intellectual Property rights: Lessons from Korea's Experience* (Geneva: International Centre for Trade and Sustainable Development, Project on IPRs and Sustainable Development, Issue Paper no. 2, 2003).
Kintisch, E. 'Scientists Say Continued Warming Warrants Closer Look at Drastic Fixes', 318 *Science* 1054 (2007).
Klass, A.B. & E.J. Wilson. 'Climate Change, Carbon Sequestration, and Property Rights', 2010 *U. Ill L Rev* 363 (2010).
Kogan, L.A. 'Brazil's IP Opportunism Threatens U.S. Private Property Rights', 38 *U Miami Inter-Am L Rev* 1 (Fall 2006).
Koh, H.H. 'Jefferson Memorial Lecture: Transnational Legal Process after September 11th', 22 *Berkeley J. Int'l L.* 337 (2004).
Koonin, S.E., et al. 'Climate Engineering Responses to Climate Emergencies', Novim 28 Jul. 2009; *Committee on Geological and Geotechnical Engineering in the New Millennium; Opportunities for Research and Technological Innovation*, Committee on Geological and Geotechnical Engineering, National Research Council (Washington, DC: National Academies Press, 2006).
Koselleck, R. *The Practice of Conceptual History. Timing History, Spacing Concepts* (Stanford: Stanford University Press, 2002).
Kosloff, L. & M. Trexler. 'State Climate Change Initiatives: Think Locally, Act Globally', 18 *Wtr. Nat. Resources & Env't* 46 (2004).
Kousky, C. & S.H. Schneider. 'Global Climate Policy: Will Cities Lead the Way?', 3 *Climate Pol'y* 1 (2003).
Kumar, S.P. 'Border Enforcement of Intellectual Property Rights against In-Transit Generic Pharmaceuticals: An Analysis of Characters and Consistency, 32 *E.I.P.R.* 506 (2010).
Ladas, S.P. *Patents, Trademarks and Related Rights: National and International Protection* (Cambridge, MA: Harvard University Press, 1975).
Lakoff, G. & M. Johnson. *Metaphors We Live By* (Chicago: University of Chicago Press, 2003).
Lamy at the High-Level Plenary Session of the Millennium Development Goals Summit, 20 Sep. 2010, <www.wto.org/english/news_e/sppl_e/sppl170_e.htm>.

Lang, A. 'Legal Regimes and Regimes of Knowledge: Governing Global Services Trade', LSE Law, Society and Economy Working Papers 15/2009.

Lasswell, H.D. & M.S. McDougal. *Jurisprudence for a Free Society: Studies in Law, Science and Policy*, vol. I (New Haven Press/Kluwer Law, 1992).

Leebron, D.W. 'Linkages', 96 *Am J Int'l L* 5 (2002).

Lemley, M.A. 'Intellectual Property Rights and Standard-Setting Organizations', 90 *Cal L Rev* 1889 (2002).

Lemley, M.A. 'Property, Intellectual Property, and Free-riding', 83 *Tex Law Rev* 1059 (2005).

Lemley, M.A. 'Ten Things to Do About Patent Holdup of Standards (And One Not To)', 48 *B C L Rev* 149 (2007).

Lenton, T.M. & N.E. Vaughan. 'The Radiative Forcing Potential of Different Climate Geoengineering Options', 9 *Atmos Chem Phys Discuss* 2559 (2009).

Lerner, J. *Patent Protection and Innovation over 150 Years* (Cambridge, MA: National Bureau of Economic Research Working Paper no. 8977, 2002).

Levit, J.K. 'A Bottom-Up Approach to International Law Making: The Tale of Three Trade Finance Instruments', 30 *Yale J. Int'l L.* 125 (2005).

Levit, J.K. 'Bottom-Up International Lawmaking: Reflections on the New Haven School of International Law', 32 *Yale J. Int'l L.* 383 (2007).

Lidgard, H.H. & J. Atik. 'Facilitating Compulsory Licensing under TRIPS in Response to the AIDS Crisis in Developing Countries', in *Corporate and Employment Perspectives in a Global Business Environment*, ed. Roger Blanpain & Boel Flodgren (The Hague: Kluwer Law International, 2006).

Lin, A.C. 'Filling in Gaps of the TRIPS Agreement: Reflections on the Taiwan-Philips CD-R Compulsory Licence Case', 3 *Asian Journal of WTO and International Health Law and Policy* 577 (2008).

Lin, A.C. 'Geoengineering Governance', 8 Issues in Legal Scholarship Art. 2 (2009).

Lin, A.C. 'International Legal Regimes & Principles Relevant to Geoengineering' (27 May 2011 draft manuscript on file with author).

Linden, G. 'Optical Storage In China: A Study in Strategic Industrial Policy', *Center for Work, Technology and Society*, University of California, Berkeley.

Litfin, K.T. *Ozone Discourses. Science and Politics in Global Environmental Cooperation* (New York: Columbia University Press, 1994).

Lobel, O. 'Setting the Agenda for New Governance Research', 89 *Minn. L. Rev.* 498 (2004).

Lobel, O. 'The Renew Deal: The Fall of Regulation and the Rise of Governance in Contemporary Legal Thought', 89 *Minn. L. Rev.* 342 (2004).

Lyman, E. '"Progress" of Bangkok Talks Shows Much Still to Be Done for 2009 Global Agreement', BNA Env't Rep, (11 Apr. 2008).

MacCracken, M. 'Beyond Mitigation: Potential Options for Counter-Balancing the Climatic and Environmental Consequences of the Rising Concentrations of Greenhouse Gases' (World Bank Pol'y Res., Working Paper No. 4938, 2009).

<www.climate.org/PDF/World-Bank_Beyond-Mitigation_MacCracken.pdf>.

Marceau, G. 'A Call for Coherence in International Law: Praises for the Prohibition against "Clinical Isolation" in WTO Dispute Settlement', 33 *J World Trade* 87 (1999).

Marceau, G. 'WTO Dispute Settlement and Human Rights', 13 *Eur J Int'l L* 753 (2002).

Maskus, K.E. *Intellectual Property Rights in the Global Economy* (Washington, DC: Institute for International Economics, 2000).

Maskus, K.E. *Encouraging International Technology Transfer.* UNCTAD-ICTSD Programmes on IPRs and Sustainable Development, Issue Paper No.7, (2004) Geneva: International Centre for Trade and Sustainable Development, <www.iprsonline.org/unctadictsd/projectoutputs.htm>.

Maskus, K.E. 'Intellectual Property Rights in the WTO Accession Package: Assessing China's Reforms', in *China and the WTO: Accession, Policy Reform, and Poverty Reduction Strategies*, ed. Deepak Bhattasali, Shantong Li & William J. Martin (Oxford: Oxford University Press, 2004).

Maskus, K.E. *Reforming US Patents: Getting the Incentives Right* (Washington, DC: Council on Foreign Relations, Special Reports series, Report no. 19, 2006).

Maskus, K.E. *Private Rights and Public Problems: The Global Economics of Intellectual Property in the 21st Century* (Washington: Peterson Institute for International Economics, 2011).

Maskus, K.E. & M. Penubarti. 'How Trade-Related Are Intellectual Property Rights?', 39 *Journal of International Economics* 227 (1995).

Maskus, K.E. & C. McDaniel. 'Impacts of the Japanese Patent System on Productivity Growth', 11 *Japan and the World Economy* 557 (1999).

Maskus, K.E. & M. Lahouel. 'Competition Policy and Intellectual Property Rights in Developing Countries', 23 *The World Economy* 595 (2000).

Maskus, K.E. & J.H. Reichman (eds), *International Public Goods and Transfer of Technology under a Globalized Intellectual Property Regime* (Cambridge, U.K.: Cambridge University Press, 2005).

Maskus, K.E. & R.L. Okediji, *Intellectual Property Rights and Technology Transfer to Address Climate Change: Risks, Opportunities and Policy Options* (Geneva: International Centre for Trade and Sustainable Development, Project on IPRs and Sustainable Development, Issue Paper no. 32, 2011).

Masoudi, G.F. 'Intellectual Property and Competition: Four Principles for Encouraging Innovation', (11 Apr. 2006) <www.usdoj.gov/atr/public/speeches/215645.pdf>.

Matthews, H.D. & K. Caldeira. 'Transient Climate-Carbon', 104 *PNAS* 9951 (2007).

McDonald, J.M. 'Domestic Regulation, International Standards, and Technical Barriers to Trade', 4:2 *World Trade Rev* 249 (2005).

McDougal, M.S. & H.D. Lasswell. 'The Identification and Appraisal of Diverse Systems of Public Order', 53 *Am. J. Int'l L.* 1 (1959).

McDougal, M.S., et al. 'The World Constitutive Process of Authoritative Decisions', 19 *J. Legal Educ.* 253 (1967).

McFetridge, D.G. 'Intellectual Property, Technology Diffusion, and Growth in the Canadian Economy', in *Competition Policy and Intellectual Property Rights in the Knowledge-Based Economy*, ed. Robert D. Anderson & Nancy T. Gallini (Calgary: University of Calgary Press, 1998).

McGoldrick, D. 'Sustainable Development and Human Rights: An Integrated Conception', 45 *Int'l Comp L Q* 796 (1996).

McGraw, D.M. 'The CBD – Key Characteristics and Implications for Implementation', 11 *Rev Eur Community and Int'l Envtl L (RECIEL)*, 24 (2002).

McKibben, B. 'Civilization's Last Chance: The Planet Is Nearing a Tipping Point on Climate Change, and It Gets Much Worse Fast', *L.A. Times*, 11 May 2008, <www.latimes.com/news/printedition/opinion/la-op-mckibben11-2008may11,0,2392815.story>.

McKinstry Jr., R.B. 'Laboratories for Local Solutions for Global Problems: State, Local and Private Leadership in Developing Strategies to Mitigate the Causes and Effects of Climate Change', 12 *Penn. St. Envtl. L. Rev.*, 15 (2004).

McManis, C.R. 'The Proposed Anti-Counterfeiting Trade Agreement (ACTA): Two Tales of a Treaty', 4 *Hous L Rev* 1237 (2009).

McPhee, J. *Encounters with the archdruid* (New York: Farrar, Straus and Giroux 1971).

Meier, M.H., B.S. Albert & S.C. Brau. 'Overview of the FTC Antitrust Actions in Health Care Services and Products', (June 2010). <www.ftc.gov/bc/0610hcupdate.pdf>.

Meurer, M.J. 'Patent Litigation, Licensing, Nonobviousness, and Antitrust', in *Antitrust Law and Economics*, ed. K.L. Hylton (Cheltenham: Edward Elgar, 2010).

Meyer, W.B., et al. 'The Local-Global Continuum', in *Geography's Inner Worlds: Pervasive Themes in Contemporary American Geography*, ed. Ronald F. Abler, Melvin G. Marcus & Judy M. Olson (Rutgers University Press 1992).

Michaels, R. 'The Re-Statement of Non-State Law: The State, Choice of Law, and the Challenge from Global Legal Pluralism', 51 *Wayne L. Rev.* 1209 (2005).

Minh, N.T. 'Legal Assistance to Ministry of Industry on Legal Issues Relating to RESPP Development in Vietnam' (July 2006), (Legal Issues Relating to RESPP 2006).

Moon, S. 'Does TRIPS Art. 66.2 Encourage Technology Transfer to LDCs? An Analysis of Country Submissions to the TRIPS Council (1999–2007)', Policy Brief No. 2, December 2008, UNCTAD-ICTSD project on IPRs and Sustainable Development, 2. <http://ictsd.org/i/publications/37159/>.

Moon, S. 'Meaningful Technology Transfer to the LDCs: A proposal for a Monitoring Mechanism for TRIPS Article 66.2', Policy Brief No 9, May 2011, <http://ictsd.org/i/publications/106434/>, *ICTSD Programme on Innovation, Technology and Intellectual Property Project on IPRs and Sustainable Development Policy Brief Number 9*, April 2011.

Moore, J.C., S. Jevrejeva & A. Grinstad. 'Efficacy of Geoengineering to Limit 21st Century Sea-Level Rise', *PNAS Early Edition* (forthcoming 2011), <www.pnas.org/cgi/doi/10.1073/pnas.1008153107>.

Morgera, E. & E. Tsioumani. 'The Evolution of Benefit Sharing: Linking Biodiversity and Community Livelihoods', 19 *Rev Eur Community & Int'l Env'l L (RECIEL)* 150 (2010).

Mortensen, T. 'An Unattainable Wedge: Four Limiting Effects on the Expansion of Nuclear Power', 5 *Envt'l & Energy L. & Pol'y J.* 60 (2010).

Moser, P. & A. Voena. 'Compulsory Licensing: Evidence from the *Trading with the Enemy Act*' (Stanford CA: Stanford University working paper, 2008).

Musungu, S. & C. Oh. 'The Use of Flexibilities in TRIPS by Developing Countries: Can They Promote Access to Medicines?', in WHO Study on IPRs, Innovation and Public Health (Geneva: WHO and South Centre, 2006), <www.who.int/intellectualproperty/studies/TRIPS_flexibilities/en/index.html>.

Nagel, T. 'The Problem of Global Justice', 33 *Philosophy & Public Affairs* 113 (2005).

National Energy Foundation. 'Fuel Consumption Stats, Primary Energy Consumption by Fuel' (1999). <www.nef1.org/ea/eastats.html>.

New, W. 'China Leads Developing Country Push for Balance in IP and Standards', *IPWATCH* 1 (24 Apr. 2007). <www.ip-watch.org/weblog/2007/04/24/china-leads-developing-country-push-for-balance-in-ip-and-standards/>.

Nguyen, A.T., et al. 'Medicine prices, availability, and affordability in Vietnam', *Southern Med Rev* (September 2009), <www.southernmedreview.org/index.php/smr/article/viewArticle/11>.

Nguyen, Q.B. 'Harmonization of Law for Economic Development in Vietnam and Impacts of the US-Vietnam Bilateral Trade Agreement toward This Process and Future US-VN Trade Relations: Part 2 BTA and WTO Negotiations', University of Louisville Legal Studies Research Paper No. 2010-06.

Nguyen, T.T. 'Competition Law and Access to Pharmaceutical Technology in the Developing World', 28 *Biotech L Rep* 693 (2009).

Nguyen, T.T. *Competition Law, Technology Transfer and the TRIPS Agreement: Implications for Developing Countries* (Cheltenham: Edward Elgar, 2010), 48–52.

Nguyen, T.T. & Ha-Duong Minh. 'CO_2 Emissions Mitigation Potential in Vietnam's Power Sector', DEPOCEN Working Paper 2009/22 (25 Nov. 2009). <depocenwp.org/index.php>.

Nguyen, T.T., et al. 'Improving the Clean Development Mechanism Post 2012', *CCLR The Carbon & Climate Law Review* 4(1) 76–85(10) (March 2010), <www.ingentaconnect.com/content/lex/cclr/2010/00000004/00000001/art00008;jsessionid=1849v0entekas.alexandra>.

OECD. 'Competition Law and Policy in Chinese Taipei' (April 2006). <www.oecd.org/dataoecd/18/22/36875564.pdf>.

OECD. 'Intellectual Property Rights', DAF/COMP(2004)24 (21 Jan. 2005), <www.oecd.org/dataoecd/61/48/34306055.pdf>.

Okafor, O.C. 'Newness, Imperialism, and International Legal Reform in Our Time: A TWAIL Perspective', 43 *Osgoode Hall L.J.* 171 (2005).

Okediji, R.L. 'Back to Bilateralism? Pendulum Swings in International Intellectual Property Protection', 1 *U Ottawa L & Tech J* 125 (2003–2004).
Okediji, R.L. 'Sustainable Access to Copyrighted Digital Information Works in Developing Countries', in *International Public Goods & Transfer of Technology Under a Globalized Intellectual Property Regime*, ed. Keith E. Maskus & Jerome H. Reichman (Cambridge: Cambridge University Press, 2005).
Okediji, R.L. *The International Copyright System: Limitations, Exceptions and Public Interest Considerations for Developing Countries* (Geneva: International Centre for Trade and Sustainable Development, Project on IPRs and Sustainable Development, Issue Paper no. 15, 2006).
Organisation for Economic Co-operation and Development. 'Intellectual Property and Competition Policy in the Biotechnology Industry', *Policy Brief*, June 2005.
Osofsky, H.M. 'Climate Change Litigation as Pluralist Legal Dialogue?', 26 *Stan. Envtl. L.J.* & 43 *Stan. J. Int'l L.* 181 (2007) (Joint Issue).
Osofsky, H.M. 'Local Approaches to Transnational Corporate Responsibility: Mapping the Role of Sub-National Climate Change Litigation', 20 *Pac. McGeorge GLOBAL Bus. & Dev. L.J.* 143 (2007).
Osofsky, H.M. & J.K. Levit. 'The Scale of Networks: Local Climate Change Coalitions', 8 *Chi. J. Int'l L.* 409 (2008).
Osofsky, H.M. 'Is Climate Change "International"?: Litigation's Diagonal Regulatory Role', 49 *Va. J. Int'l L.* 585 (2009).
Osofsky, H.M. 'The Geography of Climate Change Litigation Part II: Narratives of Massachusetts v. EPA', 8 *Chi. J. Int'l L.* 573 (2008).
Osofsky, H.M. 'Multiscalar Governance and Climate Change: Reflections on the Role of States and Cities at Copenhagen', 25 *Maryland J. Int'l L.* 64 (2010).
Osofsky, H.M. 'Multidimensional Governance and the BP Deepwater Horizon Oil Spill', __ Fla L Rev __ (forthcoming 2011).
Paasi, A. 'Place and Region: Looking through the Prism of Scale', 28 *Progress in Human Geography* 536 (2004).
Palmer, G. 'New Ways to Make International Environmental Law', 86 *Am J Int'l L*, 259 (1992).
Palmeter, D. & P.C. Mavroidis. 'The WTO Legal System: Sources of Law', 92 *Am J Int'l L* 398 (1998).
Panitchpakdi, S. Secretary-General of UNCTAD (2008–2010), UNCTAD Trade and Development Board, Forty-ninth executive session, 8 Jun. 2010. <www.unctad.org/Templates/webflyer.asp?docid=13354&intItemID=3549&lang=1>.
Park, W.G. 'International Patent Protection: 1960–2005', 37 *Res Pol'y* 761 (2008).
Park, W.G. 'Intellectual Property Rights and International Innovation', in *Intellectual Property, Growth and Trade*, ed. Keith E. Maskus (Amsterdam: Elsevier, 2008).

Park, W.G. & D. Lippoldt. 'International Licensing and the Strengthening of Intellectual Property Rights in Developing Countries during the 1990s', 40 *OECD Econ Stud* 7 (2005).

Parks, B.C. & J.T. Roberts. 'Environmental and Ecological Justice', in *Palgrave Advances in International Environmental Politics*, ed. Michele Betsill, Kathryn Hochstetler & Dimitris Stevis (Basingstoke: Palgrave Macmillan, 2006).

Patel, S.J., P. Roffe & A.A. Yusuf (eds). *International Technology Transfer: The Origins and Aftermath of the United Nations Negotiations on a Draft Code of Conduct* (The Hague: Kluwer Law International, 2000).

Patricia, B., A. Boyle & C. Redgwell. *International Law & the Environment* (Oxford: Oxford University Press, 2009).

Pauwelyn, J. 'The Role of Public International Law in the WTO: How Far Can We Go?', 95 *Am J Int'l L* 535, 577 (2001).

Peter, N. 'Democratising Biotechnology? Deliberation, Participation and Social Regulation in a Neo-liberal World', 36 *Review of International Studies* 471 (2010).

Purdy, R. 'The Legal Implications of Carbon Capture and Storage under the Sea', 7 *Sustainable Dev. L. & Pol'y* 22 (2006–2007).

Qian, Yi. 'Do National Patent Laws Stimulate Domestic Innovation in a Global Patenting Environment? A Cross-Country Analysis of Pharmaceutical Patent Protection, 1978–2002', 89 *Rev Econ Stat* 4369(2007).

Quat, P.H. *How to Comply With The TRIPS and WTO Law: The New Challenges To Vietnam's Patent Legislation from WTO Dispute Settlement Practice*, Doctoral Dissertation, Nagoya University, Japan, December 2007, 189–190. <http://iprenforcement.most.gov.vn/ckfinder/userfiles/files/How%20to%20Comply%20with%20the%20TRIPS%20and%20WTO%20Law(2).pdf>, 25 Jan. 2011.

Rajagopal, B. *International Law from Below: Development, Social Movement and Third World Resistance* (Cambridge University Press, 2003); Symposium: Critical Race Theory and International Law: Convergence and Divergence, 45 Vill. L. Rev. 827 (2000).

Rajamani, L. 'The Principle of Common but Differentiated Responsibility and the Balance of Commitments under the Climate Regime', 9 *Rev Eur Community & Int'l Env'l L (RECIEL)* 120 (2000).

Rasch, P.J., et al. 'Exploring the Geoengineerings of Climate Using Stratospheric Sulfate Aerosols: The Role of Particle Size', 33 *Geophysical Res Letters* 1 (2008).

Redgwell, C. 'Biotechnology, Biodiversity and Sustainable Development', in *Biotechnology and International Law*, ed. Francesco Francioni & Tullio Scovazzi (Oxford: Hart Publishing, 2006).

Redgwell, C. 'International Environmental Law', in *International Law*, 3rd edn, ed. Malcolm D. Evans (Oxford: Oxford University Press, 2010).

Reichmann, J.H. 'The TRIPS Agreement Comes of Age: Conflict or Cooperation with the Developing Countries?', 32 *Case Wt Res J Int'l L* 441 (2000).

Reichman, J.H. 'Intellectual Property in the Twenty-First Century: Will the Developing Countries Lead or Follow?', 4 *Hous L Rev* 115 (2009).

Reichman, J.H. & C. Hasenzahl. *Non-voluntary Licensing of Patented Inventions: Historical Perspective, Legal Framework under TRIPS, and an Overview of the Practice in Canada and the USA* (Geneva: International Centre for Trade and Sustainable Development, Project on IPRs and Sustainable Development, Issue Paper no. 5, 2003).

Reichman, J.H. & T. Lewis. 'Using Liability Rules to Stimulate Local Innovation in Developing Countries: Application to Traditional Knowledge', in *International Public Goods and Transfer of Technology under a Globalized Intellectual Property Regime*, ed. Keith E. Maskus & Jerome H. Reichman (Cambridge: Cambridge University Press, 2005).

Reichman, J.H. & P.F. Uhlir. 'A Contractually Reconstructed Research Commons for Scientific Data in a Highly Protectionist Intellectual Property Environment', Duke University Law School, 66 *L & Contemp Probs* 315 (2003).

Reisman, W.M. 'International Lawmaking: A Process of Communication', 75 *Am. Soc'y Int'l L. Proc.* 101 (1981).

Resnik, D.B. & K.A. De Ville. 'Bioterrorism and Patent Rights: Compulsory Licensure and the Case of Cipro', 2 *Am J Bioethics* 29 (2002).

Resnik, J. 'Law's Migration: American Exceptionalism, Silent Dialogues, and Federalism's Multiple Ports of Entry', 115 *Yale L. J.* 1564 (2006).

Resnik, J., J. Civin & J. Frueh. 'Ratifying Kyoto at the Local Level: Sovereigntism, Federalism, and Translocal Organizations of Government Actors (TOGAS)', 50 *Ariz. L. Rev.* 709 (2008).

Robock, A. 'Whither Geoengineering?', 320 *Science* 1166 (2008).

Robock, A., et al. 'A Test for Geoengineering?', 327 *Science* 530 (January 2010).

Rockström, J., et al. 'A Safe Operating Space for Humanity', 461 *Nature* 472 (2009).

Roffe, P. & T. Tesfachew. 'Revisiting the Technology Transfer Debate: Lessons for the New WTO Working Group', *Bridges*, 2 Feb. 2002 <www.iprsoneline.org/ictsd/docs/RoffeTesfachewBridgesYear6N2February2002.pdf>.

Roin, B.N. 'Unpatentable Drugs and the Standards of Patentability', 87 *Tex L Rev* 503 (2009).

Rosas, A. & L. Armati. *EU Constitutional Law: An Introduction* (London: Hart Publishing, 2010).

Rosch, J.T. 'The Role of Static and Dynamic Analysis in Pharmaceutical Antitrust', (18 Feb. 2010). <www.ftc.gov/speeches/rosch/100218pharmaantitrust.pdf>.

Ruhl, J.B. & J.E. Salzman. 'Climate Change, Dead Zones, and Massive Problems in the Administrative State: A Guide for Whittling Away', 98 *Cal. L. Rev.* 59 (2010).

Ruse-Khan, H.G., T. Jaeger & R. Kordic. 'The Role of Atypical Acts in EU External Trade and Intellectual Property Policy', 21 *Eur J Int'l L* 901 (2010).

Ruse-Khan, H.G. 'A Trade Agreement Creating Barriers to International Trade? ACTA Border Measures and Goods in Transit, Max Planck Institute for

Intellectual Property', Competition & Tax Law Research Paper Series No. 10-10 (forthcoming *Am U Int'l L Rev*), <http://papers.ssrn.com/sol3/papers.cfm?abstract_id=1706567&>.

Russell, B.D. & S.D. Connell. 'Letters: Honing the Geoengineering Strategy', 327 *Science* 144 (2010).

Sakakibara, M. & L. Branstetter. 'Do Stronger Patents Induce More Innovation? Evidence from the 1988 Japanese Patent Reforms', 32 *RAND J Econ* 77 (2001).

Salkin, P.E. & A.P. Ostrow. 'Cooperative Federalism and Wind: A New Framework for Achieving Sustainability', 7 *Hofstra L. Rev.* 1049 (2009).

Sands, P. 'International Law in the Field of Sustainable Development', 65 *British Yearbook of International Law* 375 (1994).

Sands, P. *Principles of International Environmental Law* (Cambridge: Cambridge University Press, 2003).

Scherer, F.M. *The Economic Effects of Compulsory Licensing* (New York: New York University, Stern School of Business, 1977).

Schnare, D.W. 'Climate Change and the Uncomfortable Middle Ground: "The Geoengineering" and "No Regrets" Policy Alternative', Paper delivered at the Thomas Jefferson Institute for Public Policy 4 March 2008 <www.heartland.org/events/NewYork08/newyork2008-ppt.html>.

Schneider, P.H. 'International Trade, Economic Growth and Intellectual Property Rights: A Panel Data Study of Developed and Developing Countries', 78 *J Devel Econ* 529 (2005).

Scott, K.N. 'The Day After Tomorrow: Ocean CO_2 Sequestration and the Future of Climate Change', 18 *Geo Int'l Envtl L Rev* 57 (2005).

Secretariat of the Convention on Biological Diversity, Handbook of the Convention on Biological Diversity Including its Cartagena Protocol on Biosafety, 3rd edn (Montreal Canada, 2005), s. VIII (Declarations). <www.cbd.int/convention/refrhandbook.shtml>.

Sell, S.K. 'The Global IP Upward Ratchet, Anti-Counterfeiting and Piracy Enforcement Efforts: The State of Play 4' (IQsensato, Occasional Papers No. 1, 2008).).

Sell, S.K. *Power and Ideas: North South Politics of Intellectual Property and Antitrust* (New York: State University of New York Press, 1998).

Senden, L. *Soft Law in European Community Law* (Oxford, England: Hart Publishing, 2004).

Senden, L. 'Soft Law and Its Implications for Institutional Balance in the EC', *Utrecht Law Review* 1, no. 2 (December 2005).

Shaffer, G. & M.A. Pollack. 'Hard vs. Soft Law: Alternatives, Complements, and Antagonists in International Governance', 94 *Minn L Rev.* 706 (2010).

Shah, A. 'UK's Implementation of the Anti-Circumvention Provisions of the EU Copyright Directive: An Analysis'. *Duke Law and Technology Review* 3 (2004), <www.law.duke.edu/journals/dltr/articles/2004dltr0003.html>.

Shepherd, John, FRS. 'Geoengineering the Climate: Science, Governance and Uncertainty', The Royal Society 1 Sep. 2009. <http://royalsociety.org/geoengineering-the-climate/>.

Slaughter, A.-M. *A New World Order* (Princeton University Press, 2004).

Smith, P.M. 'Looking for the Global Spaces in Local Politics', 17 *Pol. Geography* 35 (1998).

Staeheli, L.A. 'Globalization and the Scales of Citizenship', 19 *Geography Res. F.* 60 (1999).

Stewart, R.B. 'States and Cities as Actors in Global Climate Regulation: Unitary vs. Plural Architectures', 50 *Ariz. L. Rev.* 681 (2008).

Stiglitz, J.E. 'Competing over Competition Policy', (2001). <www.project-syndicate.org/commentary/stiglitz5>.

Suisse, C. 'Global Wealth Report 2010' (2010 Global Wealth Report), 3. <www.credit-suisse.com/news/doc/credit_suisse_global_wealth_report.pdf>.

Supachai, P. 'Opening Address for Sixth UN Conference to Review all Aspects of the Multilaterally Agreed Equitable Principles and Rules for the Control of Restrictive Business' (8 Nov. 2010). <www.unctad.info/en/6th-UN-Conference-on-Competition-Policy/Conference/Opening-Address-by-Secretary-General-of-UNCTAD/>.

Taubman, A. 'Rethinking TRIPS: "Adequate Remuneration" for Non-Voluntary Patent Licensing', 11 *J Int'l Econ L* 947 (2008).

Teubner, G. 'Self-subversive Justice: Contingency or Transcendence Formula of Law?', 72 *Mod L Rev* 1 (2009).

Tilmes, S., R. Müller & R. Salawitch. 'The Senstivity of Polar Ozone Depletion to Proposed Geoengineering Schemes', 320 *Science* 1201 (2008).

TIPO. 'TIPO Holds Public Hearing for Compulsory Licensing Draft Amendment on 18 August 2008'. <www.tipo.gov.tw/en/News_NewsContent.aspx?NewsID=3018>.

Trade and Climate Change. A report by the United Nations Environment Programme and the World Trade Organization' (Switzerland: WTO Secretariat, 2009), 42.

Trisolini, K. & J. Zasloff. 'Cities, Land Use, and the Global Commons: Genesis and the Urban Politics of Climate Change', in *Adjudicating Climate Change: Sub-National, National, and Supra-National Approaches*, ed. William C.G. Burns & Hari M. Osofsky (Cambridge University Press, forthcoming 2009). <http://papers.ssrn.com/sol3/papers.cfm?abstract_id=1267314>.

UNCTAD. 'Transfer of Technology', *UNCATD Series on Issues in International Agreements* (New York and Geneva: United Nations, 2001).

UNCTAD. *Compendium of International Arrangements on Technology Transfer: Selected Instruments* (Geneva: United Nations, 2001).

UNCTAD. *Intellectual Property in the World Trade Organization: Turning It into Developing Countries' Real Property* (New York and Geneva: United Nations, 2010), 1.

UNCTAD. *Model Law on Competition: Substantive Possible Elements for a Competition Law, Commentaries and Alternative Approaches in Existing Legislation* (New York and Geneva: United Nations, 2010).

UNCTAD. *The Least Developed Countries Report 2007: Knowledge, Technological Learning and Innovation for Development* (Geneva: UNCTAD, 2007).

UNCTAD-ICTSD. *Resource Book on TRIPS and Development* (Cambridge University Press, Cambridge, 2005).

United Nations Framework Convention on Climate Change, FCCC/INFORMAL/84, opened for signature 9 May 1992, entered into force 21 Mar. 1994 (FCCC). <http://unfccc.int/resource/docs/convkp/conveng.pdf>.

United Nations Office of the High Representative for the Least Developed Countries, Landlocked Developing Countries and the Small Island Developing States (UN-OHRLLS)', Criteria for Identification of LDCs. <www.unohrlls.org/en/ldc/related/59/>.

United Nations. *The Millennium Development Goals Report 2010* (New York: United Nations, 2010).

Updegrove, Andrew. 'Participating In Standard Setting Organizations: Value Propositions, Roles and Strategies', in *The Essential Guide to Standards*, Ch. 2. <www.consortiuminfo.org/essentialguide/>.

U.S. Department of Justice and Federal Trade Commission. 'Antitrust Enforcement and Intellectual Property Rights: Promoting Innovation and Competition' (April 2007), <www.justice.gov/atr/public/hearings/ip/222655.pdf>, 1.

U.S. Department of Justice and Federal Trade Commission. 'Antitrust Guidelines for the Licensing of Intellectual Property' (6 Apr. 1995). <www.usdoj.gov/atr/public/guidelines/0558.pdf>.

U.S. Energy Information Administration (EIA). 'Emission of Greenhouse Gases in the United States 2005', (February 2007), (EIA US Greenhouse Gases 2005), <www.eia.doe.gov/oiaf/1605/ggrpt/summary/carbon.html>.

U.S. GAO. 'International Climate Change Programs: Lessons Learned from the European Union's Emissions Trading Scheme and the Kyoto Protocol's Clean Development Mechanism', GAO-09-151, 18 Nov. 2008. <www.gao.gov/products/GAO-09-151>.

USTR Press Release. 'Statement by the Anti-Counterfeiting Trade Agreement (ACTA) negotiating partners regarding the recent round of ACTA negotiations in Lucerne, Switzerland' (5 Jul. 2010). <www.ustr.gov/about-us/press-office/press-releases/2010/june/office-us-trade-representative-releases-statement-act>.

van den Bossche, P. *The Law and Policy of the World Trade Organization: Text, Cases and Materials*, 2nd edn (Cambridge, England: Cambridge University Press, 2008).

Varney, C. 'Promoting Innovation through Patent and Antitrust Law and Policy', (26 May 2010). <www.justice.gov/atr/public/speeches/260101.pdf>.

Vaver, D. & B. McLachlin. *Intellectual Property Law: Copyrights, Patents and Trademarks (Essentials of Canadian Law)* (Toronto: Irwin Law, 1997).

Vera, I. 'Climate Change and Technology Transfer: The Need for a Regional Perspective', UN-DESA Policy Brief No. 18, <www.un.org/esa/policy/policybriefs/policybrief18.pdf>.

Vivas-Eugui, D. 'Regional and Bilateral Agreements and a TRIPS-Plus World: The Free Trade Area of the Americas (FTAA)', TRIPS Issue Papers no. 1, Quaker United Nations Office (QUNO) (August 2003), 14. <www.quno.org/economicissues/intellectual-property/intellectualLinks.htm#ISSUES>.

Wara, M. 'Measuring the Clean Development Mechanism's Performance and Potential, 55 *UCLA L Rev* 1759 (2008).

Watal, J. *Intellectual Property Rights in the WTO and Developing Countries* (The Hague: Kluwer Law International 2001).

Watchman (ed.), *Climate Change: A Guide to Carbon Law and Practice* (London: Globe Law & Business Publisher, 2008).

WHO. 'Global Strategy and Plan of Action on Public Health, Innovation and Intellectual Property' WHA61.21 (24 May 2008), revised in WHA62.16 (22 May 2009).

WHO. 'Public Health, Innovation and Intellectual Property: Global Strategy and Plan of Action – Proposed Time Frames and Estimated Funding Needs', (21 Jan. 2009) <http://apps.who.int/gb/ebwha/pdf_files/EB124/B124_16Add2-en.pdf>, para. 8.

WHO. *Public Health, Innovation and Intellectual Property Rights* (Geneva: WHO, 2006).

Wiener, J.B. 'Something Borrowed for Something Blue: Legal Transplants and the Evolution of Global Environmental Law', 27 *Ecology L Q* 1300 (2001).

Wigley, T.M.L. 'A Combined Mitigation/Geoengineering Approach to Climate Stablization', 314 *Science* 452 (2006).

Wilson, J.S. 'Standards, Regulation, and Trade: WTO Rules and Developing Country Concerns', in *Development, Trade and the WTO: A Handbook*, ed. Bernard Hoekman, Aaditya Mattoo & Philip English (World Bank, 2002).

Wood, J. & R.S. Dave, *Compulsory Licensing of Patents in the US, China, Japan, Germany, and India* <www.ipo.org/AM/CM/ContentDisplay.cfm?ContentFileID=6521>.

Wood, M.C. '"You Can't Negotiate with a Beetle": Environmental Law for a New Ecological Age', 50 *Nat Resources J* 167 (2010).

World Bank. 'Vietnam: Laying the Foundation for Steady Growth', (February 2007) <siteresources.worldbank.org/intvietnam/overview/21594788/ida-vietnam.pdf>.

World Bank. *Global Economic Prospects 2008: Technology Diffusion in the Developing World* (Washington DC: World Bank, 2008).

World Bank Statement, Ministerial Segment – COP11 – Montreal 4, <http://siteresources.worldbank.org/ESSDNETWORK/Resources/MINISTERIALSEGMENT COP11Montreal.pdf>.

WTO. 'TRIPS and Health, Frequently Asked Questions: Compulsory Licensing of Pharmaceuticals and TRIPS'. <www.wto.org/english/tratop_e/trips_e/public_health_faq_e.htm>.

WTO. *The Legal Texts: The Results of the Uruguay Round of Multinational Trade Negotiations* (Cambridge: Cambridge University Press, 2010).

WTO. TRIPS Council, 'Implementation of Article 66.2: Information from Developed Country Members – Types of Incentive Measure Notified', Note 3 May 2000 by the Secretariat, IP/C/W/169,<www.wtocenter.org.tw/SmartKMS/fileviewer?id=5256>.

WTO Director-General Pascal Lamy. Global trading system must deliver more to the least-developed, speech 9 May 2011, <www.wto.org/english/news_e/sppl_e/sppl191_e.htm>.

WTO Director-General Pascal Lamy. *The Doha Round marks a transition from the old governance of the old trade order to the new governance of a new trade order*, Speech at the occasion of the 10th anniversary of the World Trade Institute (1 Oct. 2010), <www.wto.org/english/news_e/sppl_e/sppl173_e.htm>.

WTO Secretariat. *World Trade Report 2005 – Exploring Links Between Trade, Standards and the WTO* 75 (2005), <www.wto.org/english/res_e/booksp_e/anrep_e/wtr05-0_e.pdf>.

Yamane, H. *Interpreting TRIPS: Globalisation of Intellectual Property Rights and Access to Medicines* (Oxford: Hart Publishing Ltd, 2011).

Yu, P.K. 'Access to Medicines, BRICS Alliances, and Collective Action', 34 *Am J L & Med* 345 (2008).

Yu, P.K. 'TRIPS and Its Discontents', 10 *Marq Int'l Prop L Rev* 370 (2006).

Yu, P.K. 'A Tale of Development Agendas', 35 *Ohio N U L Rev* 465 (2009).

Yu, P.K. 'The Objectives and Principles of the TRIPS Agreement', 46 *Houston L Rev* 797 (2009).

Yu, P.K. 'Six Secret (And Now Open) Fears of ACTA', (describing Argentina's, Brazil's, China's and India's, strong opposition to an EU initiative to push for further discussions of IPR enforcement issues, in the June 2006 TRIPS Council Meeting), (forthcoming 64 *SMU L Rev* 2011) (20 Oct. 2010), 10–11 (Six Open Fears). <http://papers.ssrn.com/sol3/papers.cfm?abstract_id=1624813>.

Yu, P.K. 'TRIPS Enforcement and Developing Countries' (forthcoming *Am U Int'l L Rev*) (2011). <http://papers.ssrn.com/sol3/papers.cfm?abstract_id=1736030&##>.

Index

U